Residential Construction Academy

Facilities Maintenance

Second Edition

Kevin Standiford

DELMAR
CENGAGE Learning™

Australia • Brazil • Japan • Korea • Mexico • Singapore • Spain • United Kingdom • United States

Residential Construction Academy, Facilities Maintenance, Second Edition
Kevin Standiford

Vice President, Career and Professional Editorial: Dave Garza

Director of Learning Solutions: Sandy Clark

Acquisitions Editor: Jame Devoe

Managing Editor: Larry Main

Vice President, Career and Professional Marketing: Jennifer Baker

Marketing Director: Deborah Yarnell

Marketing Coordinator: Mark Pierro

Production Director: Wendy Troeger

Production Manager: Mark Bernard

Senior Art Director: Casey Kirchmayer

Technology Project Manager: Christopher Catalina

Production Technology Analyst: Thomas Stover

For product information and technology assistance, contact us at
Cengage Learning Customer & Sales Support, 1-800-354-9706

For permission to use material from this text or product,
submit all requests online at **cengage.com/permissions**
Further permissions questions can be emailed to
permissionrequest@cengage.com

Library of Congress Control Number: 2009943120

ISBN-13: 978-1-111-31112-4

ISBN-10: 1-111-31112-9

Delmar
5 Maxwell Drive
Clifton Park, NY 12065-2919

Cengage Learning is a leading provider of customized learning solutions with office locations around the globe, including Singapore, the United Kingdom, Australia, Mexico, Brazil, and Japan. Locate your local office at:
international.cengage.com/region

Cengage Learning products are represented in Canada by Nelson Education, Ltd.

For your lifelong learning solutions, visit **delmar.cengage.com**

Visit our corporate website at **www.cengage.com**

Notice to the Reader

The publisher does not warrant or guarantee any of the products described herein or perform any independent analysis in connection with any of the product information contained herein. The publisher does not assume, and expressly disclaims, any obligation to obtain and include information other than that provided to it by the manufacturer.

The reader is expressly warned to consider and adopt all safety precautions that might be indicated by the activities herein and to avoid all potential hazards. By following the instructions contained herein, the reader willingly assumes all risks in connection with such instructions.

The publisher makes no representation or warranties of any kind, including but not limited to, the warranties of fitness for particular purpose or merchantability, nor are any such representations implied with respect to the material set forth herein, and the publisher takes no responsibility with respect to such material. The publisher shall not be liable for any special, consequential, or exemplary damages resulting, in whole or part, from the readers' use of, or reliance upon, this material.

Printed in the United States of America
2 3 4 5 6 7 13 12 11

Brief Contents

Table of Contents

Chapter 8 Surface Treatments. **235**

Preface

About the *Residential Construction Academy Series*

One of the most pressing problems confronting the construction industry today is the shortage of skilled labor. The construction industry must recruit an estimated 200,000 to 250,000 new craft workers each year to meet future needs. This shortage is expected to continue well into the next decade because of projected job growth and a decline in the number of available workers. At the same time, the training of available labor is becoming an increasing concern throughout the country. This lack of training opportunities has resulted in a shortage of 65,000 to 80,000 skilled workers per year. This challenge is affecting all construction trades and is threatening the ability of builders to construct quality homes.

These challenges led to the creation of the innovative *Residential Construction Academy* series. The *Residential Construction Academy* series is the perfect way to introduce people of all ages to the building trades while guiding them in the development of essential workplace skills, including carpentry, electrical wiring, HVAC, plumbing, masonry, and facilities maintenance. The products and services offered through the Residential Construction Academy are the result of cooperative planning and rigorous joint efforts between industry and education. The program was originally conceived by the National Association of Home Builders (NAHB)—the premier association of more than 200,000 member groups in the residential construction industry—and its workforce development arm, the Home Builders Institute (HBI).

For the first time, construction professionals and educators created national skills standards for the construction trades. In the summer of 2001, NAHB, through the HBI, began the process of developing residential craft standards in six trades: carpentry, electrical wiring, HVAC, plumbing, masonry, and facilities maintenance. Groups of employers from across the country met with an independent research and measurement organization to begin the development of new craft training standards. Care was taken to assure the representation of builders and remodelers, residential and light commercial, custom single family and high production or volume builders. The guidelines from the National Skills Standards Board were followed in developing the new standards. In addition, the process met or exceeded American Psychological Association standards for occupational credentialing.

Next, through a partnership between HBI and Cengage/Delmar Learning, learning materials—textbooks, videos, and instructor's curriculum and teaching tools—were created to teach these standards effectively. A foundational tenet of this series is that students *learn* by doing. Integrated into this colorful, highly illustrated text are Procedure sections designed to help students apply information through hands-on, active application. A constant focus of the *Residential Construction Academy* series is teaching the skills needed to be successful in the construction industry and constantly applying the learning to real-world applications.

The newest programming component to the *Residential Construction Academy* series is the industry Program Credentialing and Certification for both instructors and students by the HBI. The National Instructor Certification ensures consistency in instructor teaching/training methodologies and knowledge competency when teaching to the industry's national skills standards. Student Certification is offered for each trade area of the *Residential Construction Academy* series in the form of rigorous testing. Student Certification is tied to a national database that will provide an opportunity for easy access for potential employers to verify skills and competencies. Instructor and Student certifications serve as the basis for Program Credentialing. For more information on HBI's Program Credentialing and Instructor and Student Certification, please go to **www.hbi.org/certification**.

About This Book

A facility maintenance technician is responsible for the day-to-day maintenance and operational tasks that support a commercial facility. Duties often include but are not limited to

- Responsibility for various activities related to the repair and maintenance of the electrical, plumbing, heating, and ventilation systems
- Painting and minor repair to walls, ceilings, and floors
- Preventive/predictive maintenance per requirements and nonscheduled or emergency maintenance when required to support operations
- Recommendations to modify or replace equipment when necessary to support demand, or improve building efficiency
- Snow removal and other groundskeeping duties as assigned
- Installation and repairs of locks
- Inspection and maintenance of lights and fire extinguishers.

Facilities Maintenance, Second Edition, provides coverage for the areas in residential wiring that are required of an entry-level facility maintenance technician, including the basic hands-on skills and the more advanced theoretical knowledge needed to gain job proficiency. In addition to electrical, other topics covered include customer service skills, carpentry, surface painting, plumbing, appliance repair, pest prevention, groundskeeping, and HVAC systems. The format of the text is designed to be easy to learn and easy to teach.

New to This Edition

Most of the chapters in the Second Edition have been updated to include advances and changes in the HBI standards for facility maintenance technicians. The following chapters have had major content additions and revisions:

Chapter 2—Methods of Organizing, Troubleshooting, and Problem Solving

Chapter 3 —Applied Safety Rules
- New content on first aid, frostbite, bleeding asphyxiation, and chemical burns

Chapter 4—Fasteners, Tools, and Equipment
- Expanded coverage of fasteners

Chapter 5—Practical Electrical Theory
- Additional coverage of the structure of matter
- New coverage of movement of electrons
- Expanded coverage of electrical circuits, Ohm's Law, and characteristics of circuits

Chapter 6—Electrical Facilities Maintenance
- More coverage of electrical safety, electrical tools and test equipment, and conductor and cable types
- New content on raceways

Chapter 7—Carpentry

- Expanded coverage of carpentry safety
- Additional material on repair and replacement of interior doors, door hardware and re-key, cabinets and shelving, suspended ceiling systems, windows and screens, floor and wall tiles and fixtures, including grouting and carpet

Chapter 8—Surface Treatments

- Expanded coverage of painting safety
- More content on specialty finishes, decorative finishing (faux), types of wall covering, wall covering preparation, applying wall covering, surface textures, types of textures, and applying surface textures

Chapter 9—Plumbing

- Expanded coverage of plumbing safety and plumbing tools and equipment

Chapter 10—Heating, Ventilation, and Air Conditioning Systems

- More coverage of HVAC safety and gas furnace troubleshooting and repair

Chapter 11—Appliance Repair and Replacement

- Expanded coverage of troubleshooting

Chapter 12—Trash Compactors

- Expanded coverage of troubleshooting

Chapter 13—Elevators

- Expanded coverage of routine elevator maintenance

Chapter 14—Pest Prevention

- Expanded coverage on nonpesticidal pest control and pesticide safety

Chapter 15—Landscaping and Groundskeeping

- Expanded coverage of landscaping safety, landscaping tools, and equipment, and landscaping terminology
- Added content on introduction to botany, soil amendment prior to installation, installing a retaining wall using landscaping timbers, installing a retaining wall using bricks and blocks, installing a retaining wall using castle rock and installing paving stones

Chapter 16—Basic Math for Facilities Maintenance Technicians

- New chapter

Chapter 17—Blueprint Reading for Facility Maintenance Technicians

- New chapter

Features

This innovative series was designed with input from educators and industry professionals and informed by the curriculum and training objectives established by the National Skills Standards Committee. The following features aid learning:

Learning features such as the **Objectives, Glossary,** and **Introduction** set the stage for the coming body of knowledge and help learners identify key concepts and information. These learning features serve as a road map for continuing through the chapter. Learners also may use them as an on-the-job reference.

From Experience boxes provide tricks of the trade and mentoring wisdom that make a particular task a little easier for the novice to accomplish.

Safety is featured throughout the text to instill safety as an attitude among learners. Safe job site practices by all workers is essential; if one person acts in an unsafe manner, all workers on the job are at risk of being injured. Learners will come to know and appreciate that adherence to safety practices requires a blend of ability, skill, and knowledge that should be continuously applied to all tasks they perform in the construction industry.

Caution boxes highlight safety issues and urgent safety reminders for the trade.

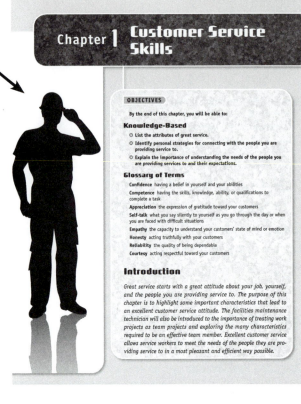

Chapter 1   Customer Service Skills

OBJECTIVES

By the end of this chapter, you will be able to:

Knowledge-Based
- List the attributes of great service.
- Identify personal strategies for connecting with the people you are providing service to.
- Explain the importance of understanding the needs of the people you are providing services to and their expectations.

Glossary of Terms

Confidence having a belief in yourself and your abilities
Competence having the skills, knowledge, ability, or qualifications to complete a task
Appreciation the expression of gratitude toward your customers
Self-talk what you say silently to yourself as you go through the day or when you are faced with difficult situations
Empathy the capacity to understand your customers' state of mind or emotion
Honesty acting truthfully with your customers
Reliability the quality of being dependable
Courtesy acting respectful toward your customers

Introduction

Great service starts with a great attitude about your job, yourself, and the people you are providing service to. The purpose of this chapter is to highlight some important characteristics that lead to an excellent customer service attitude. The facilities maintenance technician will also be introduced to the importance of treating work projects as team projects and exploring the many characteristics required to be an effective team member. Excellent customer service allows service workers to meet the needs of the people they are providing service to in a most pleasant and efficient way possible.

CHAPTER 7 *Carpentry* 161

FROM EXPERIENCE

Stretch the line tightly on nails inserted between the wall and the wall angle.

NAIL

STRETCHED LINE

WALL ANGLE

© Cut a number of hanger wires using wire cutters. The wires should be about 12 inches longer than the distance between the overhead construction and the stretched line. Attach the hanger wires to the hanger lags. Insert about 6 inches of the wire through the screw eye. Securely wrap the wire around itself three times. Pull on each wire to remove any kinks. Then make a 90° bend where it crosses the stretched line. If a laser is used, the 90° bend is done later when the main runner is installed.

JOIST

SUSPENDED CEILING LAG

TWIST HANGER WIRE AROUND ITSELF THREE TIMES

ALLOW 6" & BEND 90 DEGREES AT MAIN RUNNER LINE

STRETCHED MAIN RUNNER LINE

Procedures   Joining Plastic Pipe

- Mark the pipe at the appropriate point with a pencil.
- Ⓐ Cut the pipe using either a hacksaw or a tubing shear.
- Remove the burrs from both the inside and the outside of the pipe.
- Apply primer, if required, to both the male and female portions of the joint.
- Ⓑ Apply cement to both the male and female portions of the joint.
- Ⓒ Insert the male end of the fitting into the female end and rotate the pipe ¼ turn.
- Hold the pipe and fitting together for approximately 1 minute to prevent the pipe from pulling out of the fitting.

FROM EXPERIENCE

When working with plastic pipes, always try to dry-fit the piping arrangement before cementing. Once a joint is cemented, you don't get a second chance!

CAUTION

Follow all safety guidelines provided on the primer and cement containers. Plastic primers and cements should only be used in well-ventilated areas as the fumes from these chemicals are hazardous to your health.

Ⓐ
Photo by Bill Johnson.

Ⓑ
Photo by Bill Johnson.

Ⓒ
Photo by Bill Johnson.

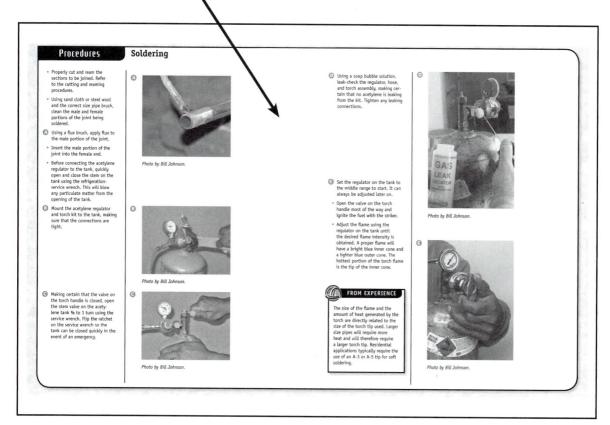

① What are the three states of water?

② What are the three principal parts of an atom?

③ State the law of charges.

④ What is a coulomb?

⑤ What is an amp?

⑥ What is electricity?

⑦ What is a watt?

CHAPTER 5 *Practical Electrical Theory* **97**

Name: _____

Date: _____

Practical
Electrical
Theory
Job Sheet

1

Practical Electrical Theory

Understanding Ohm's Law

Upon completion of this job sheet, you should have a basic understanding of Ohm's law.

① Determine the voltage of a load drawing 20 amps and having a resistance of 25 ohms.

② Determine the wattage of a device drawing 10 amps and having a resistance of 75 ohms.

Instructor's Response:

_____
_____
_____

Chapter review questions and **job sheets** enable the reader to assess the knowledge and skills obtained from reading the chapter.

Procedures provide step-by-step coverage of typical facilities maintenance tasks, explaining the work in detail.

Procedures   Soldering

- Properly cut and ream the sections to be joined. Refer to the cutting and reaming procedures.
- Using sand cloth or steel wool and the correct size pipe brush, clean the male and female portions of the joint being soldered.
- Ⓐ Using a flux brush, apply flux to the male portion of the joint.
- Insert the male portion of the joint into the female end.
- Before connecting the acetylene regulator to the tank, quickly open and close the stem on the tank using the refrigeration-service wrench. This will blow any particulate matter from the opening of the tank.
- Ⓑ Mount the acetylene regulator and torch kit to the tank, making sure that the connections are tight.

- Ⓒ Making certain that the valve on the torch handle is closed, open the stem valve on the acetylene tank ¼ to 1 turn using the service wrench. Flip the ratchet on the service wrench so the tank can be closed quickly in the event of an emergency.

Photo by Bill Johnson.

Photo by Bill Johnson.

Photo by Bill Johnson.

- Ⓓ Using a soap bubble solution, leak-check the regulator, hose, and torch assembly, making certain that no acetylene is leaking from the kit. Tighten any leaking connections.

- Ⓔ Set the regulator on the tank to the middle range to start. It can always be adjusted later on.
- Open the valve on the torch handle most of the way and ignite the fuel with the striker.
- Adjust the flame using the regulator on the tank until the desired flame intensity is obtained. A proper flame will have a bright blue inner cone and a lighter blue outer cone. The hottest portion of the torch flame is the tip of the inner cone.

Photo by Bill Johnson.

⌒ FROM EXPERIENCE

The size of the flame and the amount of heat generated by the torch are directly related to the size of the torch tip used. Larger size pipes will require more heat and will therefore require a larger torch tip. Residential applications typically require the use of an A-3 or A-5 tip for soft soldering.

Photo by Bill Johnson.

An end-of-book glossary containing definitions of terms is an important reference. In addition, key topic and skill sets are addressed in the following areas:

- Introduction to customer service skills, Chapter 1
- Introduction to task management, Chapter 2
- Introduction to OSHA safety in Chapter 3, Applied Safety Rules
- Basic concepts of electrical theory, Chapter 5

Turnkey Curriculum and Teaching Material Package

We understand that a text is only one part of a complete, turnkey educational program. We also understand that instructors want to spend their time teaching, not preparing to teach. The *Residential Construction Academy* series is committed to providing thorough curriculum and preparatory materials to aid instructors and alleviate some of their heavy preparation commitments. An integrated teaching solution is provided with the text, including the Instructor's e.resource™, a printed Instructor's Resource Guide, and Workbook.

Workbook

Designed to accompany *Residential Construction Academy Facilities Maintenance*, Second Edition, the Workbook is an extension of the core text and provides additional review questions and problems designed to challenge and reinforce the student's comprehension of the content presented in the core text.

e.resource

Delmar/Cengage Learning's e.resource is a complete guide to classroom management. The CD-ROM contains syllabi, lesson plans, chapter hints, answers to review questions, and other aids for instructors using this series. Designed as a complete and integrated package, e.resource also provides suggestions for when and how to use the accompanying PowerPoint presentations and test banks. An Instructor's Resource Guide is also available.

Features contained in the e.resource include

- Instructor Syllabus: goals, topics covered, reading materials, required text, lab materials, grading plan, and terms.
- Student Syllabus: goals, topics covered, required text, grading plans, and terms.
- Lesson Plans: goals, discussion topics, suggested reading, and suggested homework assignments. You have the option of using these lesson plans with your own course information.
- Chapter Hints: objectives and teaching hints that provide direction on how to present the material and coordinate the subject matter with student projects.
- Answers to Review Questions: solutions that enable you to grade and evaluate end-of-chapter tests and exercises.
- PowerPoint Presentations: provide the basis for a lecture outline that helps you present concepts and material. Key points and concepts can be graphically highlighted for student retention.

- Test Questions: over 150 questions of varying levels of difficulty are provided in true/false and multiple-choice formats. These questions can be used to assess student comprehension or can be made available to the student for self-evaluation.
- Correlation Grid: maps the book content to the HBI Skill Standards.

Online Companion

The Online Companion is an excellent supplement for students. It features many useful resources to support the core text. Linked from the student materials section of www.residentialacademy.com, the Online Companion includes chapter quizzes, an online glossary, product updates, related links, and more.

About the Author

Kevin Standiford, author, contributor, and consultant, has been in the technology fields of manufacturing processes, HVACR, process piping, and robotics for more than 20 years. While attending college to obtain his bachelor of science in mechanical engineering technology, he worked for McClelland Consulting Engineers as a mechanical designer, designing HVAC, complex processing piping, and cogeneration systems for commercial and industrial applications. During his college years, he became a student member of the American Society of Heating, Refrigerating and Air-Conditioning Engineers, where he developed and later wrote a paper on a computer application that enabled the user to simulate, design, and draw heating and cooling systems by using AutoCAD. The paper was entered into a student design competition and won the first prize for the region and state. After graduation, Kevin worked for Pettit and Pettit Consulting Engineers, one of the leading HVAC engineering firms in the state of Arkansas, as a mechanical design engineer. While working for Pettit and Pettit, Kevin designed and selected equipment for large commercial and government projects by using manual design techniques and computer simulations.

In addition to working at Pettit and Pettit, Kevin started teaching part-time evening engineering and design courses for Garland County Community College in Hot Springs, Arkansas.

Subsequently he stopped working full time in the engineering field and started teaching technology classes, which included heat transfer, duct design, and properties of air. It was also at this time that Kevin started writing textbooks for Cengage/Delmar Publishers. The first textbook was a descriptive geometry book, which included a section on sheet metal design. Today Kevin is a full-time consultant working for both the publishing and engineering industries, and a part-time instructor. In the publishing industry, Kevin has worked on numerous e-resource products, mapping, and custom publications for Delmar's HVAC, CAD, and plumbing titles.

Acknowledgments

The NAHB and HBI would like to thank the many individuals, members, and companies that participated in the creation of the Facilities Maintenance National Skill Standards. These standards helped guide us in the creation of this text. Thanks also go to Quantum Integrations, who helped create the original content, and Debbie Standiford for her work on Chapter 8.

In addition, we thank the following people who provided important feedback throughout the development of the book, enabling us to hone the content:

Jim Eichenlaub
Executive Director
Builders Association of Metropolitan Pittsburgh
Pittsburgh, PA

Michael Frank
Facilities Maintenance Instructor
Quentin M. Burdick Job Corps
Minot, ND

Kevin Fry
Facilities Maintenance Instructor
Excelsior Springs Job Corp
Excelsior Springs, MO

Mark Martin
Carpentry Instructor
Penobscot Job Corps
Bangor, ME

Daryl Martinez
Facilities Maintenance Instructor
Talking Leaves Job Corps Center
Tahlequah, OK

Shannon Pfeiffer
Program Administrator/Project Coordinator
Building Trades Academy
Chesapeake, VA

George Vick
Building Trades Instructor
Building Trades Academy
Chesapeake, VA

Chapter 1 Customer Service Skills

Introduction

Great service starts with a great attitude about your job, yourself, and the people you are providing service to. The purpose of this chapter is to highlight some important characteristics that lead to an excellent customer service attitude. The facilities maintenance technician will also be introduced to the importance of treating work projects as team projects and exploring the many characteristics required to be an effective team member. Excellent customer service allows service workers to meet the needs of the people they are providing service to in a most pleasant and efficient way possible.

1

An Excellent Customer Service Attitude

Attitude really is *everything* when providing service. It sets the stage for all other actions. A positive and service-oriented attitude can help you overcome many problems, dissatisfactions, or mistakes. It can also help in building long-term and loyal relationships with coworkers and customers.

Attitude is also a key factor in developing first impressions. It tells how you feel about yourself, your job, and the customers. It reveals confidence in doing an excellent job or reflects a sense of competence.

Finally, attitude is contagious, not only to other employees or contractors with whom you work, but also to the customers themselves. Remember that customers can have their own problems (including attitude problems) and you have an opportunity to improve the customer's day with your positive outlook.

Attitude constitutes several elements: *confidence, competence, appreciation, empathy, honesty, reliability, responsiveness, patience, open-mindedness,* and *courtesy*. If presented properly, these elements assure others that you are capable of doing what you say you can do. Let's look at each of these elements independently.

Confidence

Do you project confidence that you can do your job, solve a problem, or find the information necessary to achieve the customer's goals? What do the people you work with think? Do they have confidence in you that you can do the job?

A high level of **confidence** lowers the stress and anxiety in those around you. Lowered stress and anxiety leads to calm and rational thinking, better relationships, and improved patience, all of which are key ingredients to a pleasant and productive work environment.

Confidence is fostered by several behaviors:

- Know how to do your job well.
- Know that the services you provide and represent will meet the customer's needs and expectations.
- If you do not know how to do something or do not have sufficient knowledge, learn it and practice your skills often.
- Believe in your own abilities to complete the task, solve the problem, or find needed information.
- Control your self-talk so that you treat yourself (and others) in a positive way.

Competence

Are you skilled in the tasks you are asked to do? Do you have the appropriate knowledge? Are you efficient so that you can do the tasks quickly and with ease? If you have skills that need improvement, or do you know where to go for help? How will you improve your **competence** throughout your career? If you were asked to develop new skills, how would you go about doing this?

You must know how to perform well the tasks of your job. You must understand fully the services you are presented with. You must keep current and know where to find information quickly and how to effectively communicate that information to others. Being highly competent tends to increase confidence. The more you know about what you are doing, the better service you can offer your customers.

Appreciation

Do you appreciate your customers? How do you show your appreciation to others? Do you appreciate the skills and know-how you have developed in yourself?

Appreciation is a mindset established through self-talk. **Self-talk** is what you say silently to yourself as you go through the day or when you are faced with difficult situations. Self-talk includes all the things you are saying to yourself as you work with a customer and complete your tasks. What you are saying to yourself is always reflected in some type of behavior—a tone of voice, an action, or an easily perceived attitude—to your customers. You choose what you think about your customer.

> **Negative or condescending self-talk:** *She has no clue about all the things this system can do! What an idiot!*
> **Positive self-talk:** *This is a great opportunity to help her learn about all the things this system can do! Part of what she bought was the training; she is paying for it every bit as much as she paid for the TV. We owe it to her. She is going to be so glad she bought it.*

The kind of self-talk you choose is just that—your choice. Positive self-talk sets you up for a positive attitude followed by positive words and actions. Negative self-talk does just the opposite. It sets you up for a negative attitude followed by damaging words and ineffective actions.

Appreciation also includes how you feel about yourself and the efforts you have made to be successful. Self-talk applies here, too. With negative self-talk, you set yourself up for failure. Positive self-talk can provide just the right amount of confidence and motivation to succeed.

> **Negative self-talk:** *I have no clue how to fix this problem. I might as well give up now before I waste any more time.*
> **Positive self-talk:** *I haven't seen this problem before. This will be a good opportunity to test my skills. Let's see what I can find out.*

Once you make a habit of believing in yourself and your abilities, words and actions will follow accordingly. You will see that you actually find ways to succeed.

Empathy

Do you listen to your customers with empathy and truly want to help them meet their needs? How do you show empathy to those around you? What are the effects when you show empathy?

Empathy does not mean agreeing with everything your customer says. Nor does it mean promising to do everything your customer wants. **Empathy** does mean understanding what your customer's needs are and where they are coming from. Empathy is putting yourself in the other person's shoes, so to speak. Whether you agree with them, customers do have a right to express their opinions and points of view. They also have a right to be heard. Empathy is realizing this fact and acting and communicating accordingly.

The first step in empathy is to listen carefully to the customer without forming an opinion or making a judgment based on your own point of view. Taking turns to speak is not really listening. It is just being polite. Listening means that you do whatever you need to understand the customer's concerns, problems, questions, and points of view. Listening to a customer does two important things: (1) it gives you the information you need to solve the problem or meet the need and (2) it affirms the customer that he or she has been heard. This affirmation is critical and will go a long way toward building trust in you. Once the customer trusts you, you then have the opportunity to address the issue or complete your job successfully. You might be able to provide more information to help the customer fix the problem. You may change the customer's viewpoint by showing that the customer didn't really want what he or she thought in the first place. You may simplify what was initially perceived as a complicated issue. Basically, you must listen to the other person before you can expect him or her to listen to you.

Empathy is important in everyday dealings with customers, but it is especially important when the customer is frustrated, angry, or upset. When people are upset, their ability to think clearly or logically is often diminished. They often say irrational things, demand unreasonable responses, or behave inappropriately. Empathy diffuses emotion. When an angry customer experiences empathy from an employee, the typical response is to calm down and return to rational and reasonable thinking. With rational and reasonable thinking, most problems can be solved satisfactorily.

Some common phrases that show empathy include:

Wow! I would be mad too if the product broke in the first week.
This problem is definitely an inconvenience. Let's see how to fix this as quickly as possible.
I know it is frustrating when there is this much of a delay. Let me explain why our process is so important.

Be specific in your statements and try to avoid being general. Identify the emotion that you think the customer is feeling. Try to also identify the source of the emotion specifically.

A word of warning: If you try to show empathy without sincerity, you only make matters worse. Your customers are not stupid. They will see through your insincerity and feel patronized.

Honesty

Are you honest with your customers, and do your words and actions convey honesty? Do you strive to build trusting relationships with your customers? How specifically do you try to be honest with your customers? What are the consequences of dishonesty?

Honesty is an essential element to success whether you are the CEO or the lowest rung of the ladder. Also trusting relationships with customers foster customer loyalty and ongoing business.

Customers need to trust that you will be honest with them and that you are providing them with the right information. Dishonesty is quite easy to detect. It comes through clearly in words, tone, and body language, not to mention in customer dissatisfaction eventually when the dishonesty is found out—and dishonesty is *always* found out sooner or later.

To be honest, you must say the truth, follow through on promises, and state that you do not know the answer when appropriate. Customers will quickly determine that you are trustworthy and will continue to give you their business, often even when you cannot fix a problem or meet the immediate need. Also they will tell others about how they were treated.

You can get into "honesty" problems in many different ways. The following list reflects only a few examples that you might have experienced yourself.

- Promising that a product will meet a need when you know it won't
- Underestimating a wait time for service when you know it will be longer
- Stating that a product will be available on time when you know delivery has been delayed
- Underestimating the cost of a repair when you know that it will cost more

There are some basic tips that will help to ensure that the customer sees you as trustworthy:

- Overestimate wait times. If you serve the customer sooner than expected, the customer will be thrilled.
- Never promise you can deliver anything that you are not sure. If necessary, tell the customer you need to check your facts with others and get back to them (and then get back to them when you say you will).
- Don't tell a customer what he or she wants to hear, unless it is the truth. Instead, tell him the facts and work on solving any problems.
- Explain your actions thoroughly and in terms your customer can understand. Sometimes a customer may distrust you simply because he or she does not have enough information to know that you are doing what you are supposed to be doing or telling you accurate information.
- Present yourself as working on behalf of the customer. For example, address the customer's needs and offer information or make recommendations related to those needs.
- Provide advantages and disadvantages of a product or service option so that the customer can make informed decisions.

> *I have found that being honest is the best technique I can use. Right up front, tell people what you're trying to accomplish and what you're willing to sacrifice to accomplish it.*
> —Lee Iacocca

> *If you tell the truth you don't have to remember anything.*
> —Mark Twain

Reliability

How do you relate yourself in terms of reliability? Do you *always* do what you say you will do? Can customers depend on you and your products and services to meet their needs? Are you regularly late, or do you typically run on time? How important is being punctual to you?

Reliability is essential to your customers. Customers want services they can depend on. Basically, **reliability** means doing what you say you will do and when you will do it. If you have scheduled a service call for 9 AM, then show up at 9 AM (not before, unless you call to ensure it is convenient, and not after). Obviously, things happen that can be out of your control. In these cases, it is important to notify customers accordingly. It is also important to do everything in your power to control the situations. Good planning, accurate estimation of a project's time needs and travel time, and organization all help in this control.

Reliability is also reflected in availability, prompt replies, quick follow-up, and fast work. It implies accountability for actions and any potential problems. In order to be accountable, you must be available. Can customers contact you? Can your coworkers or office contact you easily? If a message must be left, do you respond quickly?

Keys to reliability are listed here. Add your own keys as they relate to your specific job.

- Ensure that customers and coworkers can contact you.
- If a message must be left, check your messages often and respond immediately.
- If you are out of touch for a specific time, let people know, so they won't be disappointed by your lack of response.
- Use an alarm to remind you of important meetings. (Some watches and cell phones have this feature.)
- Show up on time—not before or after the agreed-upon time, unless it has been approved ahead of time.
- Learn to accurately estimate how long it will take to complete a task, receive a part, schedule an appointment, etc. Don't guess; instead, wait until you have all the information before estimating time.
- Learn how to do the job right to begin with. Reliability also means being able to depend on the quality of work.

> *Quality means doing it right when no one is looking.*
> —*Henry Ford*

Responsiveness

Do you respond to your customers quickly, accurately, and with the goal of meeting their needs and answering their questions? How do you show your customers that you are responsive? How you respond to the customer typically determines how well your service is received, which ultimately translates to either customer satisfaction and loyalty or customer dissatisfaction.

There is nothing more frustrating than being passed from one person to another without getting what you need. Yet, this is often the experience of customers in many businesses. For good service, responsiveness is on the top of the list of key elements.

Patience

Do you have patience when dealing with the customer and/or solving a problem? People that are considered to be good natured or tolerance of delay or incompetence is considered to have patience. It is extremely important to exercise patience when dealing with a customer. Remember without the customer you would not have a job.

Open-mindedness

Do you have open-mindedness? People that have the ability to consider different opinions or ideals are considered to be open-minded. Being open minded does not, however, mean that all opinions or ideals are correct for a particular problem, but having the ability to listen to the customers ideals and opinions will help strengthen the technician/customer relationship.

Courtesy

What does it mean when someone is courteous? Are you courteous when you deal with others? How specifically do you demonstrate courtesy?

Courtesy sends a positive and powerful message to customers, whether they are your external paying customers or those internal customers with whom you work on a daily basis. Courtesy is also a habit that, once formed, becomes second nature.

Characteristics of a courteous employee are reflected in the following behaviors:

- Saying "Please," "Thank you," and "You're welcome"
- Responding with "Yes ma'am" or "No sir"
- Saying "I'm sorry" or "Excuse me"
- Addressing people by their names and using Mr., Ms., or Miss as appropriate (e.g., if you do not know them well)
- Saying "Yes" instead of "Yeah"
- Being friendly
- Smiling often
- Opening doors and allowing others to go through first
- Introducing yourself to new people
- Being attentive and focusing on the person in front of you without being distracted
- Using appropriate language

Courtesy also implies sincerity. Show that you sincerely appreciate your customers by thanking them for their business and being specific.

For example: *Thank you for buying our entertainment system. I know you are going to love it. We really appreciate your business.*

Remember your customers' names because you honestly feel that they are important enough to do so. Be sincere in all of your actions. Your sincerity—or lack of sincerity—will be obvious to your customers.

It is more natural for people to forget common courtesies when stressed—for example, when a customer is angry or when you are frustrated because of a problem situation. These are the times when you need to be exceptionally courteous. You should be courteous even when you feel that the person does not deserve it and when customers are not being courteous themselves. This is why developing the habit of being courteous is so important. If courtesy is a habit, you are less likely to forget about it when stressed.

Tone of Voice

Sometimes how you say something means as much as what you say. The wrong tone can cause a misinterpretation of your words. Saying "thank you" in an angry tone serves only to agitate your customer. Asking about the problem in a disinterested tone shows the customer that you are not sincere. Using sarcasm typically causes a customer to become angry and feel disrespected.

Combining a positive, friendly, and confident tone with positive and confident words such as "Absolutely," "Definitely," "Not a problem," and so forth can be very effective. Say the following phrases with a positive, friendly, and confident tone to get the point:

Absolutely. I can have this fixed in no time!
Definitely. I will order the part for you today.
Not a problem. I will reschedule the service call for that date.
Yes! I will be happy to move this for you.

Try to match your tone to the customer's needs. If the customer is in a hurry, then make your tone urgent and energetic. If the customer is frustrated, use a confident and helpful tone. If your customer is doubtful or has many questions, use a reassuring and confident tone.

Listening

There is a significant difference between listening and hearing. To listen means that you truly attempt to understand what the speaker is saying. Real listening is a highly active process. Without listening, there is no communication—but only speaking and hearing.

Effective listening does several things for the relationship:

- It shows that you sincerely care about the customer and the customer's needs.
- It demonstrates attentiveness to the customer.
- It allows you to gain critical information with which to complete your task successfully or to solve the problem.
- For frustrated customers, it reduces irritation by ensuring them that they are being heard.
- It fosters an effective and productive relationship between you and the customer.

Avoid the Words "I Can't"

Focus on what you can do rather than what you cannot do. If you cannot do exactly what the customer wants, explain what you *can* do for the customer that either comes close to the customer's request or meets the same need in a different way. Be a problem solver.

To connect with customers, develop your skills in the following strategies and then practice them consistently:

- Evaluate your body language and use it to convey the appropriate messages.
- Focus on how you say things and your tone of voice.
- Be attentive to customers.
- Develop effective listening skills and practice these with customers.
- Respond positively to customers.

Review Questions

Define the following attitudes:

1. Confidence

2. Competence

3. Appreciation

4. Empathy

5. Honesty

6. Reliability

7. Responsiveness

8. Courtesy

9. Open-mindedness

10. Patience

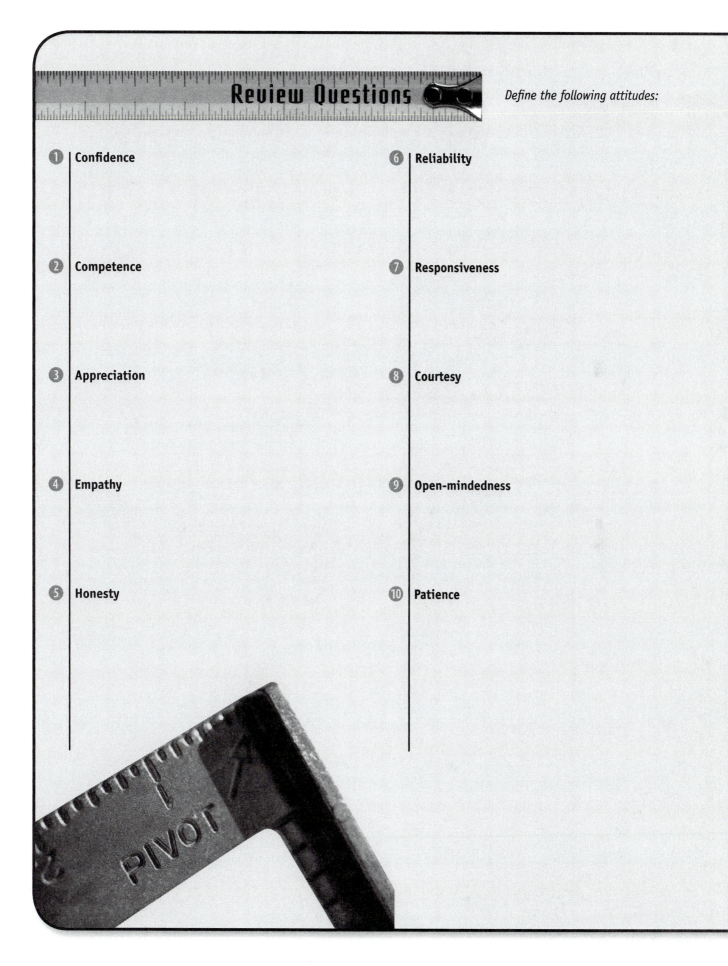

Name: _____

Date: _____

Customer Service

Customer Service Checklist

Reviewing this checklist will help you continuously improve your customer service skills and keep you focused on the only thing you can control—your own behavior.

❑ Put yourself in your client's shoes.

❑ Reserve judgment about your client and his or her problem and listen with an open mind.

❑ Listen attentively with genuine interest to everything your customer has to say.

❑ Ask questions to clarify your understanding.

❑ Tell the client what you can do and why.

❑ Be empathetic. This doesn't mean you agree with the person's feelings, but it does indicate you acknowledge him or her.

❑ Make commitments. Commitments guarantee that something will get done. It's also a way to manage the customer's expectations.

❑ Meet commitments; don't make a commitment just to get rid of a customer. Make a commitment you can keep.

Instructor's Response:

_____
_____
_____

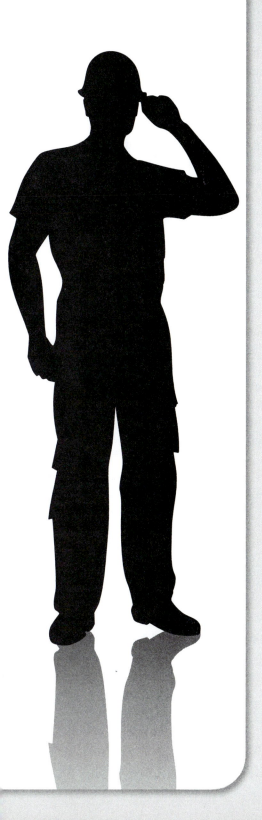

Chapter 2 | Methods of Organizing, Troubleshooting, and Problem Solving

OBJECTIVES

By the end of this chapter, you will be able to:

Skill-Based

- ✪ Establish priority of work tasks.
- ✪ Assign tasks.
- ✪ Carry out work order systems.
- ✪ Using the steps outlined in the text to properly trouble a technical issue.

Glossary of Terms

Task an activity that needs to be performed to complete a project

Priority giving a task precedence over others

Assigning tasks giving a task to someone to complete

Troubleshooting the process of performing a systematic search for a resolution to a technical problem

Diagnostics the process of determining a malfunction

Introduction

To successfully complete a project regardless of its complexity or nature, it must first be divided into tasks that can be assigned and the results measured. This is irrespective of whether it is a new installation or replacement or repair of a defective piece of equipment. The only difference is that if the project consists of replacing or repairing equipment then removal and/or troubleshooting will be included as a task to be performed.

Establish Priority of Work Tasks

Traditional wisdom about setting priorities promises you higher productivity and a greater sense of accomplishment. All you have to do is write out a to-do list, prioritize it by the order of importance and urgency (using the ubiquitous A, B, and C labels), and then tackle it, right? Then why, after a period time, is that same C item still on your to-do list? Also why, after a busy day of completing **tasks**, do you still find yourself saying, "I didn't get anything done today"? Two reasons may cause this:

1. Priorities changed during the day, but for good reason. You may not have accomplished A, B, or C on your to-do list, but you did respond appropriately to the additional tasks that you were presented with that day.

2. Don't fall into the "ACT, then THINK" method of setting priorities. To prevent yourself from falling into this method of setting **priorities**, understand three common priority-setting traps and how you can avoid them.

   a. *Whatever hits first*: Do you "choose" your priorities simply by responding to things as they happen? If so, your priorities are really choosing you. Think about how this general lack of control over your day contributes to your stress level. You need to clarify your priorities by determining each task's importance and level of urgency (i.e., "THINK, then ACT"). This means negotiating with people to respond in a time frame that's convenient to you and agreeable to them.

   b. *Path of least resistance*: When was the last time you heard yourself say, "It's just easier to do it myself"? This is not always an incorrect assumption, but if you're saying it too often, you're probably not giving the other workers enough credit or you have the wrong person working for you. Ask yourself these questions: Am I trying to avoid conflict? Does the task at hand require more expertise than the other workers have? Should time or money be invested to train someone to take on some of the lower-priority tasks I am currently performing? Answers will help you determine what alternative action you need to take.

   c. *Squeaky wheel*: In most situations, it is not hard to identify who the squeaky wheels are. Their requests are always urgent and need to be done right away. Usually, you do the work on their time frame. Unless the request is really urgent, give them a specific time or date when they can expect you to complete the task. Eventually they'll understand that their requests to complete tasks need to be prioritized with all of your own prioritized tasks.

Assign Tasks

Assigning tasks isn't just a matter of telling someone else what to do. There is a wide range of responsibilities that you can assign to a person along with a task. The more experienced and reliable the person is, the more unsupervised tasks that can be assigned to the person. The more critical the task is, the more cautious you need to be when assigning tasks. It is important that each worker understands his or her part in a job and can perform the assigned task. If a worker does not have the ability to complete an assigned task, as a supervisor or manager, you would have to assign that task to someone else who can do that.

Before assigning work to a worker, consider the following aspects of the job the worker will do:

- What hazards are in the workplace environment or around the worker?
- Are there special work situations that come up which could lead to new risks for this worker? For example, are there risks that might be encountered outside the normal work area? Just once a week? During a task to fetch materials?

- Are there occasional risks from coworkers, such as welding or machining, that could affect the workers nearby?
- In slow periods, workers might be asked to "help out" other employees. Ensure that any hazards associated with those jobs are reviewed with the worker, by both you and the coworker who will supervise those tasks.

Ensure that you communicate with the worker about the job tasks clearly and frequently, repeating and confirming this training over the first few weeks of work. Some workers are overwhelmed with instructions at first and may need to hear this information repeated more than once. Also:

- Inform workers not to perform any task until they have been properly trained.
- Inform workers that if they don't know or if they are unsure of something, they need to ask someone first. Get them to think in a safety-minded way about all their work.

Carry Out Work Order Systems

Once tasks have been assigned to the appropriate workers, the tasks will need to be completed. Following are suggestions on how to document the completed tasks:

1. Develop a progress report for the current week. This report should contain the following information:
   a. Work accomplished: Document the tasks accomplished during the previous week. They should be specific in nature.
   b. Major findings: Document any issues that were encountered when a specific task was dealt with.
   c. Worker: In the documentation, record the name of the person who completed the specific task.
   d. Estimated hours to complete: This is used to identify the worker's effectiveness in task estimation. The goal is to compare the predicted value (estimated the week before) with the actual number of hours it took for the team to accomplish a specific task.
   e. Actual hours to complete: Record the actual number of hours the worker took to complete a specific task.
2. Plan items for the following week. The report should contain the following information:
   a. Work items for next week: List all the tasks that the worker plans to accomplish for the following week.
   b. Worker: List the worker who is responsible for completing the listed task.
   c. Estimated hours to complete: For planning purposes, the estimated hours to completion should be identified to show how long it will take to actually complete the task.

The Troubleshooting Process

Some consider **troubleshooting** to be an art form; whereas, others consider it to be science. Actually it is a combination of art and science—trying to solve a problem using a pure scientific approach may not always work. In any case, some steps can be followed (especially by new technicians) to assist in the development of trouble-shooting skills. One of the most important of these steps—establishing a good rapport with the customer—was briefly discussed in Chapter 1. Often what the customer perceives as the ultimate problem is nothing more than a symptom. Often when the technician establishes a good rapport with the customer, he or she can easily get the actual information needed to determine the actual problem. However, as mentioned earlier, rapport is only a small portion of the troubleshooting process. Actually two major phases can be associated with the troubleshooting process: the identification and the repair processes. The identification process is more than just shining a flashlight into a boiler or heat exchanger and trying to spot a defect or a malfunctioning part. This process can be divided into several key phases or steps as follows:

1. Gathering information
2. Verifying the issue
3. Looking for quick fixes
4. Performing the appropriate **diagnostics** (the process of determining a malfunction)
5. Using additional resources to research the issue (if necessary)
6. Escalating the issue (if necessary)
7. Complete the repair process

Gathering Information

When starting to troubleshoot a system, the first step is to gather the information necessary to correctly identify the problem. This is often done by simply asking the customer a few simple questions. However, when questioning the customer, keep two general rules in mind. They are:

1. Start with open questions such as "What is the issue?" Open questions cannot be answered with a "yes" or "no."
2. Let the customer explain in his or her own words what he or she has experienced/is experiencing. *Never interrupt* a customer or add comments to what he or she is telling you.

Verifying the Issue

As stated earlier, the situation or problem that the customer describes is often not the actual problem. Therefore, always verify whether the problem described by the customer is the actual problem of the system and not just a symptom.

Looking for Quick Fixes

Although in many cases the actual fix is more involved than simply resetting or changing the battery in a thermostat, there are still cases in which the simplest and/or most obvious fix corrects the problem. For example, suppose that you were called to look at a customer's gas central heating unit because it would not ignite. If the customer is using an electronically controlled thermostat, it might be useful to check its battery before starting to break the furnace down.

Performing the Appropriate Diagnostics

If the quick fix does not resolve the issue, then you will need to perform a more thorough diagnostics. Often the equipment manufacturers will supply troubleshooting charts and information to help diagnose the equipment.

Using Additional Resources to Research the Issue

If you have never encountered a problem like the one that is currently before you, and you are having trouble locating the issue, don't be afraid to go to the Internet, a distributor, or even a fellow colleague to help resolve the issue.

Escalating the Issue

If you are working for a large company and continue to have trouble locating and correcting the problem, the issue can often be escalated for assistance to a service manager. If you are self-employed or working for a small company and you encounter a problem that you cannot resolve, the equipment manufacturer can often be of assistance. In any case, though, you should never escalate an issue unless you are truly stumped.

Complete the Repair Process

Once the issue has been correctly identified, the repair process can proceed. Like the identification process, the repair process also involves several steps. They are:

1. Repairing or replacing the faulty item and/or equipment
2. Testing the system thoroughly to verify that the repair actually corrected the issue
3. Educating the homeowner about the nature of the problem and the action(s) taken to correct it
4. Completing all administrative paperwork

Verifying that the Repair Actually Corrected the Issue

Verifying that the repair actually corrected the issue is one of the most critical steps in the repair process. Never leave a customer site without first testing the repair and/or installation to confirm that you actually corrected the problem.

Educating the Customer about the Nature of the Problem and Actions Taken to Correct It

Always show the customer the worn and replaced parts and explain to him or her why the old parts are defective. If a customer understands the issue and the corrective action taken, he or she is less likely to become dissatisfied with the repair job.

Completing All Administrative Paperwork

Completing the paperwork is especially important when dealing with warranty work. If the necessary paperwork is not completed correctly and on time, then there will be a delay in the service company receiving its payment. In some cases, the claim may even be denied.

Solving a Technical Problem

When solving a technical problem, the facility maintenance technician doesn't start by pulling out a calculator and entering numbers; there is a systemic approach that must be taken. This approach starts by defining and then researching the problem. Once sufficient information has been ascertained, the facility maintenance technician can start determining all possible solutions to the task. The steps required for successful problem solving are listed next and can be executed in order with the exception of steps 1, 2, 7, and 8.

Step 1. State the problem
Step 2. List unknown variables
Step 3. List what is given
Step 4. Create diagrams
Step 5. List all formulas
Step 6. List assumptions
Step 7. Perform all necessary calculations
Step 8. Check answers

The following example illustrates each step of this process. Note that not every step may apply to all situations; however, the overall concept is still the same. In this example you will calculate the amount of concrete necessary to pour a slab for the condenser shown in Figure 2-1. The cooling tower is 8 ft × 5 ft and requires a 1 ft × 6 in. overhang around the perimeter.

Step 1 State the problem

In this step a statement of the problem is created. The statement should be kept as simple and direct as possible. There is no need to list every detail about the problem, but just the key points. Additional information can be added later if necessary. A statement for the problem shown in Figure 2-1 might read:

Statement:
For a 8 ft × 5 ft condensing unit, calculate the amount of concrete required for a slab having a 1 ft × 6 in. overhang and 6 in. thick.

Step 2 List unknown variables

Create a list of all the unknown attributes for the stated problem. Even though the problem is in its infancy, some calculations might have to be performed to determine some of the missing information. Be sure to provide ample space for any future elements that might require calculations. Therefore, the list should look like:

Find:
Amount of concrete needed in cubic feet.
Amount of concrete needed in cubic yards.

Figure 2-1: **Air conditioning unit to be set**
(Courtesy Heatcraft, Inc., Refrigeration Products Division)

Step 3 List what is given

A list of all known parameters associated with the stated problem is next created:

Given:

Condensing unit width = 5 ft
Condensing unit length = 8 ft
Required overhang = 1 ft × 6 in.

Step 4 Create diagrams

Often the best way to determine exactly what is going on in a problem is to make a simple sketch. The sketch should be void of any unnecessary details that might hinder the interpretation of the actual problem (see Figure 2-2).

Step 5 List all formulas

In this step, a list is created of all the formulas that will be used to solve the problem, as well as a source reference for each formula. This list will serve as a guide that will facilitate checking the final answer. When a technical solution is validated, all stages of the solution are verified including formulas and their connotation. To solve the example problem the following formulas will be used:

Formulas:

Feet = inches/12
Volume = length × width × thickness (height)
Cubic yards = cubic feet/27

Step 6 List assumptions

Often when solving a technical problem it is necessary to assume some of the details of the project. Suppose that an engineer is calculating the amount of heat that is transferred via conduction through the exterior walls of a building to its surroundings. Although the average outside temperature can be obtained from the ASHRAE Fundamentals Handbook for all major cities in the United States, it may be necessary to assume a temperature if the area where the building is located is not listed. In the case of the example problem we do not have any assumptions. Therefore we would list N/A for our assumptions or leave it completely off.

Assumptions:

N/A

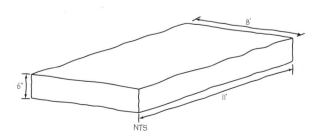

Figure 2-2: **Rough hand sketch of concrete pad**

Step 7 Perform all necessary calculations

The necessary calculations are carried out. Solving the example problem would yield the following calculations.

Converting the thickness from inches to feet

Feet = inches/12
Feet = 6 in./12
Feet = 0.5

Determining the volume of the slab (cubic feet)

$\text{Volume}_{\text{cubic feet}}$ = length × width × thickness
$\text{Volume}_{\text{cubic feet}}$ = 11 ft × 8 ft × 0.5 ft
$\text{Volume}_{\text{cubic feet}}$ = 44 cubic feet

Converting the volume from cubic feet into cubic yards

$\text{Volume}_{\text{cubic yards}}$ = $\text{volume}_{\text{cubic feet}}$/27
$\text{Volume}_{\text{cubic yards}}$ = $44_{\text{cubic feet}}$/27
$\text{Volume}_{\text{cubic yards}}$ = 1.63

Step 8 Check answers

Any time a calculation is made it must be checked for accuracy.

Check:

To check our work for this example problem we will work the problem in reverse.

Convert from cubic yards to cubic feet

$1.62_{\text{cubic yards}}$ × 27 = $\text{volume}_{\text{cubic feet}}$
44 = volume cubic feet

Finding the thickness of the slab

If we use two of the known dimensions of the slab, we should be able to find the remaining dimension. For example, if we divide the $\text{volume}_{\text{cubic feet}}$ by the length and then again by the width, the remaining portion should be the thickness of the slab.

First
Unknown = total cubic feet/length of slab
Unknown = 44 cubic feet/11 feet
Unknown = 4 square feet

Second
Thickness = unknown (from step #1)/width of slab
Thickness = 4 square feet/8 ft
Thickness = 0.5 ft

Or
Thickness (in.) = 0.5 ft × 12 in.
Thickness (in.) = 6 in.

Review Questions

1 Why is it important to assign priority to work tasks?

2 How is priority assigned to a task?

3 List four aspects that should be considered before a task is assigned.

4 Why is it important to estimate the amount of time to complete a task before starting it?

5 When the amount of time to complete a task spans across multiple days or weeks, why is it important to plan the event of consecutive days?

Name: _____

Date: _____

Assigning a Task

Assigning a Task Checklist

Completing this checklist will help you continuously improve your skills in assigning a task. As you are driving a project into task to be assigned and completed, use the comments section to record special notes and/or concerns related to that task.

Considerations	Yes/No	Comment
Are there hazards in the workplace environment or around the worker?		
Are there special work situations that come up, which could lead to new risks for this worker? For example: are there risks that might be encountered outside the normal work area? Just once a week? Can the materials needed to complete the project?		
Are there occasional risks from co-workers, such as welding or machining, which could affect the workers nearby?		
If hazards are associated with a task, have those hazards been reviewed with the worker?		

Instructor's Response:

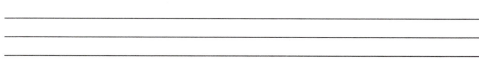

Chapter 3 Applied Safety Rules

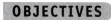

OBJECTIVES

By the end of this chapter, you will be able to:

Knowledge-Based
- ⊗ Explain the purpose of OSHA.
- ⊗ Explain the basic safety guidelines and rules for general workplace safety.
- ⊗ Explain the basic safety guidelines and rules for working with and around an electrical power tool and circuit.

Skill-Based
- ⊗ Create a basic fall protection plan.
- ⊗ Work safely with ladders and extension ladders.
- ⊗ Correctly identify and select the proper fire extinguisher for a particular application.

Glossary of Terms

Class A fire extinguishers fire extinguishers used on fires that result from burning wood, paper, or other ordinary combustibles

Class B fire extinguishers fire extinguishers used on fires that involve flammable liquids such as grease, gasoline, or oil

Class C fire extinguishers fire extinguishers used on electrically energized fires

Class D fire extinguishers fire extinguishers typically used on flammable metals

Ground Fault Circuit Interrupter (GFCI) electrical device designed to sense small current leaks to ground and de-energize the circuit before injury can result

Occupational Safety and Health Administration (OSHA) branch of the U.S. Department of Labor that strives to reduce injuries and deaths in the workplace

Asphyxiation loss of consciousness caused by a lack of oxygen or excessive carbon dioxide in the blood

Cardiopulmonary resuscitation (CPR) an emergency first aid procedure used to maintain circulation of blood to the brain

Frostnip The first stage of frostbite, which causes whitening of the skin, itching, tingling, and loss of feeling

Frostbite injury to the skin resulting from prolonged exposure to freezing temperatures

Personal protective equipment (PPE) any equipment that will provide personal protection from a possible injury

Introduction

The Occupational Safety and Health Act (OSHA) of 1970 was passed by Congress "to assure so far as possible every working man and woman in the Nation safe and healthful working conditions and to preserve our human resources." Under the act, OSHA was established within the Department of Labor and was authorized to regulate health and safety conditions for all employers with few exceptions. This chapter is designed to provide the facilities technicians with the knowledge to ensure safety for themselves and their coworkers when performing maintenance duties at the facilities where they work.

Purpose of OSHA

OSHA was created to:

- Encourage employers and employees to reduce workplace hazards and implement new or improve existing safety and health standards.
- Provide for research in occupational safety and health and develop innovative ways of dealing with occupational safety and health problems.
- Establish "separate but dependent responsibilities and rights" for employers and employees for achieving better safety and health conditions.
- Maintain a reporting and recordkeeping system to monitor job-related injuries and illnesses.
- Establish training programs to increase the number and competence of occupational safety and health personnel.
- Develop mandatory job safety and health standards and enforce them effectively.

Basic Fall Protection Safety Procedures

Maintaining written fall protection procedures protects not only workers from falls but also management from charges of incompetence. Having individual workers or supervisors decide as to when fall protection is required and what kinds of fall protection equipment to use is an acceptable practice only where workers are routinely exposed to simple hazards, such as homebuilders on a roof. However, when workers are involved with lots of nonroutine jobs, such as removing a branch from a roof, safety is enhanced if management puts in writing the fall protection and rescue procedures that employees are required to use.

The written plan must describe how workers will be protected when working 10 feet or more above the ground, other work surfaces, or water.

The plan should:

1. Identify all fall hazards in the work area.
2. Describe the method of fall arrest or fall restraint to be provided.
3. Outline the correct procedures for assembly, maintenance, inspection, and disassembly of the fall protection system to be used.
4. Explain the method of providing overhead protection for workers who may be in or pass through the area below the work site.
5. Communicate the method for prompt and safe removal of injured workers.

Before a fall protection plan can be developed, understand two important definitions:

- **Fall arrest system**—equipment that protects someone from falling more than 6 feet or from striking a lower object in the event of a fall, whichever distance is less. This equipment includes approved full-body harnesses and lanyards properly secured to anchorage points or to lifelines, safety nets, or catch platforms.
- **Fall restraint system**—apparatus that keeps a person from reaching a fall point; for example, it allows someone to work up to the edge of a roof but not fall. This equipment includes standard guardrails, a warning line system, a warning line and monitor system, and approved safety belts (or harnesses) and lanyards attached to secure anchorage points.

Developing a Fall Protection Work Plan

To develop a fall protection work plan, you must identify the responsibilities of your company and the work areas to which the plan applies. This information should be listed as the first item in your plan. After listing your company responsibilities and the work areas in the plan:

1. Identify all fall hazards in the work area. To determine fall hazards, you must review all jobs and tasks to be done. After all fall hazards have been identified, list the employees required to work 10 feet or more above the ground, other work surface, or water.
2. Determine the method of fall arrest or fall restraint to be provided for each job and task that is to be done 10 feet or more above the ground, another work surface, or water.
3. Describe the procedures for assembly, maintenance, inspection, and disassembly of the fall protection system to be used.
4. Describe the correct procedures for handling, storage, and security of tools and materials.
5. Describe the method of providing overhead protection for workers who may be in or pass through the area below the work site.
6. Describe the method for prompt and safe removal of injured workers.

7. Identify where a copy of this plan has to be posted.

8. Train and instruct all personnel in all of these items.

9. Keep a record of employee training and maintain it on the job.

First Aid

No matter how careful the service technicians are, accidents and mishaps do happen, which require immediate medical attention. Although not all injuries require a visit to the doctor or hospital, some treatment is at least necessary to prevent further injury or infection. All service vehicles should be equipped with a first aid kit, which contains the basic medical supplies, such as burn cream, bandages, alcohol pads, eye wash, eye pads, tweezers, antiseptic spray, gauze bandages, and **Cardiopulmonary Resuscitation (CPR)** face shields. Note that the following sections are not intended to provide medical advice, but to provide basic information regarding immediate treatment for a number of situations commonly encountered in the field.

Bleeding

If a cut results in bleeding, place a clean folded cloth over the area and apply firm pressure. If blood soaks through the cloth, do not remove it. Simply cover the cloth with another and continue to apply pressure until the bleeding stops. If at all possible, elevate the cut area to a level above the heart to help stop the bleeding. If the cut is relatively small, the injury can be washed with soap and warm water and then bandaged.

Asphyxiation

Asphyxiation is loss of consciousness caused by a lack of oxygen or excessive carbon dioxide in the blood. An oxygen level below 19 percent may result in unconsciousness. As a result of electric shock or inhalation of refrigerant, the victim may stop breathing. When a victim's respiratory system fails, the flow of oxygen through the body may stop within a matter of minutes. If the victim stops breathing, (CPR) should be administered. CPR is an emergency first aid procedure used to maintain circulation of blood to the brain.

Exposure

Exposure results when a refrigerant comes in contact with the skin. Frostbite can occur any time a technician is exposed to prolonged freezing temperatures. The two main stages of exposure are **frostnip** and **frostbite**. The first stage of exposure, called frostnip, causes whitening of the skin, itching, tingling, and loss of feeling. In the final stage, called frostbite, the skin turns purple and blisters are formed on the skin. In rare situations the exposure can result in gangrene and requires amputation of the affected area. It can be treated by covering the area with something warm and dry and then obtaining professional medical attention. Never rub, massage, poke, or squeeze the affected area as this can result in tissue damage. A warm bottle of water can be placed gently against the affected area to warm it slightly.

Environmental Protection Agency (EPA) and Department of Transportation (DOT) Hazardous Materials Safety Procedures

Hazardous materials or chemicals are those substances regulated by federal, state, and local laws, regulations, and ordinances.

Safety Procedures

When dealing with hazardous materials or chemicals, be sure to follow these general guidelines:

- Make sure that the names on container labels match the substance names on the corresponding material safety data sheets (MSDSs). If a label is missing or the MSDS is unavailable, notify your supervisor; do not use the chemical until the correct MSDS is obtained. Never remove a manufacturer-affixed label from any container.
- Be familiar with the hazards associated with the chemicals intended to be used and ensure that all required hazard controls are in place.
- Handle and store hazardous materials only in the areas designated by your supervisor.
- Use an appropriate fume hood or other containment device for procedures that involve the generation of aerosols, gases, or vapors containing hazardous substances.
- When working with materials of high or unknown toxicity, remain in visual and auditory contact with a second person who understands the work being performed and all pertinent emergency procedures.
- Avoid skin contact by wearing gloves, long sleeves, and other protective apparel as appropriate. Upon leaving the work area, remove any protective apparel; place it in an appropriate labeled container; and thoroughly wash your hands, forearms, face, and neck.
- Be prepared for accidents and spills. If a major spill occurs, evacuate the area and dial 911.

Electrical Safety Procedures

Electrical accidents can occur when electricity is present in faulty wiring and equipment or when poor work practices are followed. Accidents involving electricity can lead to burns and tissue damage and, in some cases, cardiac arrest and death when the body forms part of the electric circuit. Electric shock can be unsettling to the victim even if there is no apparent injury.

Other possible consequences of electrical accidents are fire and explosion (as sparking can be a source of ignition) and damage to equipment. Many of the accidents can be traced back to faults such as frayed or broken insulation or practices such as inappropriate work on live equipment.

General Safety Precautions

- Never work on "hot" or energized equipment unless it is necessary to conduct equipment troubleshooting.
- Do not connect too many pieces of equipment to the same circuit or outlet, as the circuit or outlet could become overloaded.
- Be sure that **ground fault circuit interrupters (GFCIs)** are used in high-risk areas such as wet locations. (GFCIs are electrical devices designed to sense small current leaks to ground and de-energize the circuit within as little as 1/40 of a second before injury can result.)
- Test the meter on a known live circuit to make sure that it is operating.
- Test the circuit that is to become the de-energized circuit with the meter.
- Inspect all equipment periodically for defects or damage.
- Replace all cords that are worn, frayed, abraded, corroded, or otherwise damaged.
- Always follow the manufacturer's instructions for use and maintenance of all electrical tools and appliances.
- Keep equipment operating instructions on file.
- Always unplug electrical appliances before attempting any repair or maintenance.
- All electrical equipment used on campus should be UL or FM approved.
- Keep cords out of the way of foot traffic so that they don't become tripping hazards or be damaged by traffic.
- Never use electrical equipment in wet areas or run cords across wet floors.

Safety and Maintenance Procedures for Power Tools and Cords

Hand and power tools are a common part of our everyday lives and are present in nearly every industry. These tools help us easily perform tasks that would otherwise be difficult or impossible. However, these simple tools can be hazardous and have the potential for causing severe injuries when used or maintained improperly. Special attention toward hand and power tool safety is necessary to reduce or eliminate these hazards.

- Do not use electric-powered tools in damp or wet locations.
- Keep guards in place, in working order, and properly adjusted. Safety guards must never be removed while using the tool.
- Avoid accidental starting. Do not hold a finger on the switch button while carrying a power tool.
- Safety switches must be kept in working order and must not be modified. If you feel it necessary to modify a safety switch for a job you're doing, use another tool.
- Work areas should have adequate lighting and be free of clutter.
- Observers should remain a safe distance away from the work area.
- Be sure to keep good footing and maintain good balance.
- Do not wear loose clothing, ties, or jewelry when operating tools.
- Wear appropriate gloves and footwear while using tools.

Ladder Safety and Maintenance Procedures

Ladders can be divided into two main types: straight and step. Straight ladders are constructed by placing rungs between two parallel rails. They generally contain safety feet on one end that help prevent the ladder from slipping (see Figure 3-1).

Figure 3-1: **Straight ladder**

Step ladders are self-supporting, constructed of two sections hinged at the top. The front section has two tails and steps; the rear portion has two rails and braces (see Figure 3-2).

Safe Ladder Placement

- Ladders, including step ladders, shall be placed in such a way that each side rail (or stile) is on a level and firm footing and that the ladder is rigid, stable, and secure.
- The side rails (or stiles) shall not be supported by boxes, loose bricks, or other loose packing.
- No ladder shall be placed in front of a door opening toward the ladder unless the door is fastened open, locked, or guarded.
- According to OSHA Standard CFR 1926.1053(b)(5)(i) states that "Non-self-supporting ladders shall be used at an angle such that the horizontal distance from the top support to the foot of the ladder approximately one-quarter of the working length of the ladder (the distance along the ladder between the foot and the top support)." In other words, for every four feet of working length the ladder is extended upward, the base must be moved out 1 foot (as indicated in Figure 3-3).

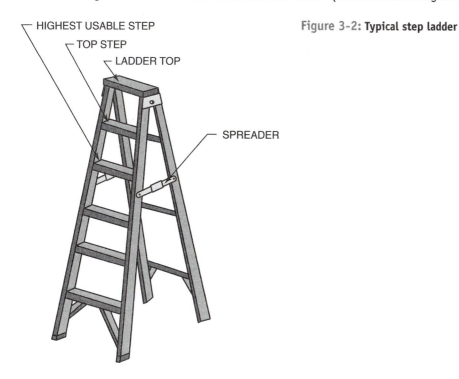

HIGHEST USABLE STEP

TOP STEP

LADDER TOP

SPREADER

Figure 3-2: **Typical step ladder**

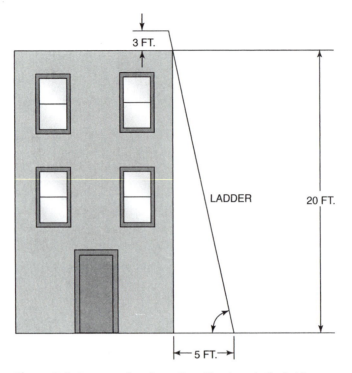

Figure 3-3: For every four feet of working length the ladder is extended upward, the base must be moved out 1 foot

- Where a ladder passes through an opening in the floor of a landing place, the opening shall be as small as is reasonably practicable.
- A ladder placed in such a way that its top end rests against a window frame shall have a board fixed to its top end. The size and position of this board shall ensure that the load to be carried by the ladder is evenly distributed over the window frame.

Safely Securing Ladders

- Ladder shall be securely fixed at the top and foot so that it cannot move either from its top or from its bottom points of rest. If this is not possible, then it shall be securely fixed at the base. If this is also not possible, then a person should stand at the base of the ladder and secure it manually against slipping.
- Ladders set up in public thoroughfares or other places (where there is potential for accidental collision with them) must be provided with effective means to prevent the displacement of the ladders due to collisions, for example, use of barricades.

Safe Use of Ladders

- Only one person at a time may use or work from a single ladder.
- Always face the ladder when ascending or descending it.
- Carry tools in a tool belt, pouch, or holster, not in your hands, so you can keep hold of the ladder.
- Wear fully enclosed slip-resistant footwear when using the ladder.
- Do not climb higher than the third rung from the top of the ladder.
- Do not use ladders made by fastening cleats across a single rail or stile.
- When there is significant traffic on ladders used for building work, separate ladders for ascent and descent shall be provided, designated, and used.

- Make sure the weight your ladder is supporting does not exceed its maximum load rating (user and materials). There should be only one person on the ladder at one time.
- Use a ladder that has proper length for the job. Proper length is a minimum of 3 feet extending over the roofline or working surface. The three top rungs of a straight, single, or extension ladder should not be stood on.
- Set up straight, single, or extension ladders for every four feet of working length the ladder is extended upward, the base must be moved out 1 foot.
- Metal ladders will conduct electricity. Use a wooden or fiberglass ladder in the vicinity of power lines or electrical equipment. Do not let a ladder, made from any material, contact live electric wires.
- Be sure all locks on extension ladders are properly engaged.
- Make sure that the ground under the ladder should be level and firm. Large flat wooden boards braced under the ladder can level a ladder on uneven or soft ground. A good practice is to have a helper hold the bottom of the ladder.
- Don't stand on the two top rungs of a step ladder.
- Follow the instruction labels on ladders.

Figures 3-4 and 3-5 illustrate these safety guidelines.

Appropriate Personal Protective Equipment

Personal protective equipment (PPE) is defined as all equipment, including clothing for shielding against weather, intended to be worn or held by people at work and

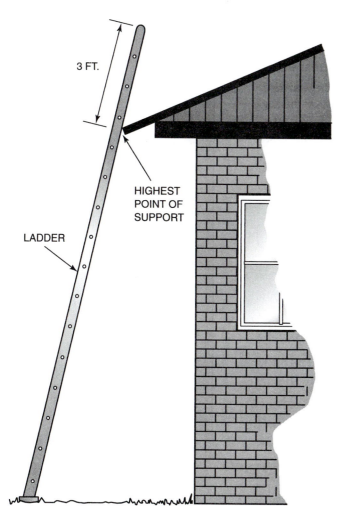

3 FT.

HIGHEST
POINT OF
SUPPORT

LADDER

Figure 3-4: **The ladders should extend at least 3 feet above the top support**

SAFETY PRACTICES FOR STEP LADDERS

CORRECT   INCORRECT

Figure 3-5: **Safe practices for using step ladders**

that protects them against one or more risks to their health or safety. This equipment includes, but is not limited to, the following:

- Helmets
- Gloves
- Eye protection
- High-visibility clothing
- Safety footwear
- Safety harnesses

To choose the right type of PPE, carefully consider the different hazards in the workplace. This will enable you to assess which types of PPE are suitable to protect against the hazard and for the job to be done. Figures 3-6 through 3-11 show examples of PPE.

Figure 3-6: **Typical electrician's hard hat with attached safety goggles**

Figure 3-7: **Safety glasses provide side protection**

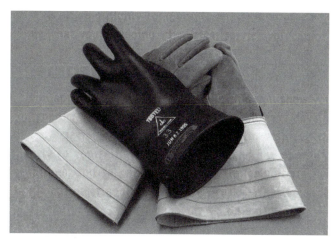

Figure 3-8: **Leather gloves with rubber inserts**

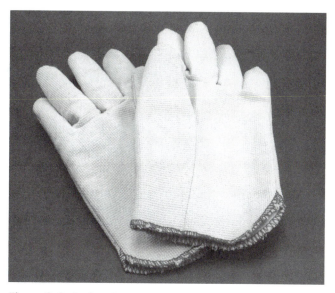

Figure 3-9: **Kevlar gloves protect against cuts**

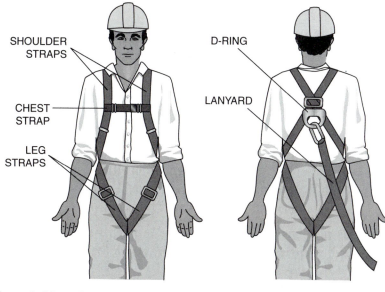

SHOULDER STRAPS

CHEST STRAP

LEG STRAPS

D-RING

LANYARD

Figure 3-10: **Typical safety harness**

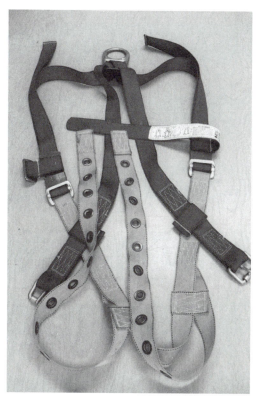

Figure 3-11: **Safety harness**

Safe Methods for Lifting and Moving Materials and Equipment to Prevent Personal Injury and Property Damage

General safety principles can help reduce workplace accidents. These include work practices, ergonomic principles, and training and education. Whether moving materials manually or mechanically, employees should be aware of the potential hazards associated with the task at hand and know how to exercise control over their workplaces to minimize the danger.

Proper methods of lifting and handling protect against injury and make work easier. You need to "think" about what you are going to do before bending to pick up an object. Over time, safe lifting technique should become a habit.

Learn the correct way to lift: Get solid footing, stand close to the load, bend your knees, and lift with your legs, not your back (see Figure 3-12).

1
APPROACH THE LOAD AND SIZE IT UP AS TO WEIGHT, SIZE, AND SHAPE. CONSIDER YOUR PHYSICAL ABILITY TO HANDLE THE LOAD.

2
PLACE FEET CLOSE TO THE OBJECT TO BE LIFTED AND 8 TO 12 INCHES APART FOR GOOD BALANCE.

3
BEND THE KNEES TO THE DEGREE THAT IS COMFORTABLE AND GET A HANDHOLD. THEN USING BOTH LEG AND BACK MUSCLES . . .

4
LIFT THE LOAD STRAIGHT UP, SMOOTHLY AND EVENLY. PUSH WITH YOUR LEGS AND KEEP THE LOAD CLOSE TO YOUR BODY.

5
LIFT THE OBJECT INTO CARRYING POSITION, MAKING NO TURNING OR TWISTING MOVEMENTS UNTIL THE LIFT IS COMPLETED.

6
TURN YOUR BODY WITH CHANGES OF FOOT POSITION AFTER LOOKING OVER YOUR PATH OF TRAVEL, MAKING SURE IT IS CLEAR.

7
SETTING THE LOAD DOWN IS JUST AS IMPORTANT AS PICKING IT UP. USING LEG AND BACK MUSCLES, COMFORTABLY LOWER LOAD BY BENDING YOUR KNEES. WHEN LOAD IS SECURELY POSITIONED, RELEASE YOUR GRIP.

Figure 3-12: **How to lift safely**

Procedures to Prevent and Respond to Fires and Other Hazards

For a fire to burn, three things are needed: fuel, heat, and oxygen. Fuel is anything that can burn, including materials such as wood, paper, cloth, combustible dusts, and even some metals. Fires are divided into four classes: A, B, C, and D (see Figure 3-13).

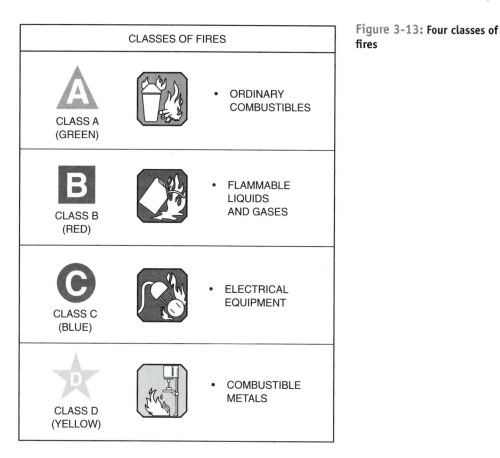

Figure 3-13: **Four classes of fires**

Class A Fires

This class involves common combustible materials such as wood or paper. **Class A fire extinguishers** often use water to extinguish a fire (see Figure 3-14).

Class B Fires

This class involves fuels such as grease, combustible liquids, or gases. **Class B fire extinguishers** generally employ carbon dioxide (CO_2).

Class C Fires

This class involves energized electrical equipment. **Class C fire extinguishers** usually use a dry powder to smother the fire.

Class D Fires

This class consists of burning metal. **Class D extinguishers** place a powder on top of the burning metal that forms a crust to cut off the oxygen supply to the metal.

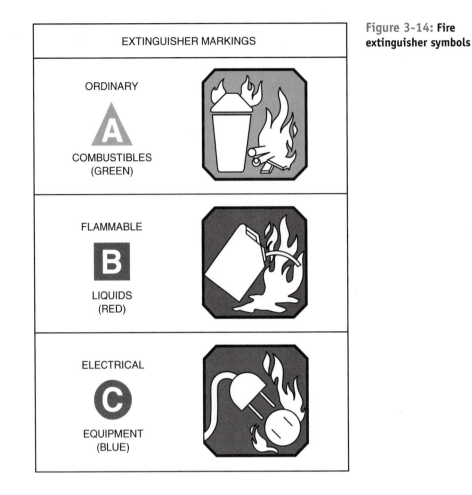

Figure 3-14: **Fire extinguisher symbols**

Prevent a Fire from Starting in Your Home

The most common causes of residential fires are careless cooking and faulty heating equipment. When cooking, never leave food on a stove or in an oven unattended. Avoid wearing clothes with long, loose-fitting sleeves. Have your heating system checked annually, and follow manufacturer's instructions when using portable heaters.

Smoking is the leading cause of fire deaths and the second-most common cause of residential fires. If you are a smoker, do not smoke in bed, never leave burning cigarettes unattended, do not empty smoldering ashes in the trash, and keep ashtrays away from upholstered furniture and curtains. In addition, keep matches and lighters away from children. Safely store flammable substances used throughout the home. And finally, never leave burning candles unattended.

Procedure to Prevent Uncontrolled Chemical Reactions

A chemical reactivity hazard is a situation with the potential for an uncontrolled chemical reaction that can result directly or indirectly in serious harm to people, property, or the environment.

To maintain a safe and healthful working environment, the Department of Energy recommends the following practices wherever chemicals are stored. These practices are based on regulations, rules, and guidelines designed to reduce or eliminate hazardous incidents associated with the improper storage of chemicals.

1. Adhere to the manufacturer's recommendations for each chemical stored, noting any precautions on the label.
2. Label all chemicals. The name and address of the manufacturer or other responsible party must be listed on the label. Chemicals with a shelf life should be labeled with the date received.
3. Store chemicals in the locations recommended (i.e., where the temperature range, vibration, or amount of light does not exceed the manufacturer's recommendations).
4. Inspect annually all chemicals in stock and storage. Hazardous chemicals should be inspected every six months. Some hazardous chemicals may require more frequent inspections. Any outdated materials should be properly disposed of or replaced if necessary.
5. Ensure that provisions are made for liaison with local planning committees, the state emergency planning commission, and local fire departments in the event of a chemical emergency.
6. Keep only enough inventory necessary for uninterrupted operation. Chemical inventory should be maintained at a minimum to reduce fire, exposure, and disposal hazards.
7. Rotate new shipments of chemicals with existing stock so that the oldest stock is available first.

Review Questions

1 What is the most common cause of residential fires?

2 What is the leading cause of fire deaths in residential buildings?

3 Define the following:

Class A fires
Class B fires
Class C fires
Class D fires

4 Define the term *personal protective equipment (PPE)*.

5 True or false? Ladders shall be securely fixed at the top and foot in such a way that they cannot move either from their top or from their bottom points of rest.

6 True or false? If it is not possible to secure a ladder at both the top and the bottom, then it shall be securely fixed at the top.

7 What is the purpose of OSHA?

8 True or false? To develop a fall protection work plan, you must identify the responsibilities of your company and the work areas to which the plan applies.

9 List two items that should be addressed in a fall protection plan.

10 True or false? Always follow the manufacturer's instructions for use and maintenance of all electrical tools and appliances.

Name: _____

Date: _____

Applied Safety Rules

Workplace Safety

Upon completion of this job sheet, you should be able to demonstrate your awareness of work area safety items. As you survey your work area and answer the following questions, you should learn how to evaluate the safety of your area.

Evaluate your work area and how you fit into it.

1 Are you properly dressed for work?
a. If yes, describe how you are dressed.

_____

_____

b. If no, explain why you are not properly dressed.

_____

_____

2 Are your safety glasses OSHA approved? **YES NO**
a. Do they have side shields? **YES NO**

3 Carefully inspect your work area, note any potential hazards.

4 Where are tools stored at your facility?
a. Are they clean and neatly stored? **YES NO**

5 Explain how you could improve tool storage.

6 Where is the first aid kit at your facility?

Instructor's Response:

_____

_____

_____

Name: _____

Date: _____

Applied Safety Rules

Identifying and Handling Hazardous Materials

Upon completion of this job sheet, you should be able to demonstrate your ability to identify hazardous materials and explain how to handle them.

1 Inspect your facility. Identify and list all hazardous materials found.
   a. Solvents

   _____

   _____

   b. Gasoline

   _____

   _____

   c. Cleaners

   _____

   _____

   d. Others

   _____

   _____

2 Check the containers in which hazardous materials are stored. Are they clearly marked? **YES NO**

3 Check to see if your facility has a material safety data sheet (MSDS) file. Is it located near the hazardous materials? **YES NO**

4 Make sure your facility has an MSDS list posted on a bulletin board where everyone can read it.

5 Read the MSDS bulletins on each of the materials you find at the facility and explain to the instructor how you would handle a spill of each material.

Instructor's Response:

   _____

   _____

   _____

Name: _____

Date: _____

Applied Safety Rules

Basic Fall Protection Safety

Upon completion of this job sheet, you should be able to demonstrate your understanding of basic fall protection safety procedures.

1. Does your facility have written fall protection and rescue procedures?
   a. If yes, are the procedures readily accessible to the workers?

   _____

   _____

   b. If no, explain why this should be brought to the attention of your supervisor.

   _____

   _____

2. When working higher than 6 feet off the ground, do workers use a fall arrest system?
   a. If no, should this be brought to the attention of your supervisor?

   _____

   _____

3. Identify all fall hazards at your facility.

4. Explain the method of providing overhead protection for workers who may be in or pass through the area below a work site.

   _____

   _____

Instructor's Response:

   _____

   _____

   _____

Name: _____

Date: _____

Applied Safety Rules

Ladder Safety and Maintenance Procedures

Upon completion of this job sheet, you should be able to demonstrate your understanding of ladder safety and maintenance procedures.

1 Name the two main types of ladders and explain the difference between the two.

_____

_____

2 Explain why it is important for straight, single, or extension ladders to be set up for every four feet of working length the ladder is extended upward, the base must be moved out 1 foot.

_____

_____

3 Explain why it is important for the ground under the ladder be level and firm.

_____

_____

4 List the ladders at your facility and write down the weight limitations for each.

_____

_____

Instructor's Response:

_____

_____

_____

Chapter 4 Fasteners, Tools, and Equipment

OBJECTIVES

By the end of this chapter, you will be able to:

Knowledge-Based

- Describe the safe use of tools, including power tools used by facilities maintenance technicians.
- Describe the proper anchors, fasteners, and adhesives necessary for a specific project.

Skill-Based

- Select, and install the proper anchors, fasteners, and adhesives necessary for a specific project.
- Select and properly use the appropriate hand tool for a specific project.
- Select and properly use the appropriate power or stationary tool for a specific project.

Glossary of Terms

Box nail a thin nail with a head, usually coated with a material to increase its holding power

Finishing nail a thin nail with a small head designed for setting below the surface of finished materials

Duplex nail a double-headed nail used for temporary fastening such as in the construction of wood scaffolds

Brad a thin, short, finishing nail

Power tool a tool that contains a motor.

Screws used when stronger joining power is needed

Caulk used to fill outside wall and foundation cracks

Introduction

Choosing the appropriate tool or equipment for a project is an important element in maintaining any facility. Many types of tools are available for handling many of the problems that may occur. There are times when using tools is not the appropriate method for repair, but other equipment may be necessary. This chapter introduces a variety of fasteners, solvents, and adhesives.

Fasteners

A fastener is a mechanical device that is used to mechanically join two or more mating surfaces or objects. Fasteners are now on the market for just about anything. They can be used where wood meets wood, concrete, or brick, and most are approved by the Uniform Building Code requirements. However, you should always consult your local building code before selecting a particular type of fastener to incorporate.

Nails

There is a huge difference in nails; some are specialty nails.

- **Common nails**—most often used nails. Used for most applications in which the special features of the other nail types are not needed (see Figure 4-1)
- **Box nails**—used for boxes and crates
- **Finishing nails**—can be driven below the surface of the wood and concealed with putty so that it is completely hidden
- **Casing nails**—used for installing exterior doors and windows
- **Duplex nails**—used for temporary structures, such as locally built scaffolds
- **Roofing nails**—used for installing asphalt and fiberglass roofing shingles
- **Masonry nails**—used when nailing into concrete or masonry
- **Brad**—a thin, short, finishing nail used for nailing trim

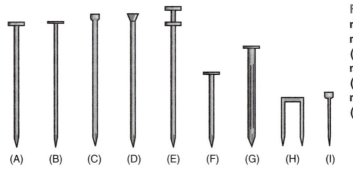

(A)   (B)   (C)   (D)   (E)   (F)   (G)   (H)   (I)

Figure 4-1: **Some of the most common kinds of nails: (A) common nail, (B) box nail, (C) finishing nail, (D) casing nail, (E) duplex nail, (F) roofing nail, (G) masonry nail, (H) staple, and (I) brad**

Screws

Screws are used when stronger joining power is needed, or when other materials must be fastened to wood. The screw is tapered to help draw the wood together as the screw is inserted. Screw heads are usually flat, oval, or round, and each has a specific purpose for final seating and appearance (see Figure 4-2).

Types of screws include the following:

- **Drywall screws**—used to attach drywall to wall studs.
- **Sheet metal screws**—used to fasten metal to wood, metal, plastic, or other materials. They are threaded completely from the point to the head, and the threads are sharper than those of wood screws. Machine screws are for joining metal parts such as hinges to metal door jambs.

- **Particleboard and deck screws**—corrosion-resistant screws used for installing deck materials and/or particleboard.
- **Lag screws**—used for heavy holding and are driven in with a wrench rather than a screwdriver.
- **Wood Screws**—are similar in function to lag screws and are available in a wide variety of sizes, head styles, and materials.

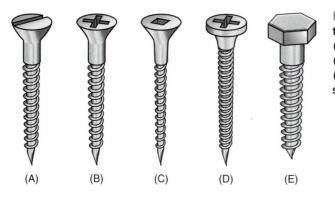

(A)　(B)　(C)　(D)　(E)

Figure 4-2: **Common screw types: (A) wood screw, (B) twinfast drywall screw, (C) particle board screw, (D) panhead sheet metal screw, and (E) lag screw**

Screw length should penetrate two-thirds of the combined thickness of the materials being joined. Use galvanized or other rust-resistant screws where rust could be a problem.

Screw head shapes are usually determined by the screw types and the pitch and/or depth of the threads. The most common shapes are oval, pan, bugle, flat, round, and hex (Figure 4-3). Different types of slots are also available for these screws (Figure 4-4).

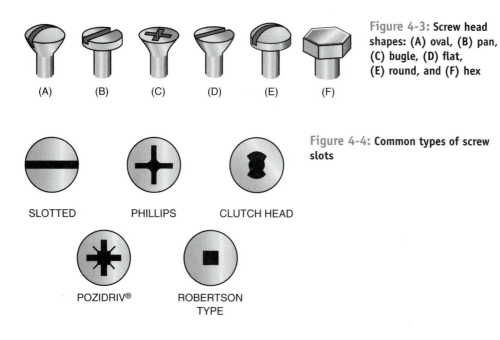

(A)　(B)　(C)　(D)　(E)　(F)

Figure 4-3: **Screw head shapes: (A) oval, (B) pan, (C) bugle, (D) flat, (E) round, and (F) hex**

SLOTTED　PHILLIPS　CLUTCH HEAD

POZIDRIV®　ROBERTSON TYPE

Figure 4-4: **Common types of screw slots**

Nuts and Bolts

Nuts and bolts are usually used at the same time: The bolt is inserted through a hole drilled in each item to be fastened together, and then the nut is threaded onto the bolt from the other side and tightened to give a strong connection. Using nuts and bolts also allows for the disassembly of parts.

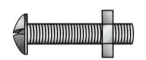

Figure 4-5: **Cap screw**

Types of bolts include the following:

- **Cap screws**—available with hex heads, slotted heads, Phillips head, and Allen drive (Figure 4-5).
- **Stove bolts**—either round or flat heads and threaded all the way to the head. Used to join sheet metal parts (Figure 4-6).
- **Carriage bolts**—a round-headed bolt for timber and threaded along part of the shank; inserted into holes already drilled (Figure 4-7).

See Appendix A, page 487 for a complete list of screws, bolts, washers, nuts, nails, and other fasteners.

Figure 4-6: **Stove bolt**   Figure 4-7: **Carriage bolt**

Adhesives

Sometimes when nails and screws just aren't holding it together by themselves, adhesives are needed.

Types of adhesives include the following:

- **Carpenter's wood glue**—a white, creamy glue, usually available in convenient plastic bottles. It is mainly used for furniture, craft, or woodworking projects. Polyvinyl sets in an hour, dries clear, and will not stain. However, it is vulnerable to moisture.
- **Epoxy**—the only adhesive with a strength greater than the material it bonds. It resists almost anything from water to solvents. Epoxy can be used to fill cavities that would otherwise be difficult to bond. Use it in warm temperatures but read the manufacturer's instructions carefully, since drying times vary, and mixing the resin and hardener must be exact.

Tips for using epoxy include the following:

- One gallon of epoxy will cover 12.8 square feet at 1/8 in. thickness or 6.4 square feet at 1/4 in. thickness.
- Although epoxies are generally able to withstand high temperatures for short periods, we do not recommend using them near or in temperatures above 200°F.
- To achieve maximum adhesion, remove oil, dirt, rust, paint, and water. Use a degreasing solvent (alcohol or acetone) to remove oil and grease. Sand or wipe away paint, dirt, or rust. Roughing up the surface increases surface area for a better bond.
- Epoxy cures quickly enough that there is significant strength after 3 hours. At 70°F the working time is 15 minutes. It is still possible to reposition work up to 45 minutes after application.
- Uncured epoxy will clean up with soap and water or denatured alcohol. Wash contaminated clothing. Cured epoxy can be removed by scraping, cutting, or removing in layers with a good paint remover.

Contact Cements

Contact cement is used to bond veneers or plastic laminates to wood for table tops and counters. Apply a thin coat on both surfaces and allow to dry somewhat before bonding. Align the surfaces perfectly before pressing together, because this adhesive will not pull apart. Use in a well-ventilated area.

Caulks

Caulk is used around outside window and door frames, and to fill outside wall and foundation cracks.

Types of caulk include:

- **Painter's caulk**—inexpensive latex caulk often used by painters to plug holes and cracks prior to painting. It can also be used to provide a smooth joint in a corner where textured materials meet. This allows the painter to paint a straight line in the corner when using contrasting paint colors.
- **Acrylic latex**—paintable, acrylic fortified caulk used for both interior and exterior applications. This caulk cleans up with water.
- **Siliconized latex**—very durable, latex caulk with silicone. It is available in colors and cleans up with water.
- **100 percent silicone**—is great for nonporous substances. It is the best choice for sealing ceramic tiles, glasses, and metal surfaces, but is less appropriate for porous surfaces such as wood and masonry. Silicone caulk remains flexible and is impervious to water. It cannot be painted and must be cleaned up with solvent. It also has a sharp odor when curing. It requires adequate ventilation and is usually colorless or white.
- **Tub and tile**—acrylic sealant that gives a flexible, watertight seal. It is mildew resistant with water cleanup.
- **100 percent silicone kitchen and bath sealant**—has the same characteristics as plain 100 percent silicone sealant.
- **Gutter and foundation sealant (butyl rubber)**—can be used on metal, wood, or concrete. It is appropriate for use in areas that experience extreme temperature variations. It requires solvent cleanup and is often used on metal flashing and around skylights.
- **Roof repair caulk**—convenient butyl rubber/asphalt formulation for sealing flashing, roofing, skylights, and so on. It cleans up with mineral spirits.
- **Adhesive caulk**—used as an adhesive during the installation of sinks, countertops, and so on. It dries harder than other caulks and so is less flexible.
- **Concrete and mortar repair**—retains some elasticity to remain in cracks in mortar and concrete. It cleans up with water.

If you're caulking around your bathtub, do it right. Fill the tub with water before you start. Tubs tend to sink ever so slightly when they are full. So, when you caulk an empty tub, you may not apply enough caulk to compensate for the sinking, which means you'll end up with cracked caulk the next time someone takes a bath.

Applying Caulk

Step 1. Caulk comes either in a squeeze tube or as a cartridge (see Figure 4-8). If caulk with a cartridge and caulk gun is applied, cut the tapered cartridge nozzle at a 45° angle, with the diameter of the opening equal to the size of the gap. Poke a hole through the tip with a piece of wire to break the internal seal.

Figure 4-8: **Caulk is made of various materials**

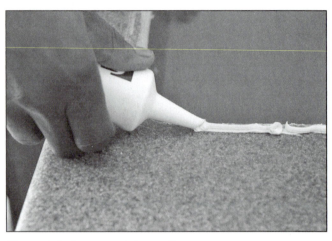

Figure 4-9: **Person applying caulk**

With the exception of epoxy, too much adhesive will weaken the hold of the materials you are bonding. Rough up smooth surfaces slightly before applying adhesives, so they will grip more securely. Apply a thin coat of glue, clamp securely, and allow to dry the recommended amount of time. Wipe away excess glue immediately after clamping.

Step 2. Apply caulk holding the tube or gun at a 45° angle. Use even pressure to squeeze the tube or trigger (Figure 4-9).

Step 3. Once it begins flowing, move the tip at an even pace along the joint. Remember that less is better. It's easier to apply more caulk than to remove too much.

Step 4. Once you have applied all of your caulk, use a caulk smoother to even out the finish. If you don't have a caulk smoother, then just wet your finger or a Popsicle stick to smooth out the bead of caulk.

Step 5. When application of all of your caulk is done, give it time to thoroughly set. It should indicate on the package how much time is required for the caulk to set. If you plan to paint over the caulk, it is important that the caulk has completely set before you start painting.

Solvents

Solvents are used in hundreds of products to improve cleaning efficiency. The solvents in cleaners help make counters, showers, toilets, tubs, carpets, and other items clean.

- **All purpose cleaner**—nonhazardous, biodegradable, nontoxic, super-concentrated cleaner. Good for cleaning everyday spills and messes.
- **Spot remover**—removes chewing gum, tar, grease, makeup, crayon, candle wax, and adhesives from most surfaces
- **Basin, tub, and tile cleaner**—cleans and removes soap scum, lime scale, body oils, and other deposits from shower walls and doors, sinks, ceramic tile, fiberglass, vinyl, porcelain, and stainless steel

Hand Tools

Hand tools, such as hammers, screwdrivers, levels, and tapes, do not use a motor.

Measuring, Marking, Leveling, and Layout Tools

Accuracy and care in measuring are all-important. They can mean the difference between a well-put-together project and a sloppy one. Tools that fall in this category are tape measures, squares, combination squares, chalk line reels, levels, and plumb bobs.

Tape Measures

Tape measures are available in lengths ranging from 6 to 100 feet. A blade that is 1 in. (or more) wide will be safer and easier to use. Those with cushioned bumpers protect the hook of the tape from damage, which is likely to occur when the tape retracts back into the case. The play in the hook (see Figure 4-10) allows you to make either inside or outside measurements without compensating for the hook (see Figure 4-11). Its flexibility allows it to measure round, contour, and other odd shapes. When making inside measurements, add the measurement of the tape case, usually marked on the case (see Figure 4-12).

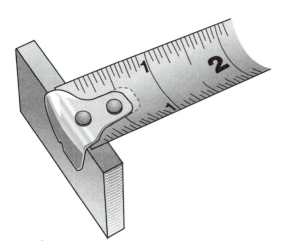

Figure 4-10: **The hook or fitting on the end of a tape measure will slide to adjust for the thickness of the fitting**

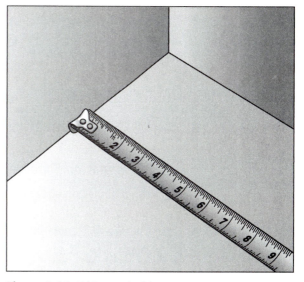

Figure 4-11: **Taking an inside measurement**

Figure 4-12: **Tape measure**

Squares

Squares are used for laying out work, checking for squareness during assembly, and marking angles. The carpenter's square, also called a framing square, is used for marking true perpendicular lines to be cut on boards and for squaring some corners, among other things (see Figure 4-13).

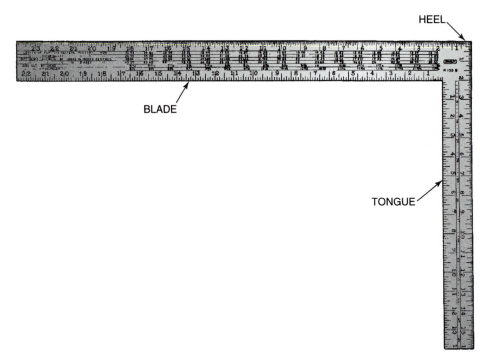

Figure 4-13: **Framing square**

Combination Squares

A combination square (Figure 4-14) is a most versatile tool. It has a movable handle that can lock in place on the 12-in. steel rule. It is used to square the end of a board, mark a 45° angle for mitering, and even make quick level checks with the built-in spirit level. It can also be used as a scribing tool to mark a constant distance along the length of a board.

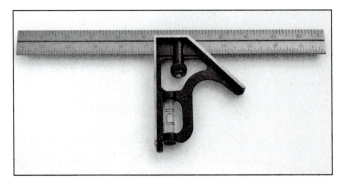

Figure 4-14: **Combination square**

Chalk line Reels

A chalk line is a string or line coated with colored chalk and is used to transfer a straight line to a working surface easily and accurately (see Figure 4-15). To use it, pull the line out and hold it tight between the two points of measurement, then snap it to leave a mark (shown in Figure 4-16). Some have a pointed case to double as a plumb bob.

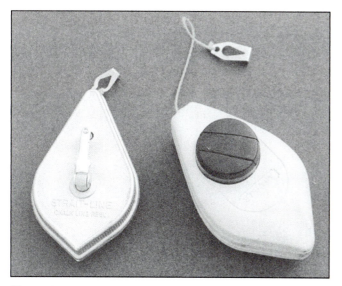

Figure 4-15: **Chalk line reel**

Figure 4-16: **Snapping a chalk line**

Levels

A level (Figure 4-17) is used to make sure your work is true horizontal (level) or true vertical (plumb).

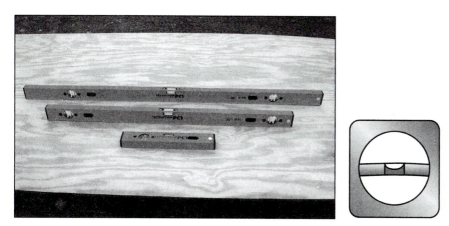

Figure 4-17: **A spirit level has one or more transparent vials containing a liquid and a bubble**

Plumb Bobs

A plumb bob (Figure 4-18) is a heavy, balanced weight on a string, which you drop from a specific point to locate another point exactly below it, or to determine true vertical.

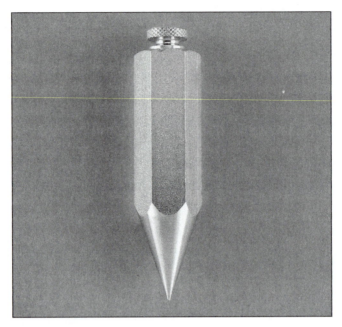

Figure 4-18: **Plumb bob**

Boring and Cutting Tools

Boring and cutting tools are a category of tools used for making holes or cutting wood, metals, and plastic. They include drill bits, saws, and so on.

Drill Bits

To drill a satisfactory hole in any material, use the correct type of drill bit; it must be used correctly and be sharpened as appropriate. A set of high-speed steel (HSS) twist drills and some masonry bits will probably be sufficient for the average facilities technician.

Types of drill bits include:

- **Twisted bit**—the most common drilling tool used with either a hand or electric drill (see Figure 4-19). It can be used on timber, metal, plastic, and similar materials. Most twist bits are made from either of the following:
  - **HSS**—suitable for drilling most types of material. When drilling metal, the HSS stands up to the high temperatures.
  - **Carbon steel**—specially ground for drilling wood and should not be used for drilling metals. These bits tend to be more brittle and less flexible than HSS bits.
- **Masonry bit**—designed for drilling into brick, block, stone, quarry tiles, or concrete. These bits are normally used in power drills (see Figure 4-19).
- **Spade bit**—used to drill large, flat-bottomed holes. A spade bit is inexpensive and suitable for general work; however, it does not have good chip clearing ability and tend to split thin material (see Figure 4-19).

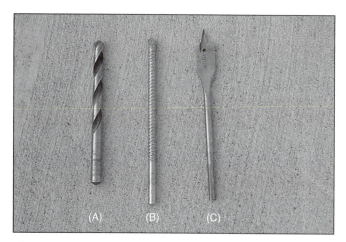

Figure 4-19: **Common drill bits: (A) twist drill bit, (B) masonry bit, and (C) spade bit**

Figure 4-20: **Hole saw**

- **Hole saw**—used for cutting large, fixed, diameter holes in wood or plastic (see Figure 4-20).

  Tips for using a drill:

- Always wear eye protection.
- Don't apply too much pressure on small drill bits.
- Ease up on pressure when drill breaks through material.
- The larger the drill bit, the slower the speed of the drill.
- Use a vise or clamp to hold the material to prevent it from spinning if the drill bit catches.

Saws

Various types of saws, such as the rip, crosscut, and Wallboard saws, are available. All of them look basically the same and their primary purpose is cutting of timber from boards, and sometimes making larger joints.

Types of saws include:

- **Crosscut saw**—available in many sizes and configurations. The crosscut saw is a good general-purpose saw typically used for cutting wood across its grain (see Figure 4-21). The kerf, or the actual cut made by the saw, is as wide as the set of teeth in the saw blade.

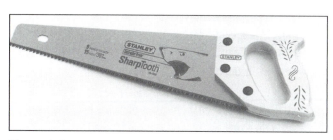

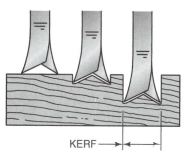

Figure 4-21: **Crosscut saw**

- **Ripsaw**—used to cut with the grain of the wood (see Figure 4-22). Ripsaw teeth are filed straight across the blade, and so each tooth is shaped like a little chisel.

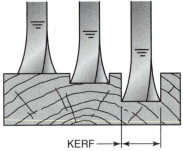

KERF

CROSS SECTION OF RIP TEETH

Figure 4-22: **Ripsaw**

- **Hacksaw**—basic hand saw used mostly for cutting metals (see Figure 4-23). Some have pistol grips, which make the hacksaw easy to grip. Hacksaws cut in straight lines.
- **Wallboard saw**—used for cutting electrical outlet holes and other small rough sawing where a powered saber saw will not fit (see Figure 4-24). A self-starting keyhole saw (a type of wallboard saw) is very handy and comfortable to use.
- **Coping saw**—used to cut the profile of one piece of molding on the end of molding (see Figure 4-25).

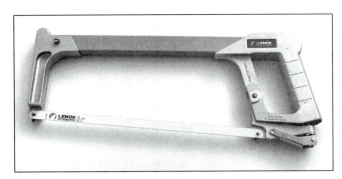

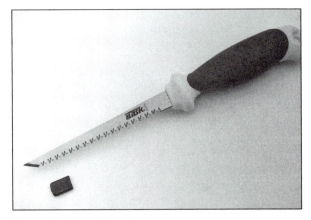

Figure 4-23: **Hacksaw**

Figure 4-24: **Wallboard saw**

Figure 4-25: **Coping saw**

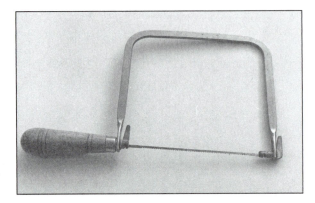

Hammers

Hammers are used for driving nails, fitting parts, and breaking up objects. Various types of hammers include:

- **Sledge hammer**—mainly used on outdoor projects. These hammers are designed to deliver heavy force (see Figure 4-26).
- **Mason's hammer**—used for working on brick, concrete, or mortar (see Figure 4-27). This hammer is often used for cutting and setting brick.
- **Claw hammer**—a general use hammer (see Figure 4-28).

Tips for using a hammer:

- If working with hard wood, drill a pilot hole in the material to prevent splitting.
- To begin hammering, grip the hammer firmly in the middle of the handle and shake hands with your hammer.
- Don't hold the hammer too tightly.
- Hold the nail between the thumb and the forefinger of the other hand and place the nail where it is to be driven.

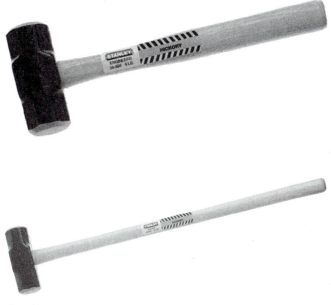

Figure 4-26: **Sledge hammers**

Figure 4-27: **Mason's hammer**

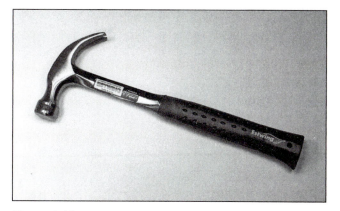

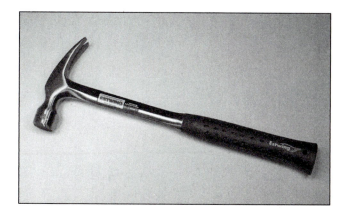

Figure 4-28: **Curved-claw and framing hammers**

- Using the center of the hammer face, drive the nail in with firm, smooth blows.
- Make sure that the striking face should always be parallel with the surface that's being hit.
- Avoid sideways or glancing blows. Take care not to mark the work surface.

Screwdrivers

Screwdrivers (Figure 4-29) are used to insert and tighten, or to loosen and remove, screws. To select the proper screwdriver for a particular job, match the appropriate screwdriver and size to the screw head (see Figure 4-30).

Types of screwdrivers include:

- Slotted
- Torx
- Phillips
- Square

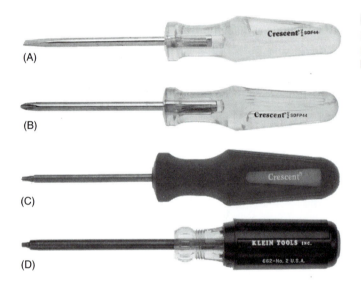
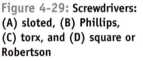
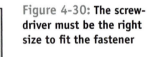

Figure 4-29: Screwdrivers: (A) sloted, (B) Phillips, (C) torx, and (D) square or Robertson

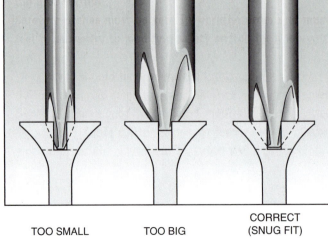

Figure 4-30: The screwdriver must be the right size to fit the fastener

TOO SMALL   TOO BIG   CORRECT (SNUG FIT)

Pliers

Pliers are used primarily for gripping objects that utilize leverage. They are designed for numerous purposes and require different jaw configurations to grip, turn, pull, or crimp various things (see Figure 4-31).

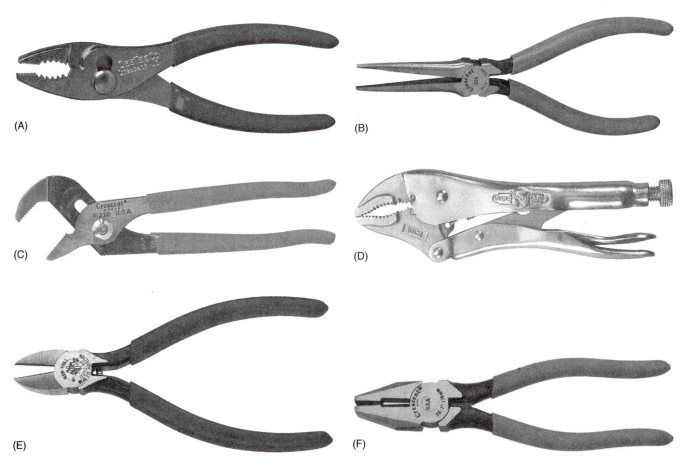

Figure 4-31: **Pliers: (A) common slip-joint, (B) needle nose, (C) channel-lock, (D) vise grip, (E) side-cutting, and (F) electrician's**

Wrenches

Wrenches are used to turn bolts, nuts, or other hard-to-turn items.

Some common types of wrenches include:

- **Socket wrench**—has a ratchet handle, which allows the user to move the handle back and forth without having to take the socket off the nut and reposition it. This wrench uses separate, removable sockets to fit many different sizes of nuts (see Figure 4-32).
- **Open-end wrench**—usually has different sizes at each end (see Figure 4-33).
- **Box-end wrench**—usually has different sizes at each end that form a complete circle and either six or twelve points (see Figure 4-34).
- **Adjustable wrench**—has jaws that can be adjusted by turning the adjusting screw (see Figure 4-35).

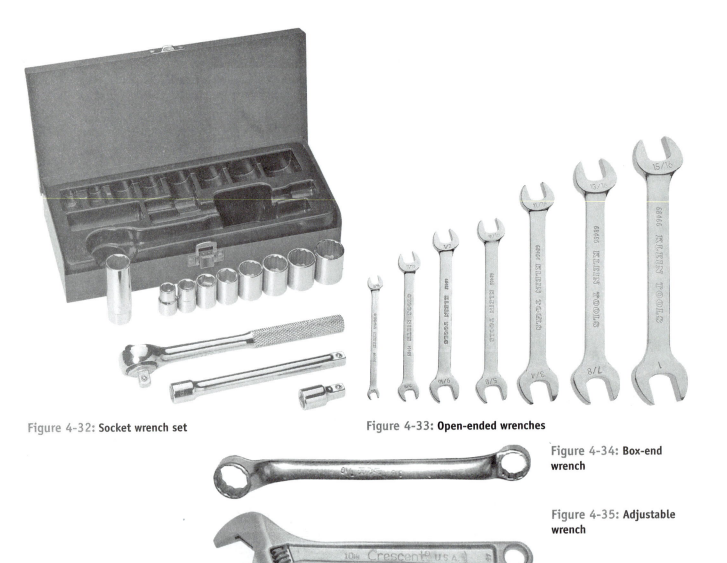

Figure 4-32: Socket wrench set

Figure 4-33: **Open-ended wrenches**

Figure 4-34: **Box-end wrench**

Figure 4-35: **Adjustable wrench**

Portable Power Tools

The term **power tool** is used to describe a tool that contains a motor. Power tools are used for some specific purposes or operations that one cannot perform manually. Portable tools are easy to move from one location to another.

Figure 4-36: Circular saw

Power Saws

Power saws are more commonly used in construction because they provide a means of cutting a material quickly and more efficiently than a hand saw.

Types of power saws include:

- **Circular saw**—the most basic portable power saw (see Figure 4-36). Most of these saws can be equipped with a rip guide to maintain a uniform width of cut on long passes.

- **Reciprocating saw**—similar to a nonpowered hand saw except that the blade moves back and forth under power (see Figure 4-37). It is good for ripping and crosscutting, but lacks the control achieved from the platform design of the saber saw.
- **Saber saw**—has a small, thin blade that cuts with an up-and-down motion, making it ideal for irregularities and scroll work, as well as for ripping and crosscutting (see Figure 4-38).

    Tips for using saws:
    - Read the owner's manual carefully before operating your power saw.
    - Observe all of the safety precautions discussed in the owner's manual.
    - Use the safety devices that are available with your saw.
    - Keep your hands away from the blade.
    - Stand to one side of the blade while it is in motion in case wood is kicked back.
    - Unplug the saw before changing blades or making adjustments.
    - Be sure that the blades you use are sharp and clean to avoid binding and burning.

Figure 4-37: **Reciprocating saw**

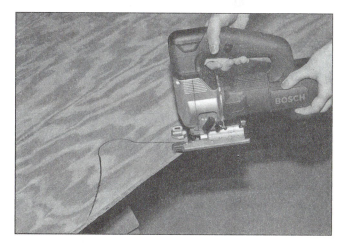

Figure 4-38: **Saber saw**

Drills and Drivers

The rotary drill, the most basic of the electric drills, is used mainly for boring holes in various materials. These drills can be either corded or cordless. Drivers are similar to the rotary drill but with a greater torque, allowing the user to drive and remove screws as well as drill through materials at a more rapid rate (see Figure 4-39).

Planes

Planes are used for planing rough boards and are ideal for trimming doors and frames for a perfect fit (see Figure 4-40).

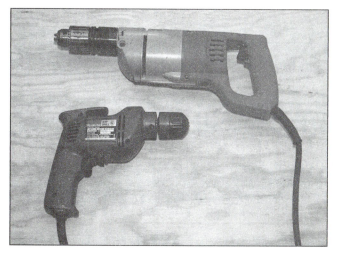

Figure 4-39: **Portable power drills**

Figure 4-40: **Portable electric plane**

Tips on how to plane:

- As with all power tools, follow the manufacturer's safety instructions carefully.
- Place the front of the plane on the end of the wood, but make sure that the cutter block is not touching the timber. Gently press down on the front handle and turn on the planer. Move the tool at an even rate along the wood. When you reach the end, transfer pressure to the rear handle and glide off the wood to avoid taking a deep gouge out of the last few millimeters of the work. Choose a portable planer that can be inverted in its own accessory stand if you want to plane several small pieces.
- To even out a wide surface, set the planer to its finest cut and plane diagonally to the grain, in overlapping strokes. You will still need to use a hand plane to get rid of slight machining marks.
- Make several light passes rather than taking off a lot of timber in one go.

Figure 4-41: **Portable electric router**

Routers

Routers can be used to make raised panel doors, round the edge of a coat rack, or create your own baseboard molding. The router can also trim plastic laminate.

There are two types of routers: the plunge and fixed (or standard) routers. Both types offer the same results, although each type is better for particular jobs.

1. **Fixed router**—used to make many different cuts including grooves, dadoes, rabbets, and dovetails. It is also used to shape edges and make cutouts (see Figure 4-41).

2. **Plunge router**—used for various applications in which a fixed router cannot be used. A typical application for a plunge router in construction is to trim the edges of plastic laminates (see Figure 4-42).

There are special router blades for finishing plastic laminates. The most popular are a flush cut blade and a beveled blade.

Four basic types of router bits (Figure 4-44) are:

1. **Grooving bits**—used to make a groove in the piece of wood. This type of bit is commonly used for street address signs for homes. Different types of grooving bits include the V-groove, the round-nose, and the straight bit.

2. **Joinery bits**—used to make several different types of joints. This type of router bit includes the finger joint, the drawer lock, the rile and stile, and the dovetail bits.

3. **Edge bits**—used to create different-shaped edges in woodwork. Examples of these types of bits include the beading, flush, and round-over bits.

4. **Specialized bits**—do not fit into one of the previous categories and have more specialized purposes, including the key hole, raised panel, and T-slot bits.

A tip for using the router: The edge guide has a straight face that can be adjusted at different distances from the bit. You adjust it so that it runs against a straight edge of the workpiece, and the edge guide keeps the router going in a straight line (see Figure 4-43).

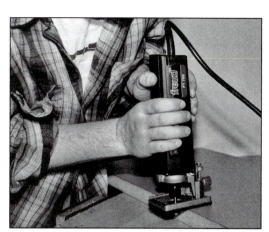

Figure 4-42: **Laminate router**

Figure 4-43: **A guide attached to the base of the router rides along the edge of the stock and controls the sideways motion of the router**

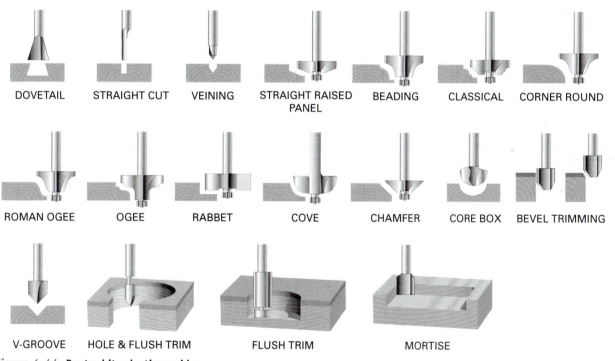

DOVETAIL STRAIGHT CUT VEINING STRAIGHT RAISED PANEL BEADING CLASSICAL CORNER ROUND

ROMAN OGEE OGEE RABBET COVE CHAMFER CORE BOX BEVEL TRIMMING

V-GROOVE HOLE & FLUSH TRIM FLUSH TRIM MORTISE

Figure 4-44: **Router bit selection guide**

Sanders

Sanders can remove large amounts of stock or finishing material quickly and put a glass-smooth surface on your projects.

Types of sanders include:

- **Detail sanders**—small handheld sanders designed for sanding around odd shapes and small nooks in woodwork (see Figure 4-45).
- **Random orbit sanders**—random motion lets the operator move the sander in any direction without scarring the work surface. These sanders give you a

Figure 4-45: **Detail sander (Image copyright Alena Brozova, 2009. Used under license from Shutterstock.com)**

Figure 4-46: **Random orbit sander (Image copyright Eimantas Buzas, 2009. Used under license from Shutterstock.com)**

superfine finish and leave a minimum amount of marks (see Figure 4-46).

- **Belt sanders**—used for removing paint or varnish from large areas, smoothing out rough wood, and preparing surfaces for finishing and thinning out thick wood (see Figure 4-47).
- **Disk sanders**—bench-mounted tools with a circular pad that accepts specially made sanding sheets. Most disk sanders also have a belt mounted vertically or horizontally on their frame (see Figure 4-48).
- **Spindle sanders**—bench-mounted tools with a cylindrical spindle located in the center of a large worktable. The spindle holds special sanding tubes of various grit sandpapers. Some spindle sanders have an oscillating feature that raises and lowers the spindle as it rotates. The oscillating feature increases the rate at which the sander removes stock. Spindle sanders are good for edge sanding, especially around curves and circles (see Figure 4-49).

Sandpaper is a sheet abrasive composed of particles of flint, garnet, emery, aluminum oxide, or silicon carbide. These particles are mounted on paper or cloth in "open coat" or "closed coat" density (see Figure 4-50). Types of sandpaper include:

- **Flint**—the least expensive sandpaper sold; this is a gray material that wears down quickly.
- **Garnet**—a much harder grit than flint and more suitable for woodworking, costs slightly more than flint paper.
- **Emery**—has a distinctive black color and is generally used on metal.
- **Aluminum oxide**—a reddish-colored, very sharp grit and is used on either wood or metal.
- **Silicon carbide**—this bluish-black material is the hardest of all, and is commonly used for finishing metal or glass.

Sandpaper grit is identified by numbers from 1500 to 12; the smaller the number, the coarser the grit.

Figure 4-47: **Belt sander**

Figure 4-48: **Disk sander**

Figure 4-49: **Spindle sander**

Grit	Common Name	Uses
40–60	Coarse	Heavy sanding and stripping, roughing up the surface
80–120	Medium	Smoothing of the surface, removing smaller imperfections and marks
150–180	Fine	Final sanding pass before finishing the wood
220–240	Very Fine	Sanding between coats of stain or sealer
280–320	Extra Fine	Removing dust spots or marks between finish coats
360–600	Super Fine	Fine sanding of the finish to remove some luster or surface blemishes and scratches

Figure 4-50: **Sample sandpaper grit table. The grit range may vary depending on manufacturers**

Stationary Tools

Stationary tools tend to be larger, more powerful, and more complex than hand tools.

- **Miter**—replaces the miter block used with a hand saw and the various miter saw mechanisms that attempt to support a manual saw in a framework. It excels at any cut in which a precise angle is required. Typical applications include cutting studwork, cutting miters for picture frames, dados, skirting, architraves, and coves. Also used in all sorts of generalpurpose carpentry and joinery (see Figure 4-51).
- **Chop**—a lightweight circular saw mounted on a spring-loaded pivoting arm and supported by a metal base. Chop saws are considered the best saw to get exact, square cuts (see Figure 4-52).
- **Band saw**—capable of performing a whole range of cuts, such as ripping, crosscutting, beveled cuts, and curves. The band saw is also capable of re-sawing and cutting a thick board into several thinner boards (see Figure 4-53).

Figure 4-51: **Miter saw**

Figure 4-52: **Chop saw**

Figure 4-53: **Band saw**

Figure 4-54: **Table saw**

Tips for using the band saw:

- Always stand to the left of the band saw. In the event of a broken blade, the blade will fly off to the right. If the blade breaks, shut off the power and stay away from the saw until it stops.
- Use gloves when uncoiling, removing, and installing the band saw blade.
- Keep your hands and fingers away from exposed parts of the blade.
- Follow the manufacturer's guidelines for adjustment of the sliding bar or post. If the guide is too high, the blade will not have the proper support.
- Avoid backing out of the cut. This could push the blade off the wheels.

- **Table saws**—the superstar of stationary woodworking power tools. It rips, miters, bevels long edges, cross-cuts, and mills nearly every imaginable woodworking joint. The table saw size is determined by the diameter of the largest blade that can be used on the saw. A 10-in. table saw can use up to a 10-in. saw blade (see Figure 4-54).

Tips for using the table saw include:
- Use the saw guard at all times. No operation should be done with the guards removed.
- Never reach over the saw blade to remove scraps or provide support to the workpiece.
- Always stand to the side of the saw, and never directly in line with the blade. If the saw catches the material you are working on, the saw will throw it in line with the blade.

- When cutting, NEVER PULL the workpiece through the saw. Start and finish the cut from the front of the saw.
- When crosscutting, hold the workpiece firmly against the miter gauge. Make sure that the miter gauge works freely in the slot and that it will clear both sides of the blade when tilted. Note that on some saws the miter gauge can be used only on one side when the blade is tilted.
- Use a push stick according to the manufacturer's guidelines.

- **Drill press**—more accurate than any portable drill. A drill press uses a drilling head positioned above an adjustable workbench, both being fixed to a sturdy base. Most models include a clamp and a guide, allowing the user greater control when drilling (see Figure 4-55).

Figure 4-55: **Drill press (Image copyright Eimantas Buzas, 2009. Used under license from Shutterstock.com)**

Tips for using the drill press include:

- Always secure the material being drilled.
- Use bits designed only for the drill press.
- Never try to stop the machine by taking hold of the chuck after the power is off.
- Check to make sure that the chuck is secured before turning the drill press on.

Safety tips for using stationary tools include:

- Safety devices and guards must always be in place. These devices were designed by the manufacturer to be used with the tool.
- Always keep blades and cutting edges sharp.
- Perform maintenance, accessory changes, and adjustments only when the tool is off and unplugged.

- Don't wear loose fitting clothing. High-powered stationary tools can catch clothing and draw the operator's body into the tool.
- When using any type of stationary saw, never use gloves. They can get caught in the saw.
- Never put your fingers and hands in front of the saw blades and other cutting tools.
- Read the owner's manual and safety precautions before using the tools.
- Know the location of emergency on and off switches and the tools power switch.
- Make sure that blades, bits, and accessories are properly mounted. In addition, make sure that all locking handles and clamps are tight before using a tool.

Tools, whether hand or power tools, can be dangerous if proper safety precautions are not followed. When working on projects that require the use of these tools, kindly ask the client to stay a safe distance from you while you are working and use all of the possible safety precautions.

Review Questions

1. What are fasteners commonly used for?

2. Define the following:

   Common nail
   Box nail
   Finishing nail

3. When is a screw used?

4. When are adhesives used?

5. What is a solvent?

6. List three types of saws commonly used today in facility maintenance.

7. How does a claw hammer differ from a mason's hammer?

8. How does an open-end wrench differ from a box-end wrench?

9. List two types of routers commonly used today.

10. List the four basic types of router bits commonly used today.

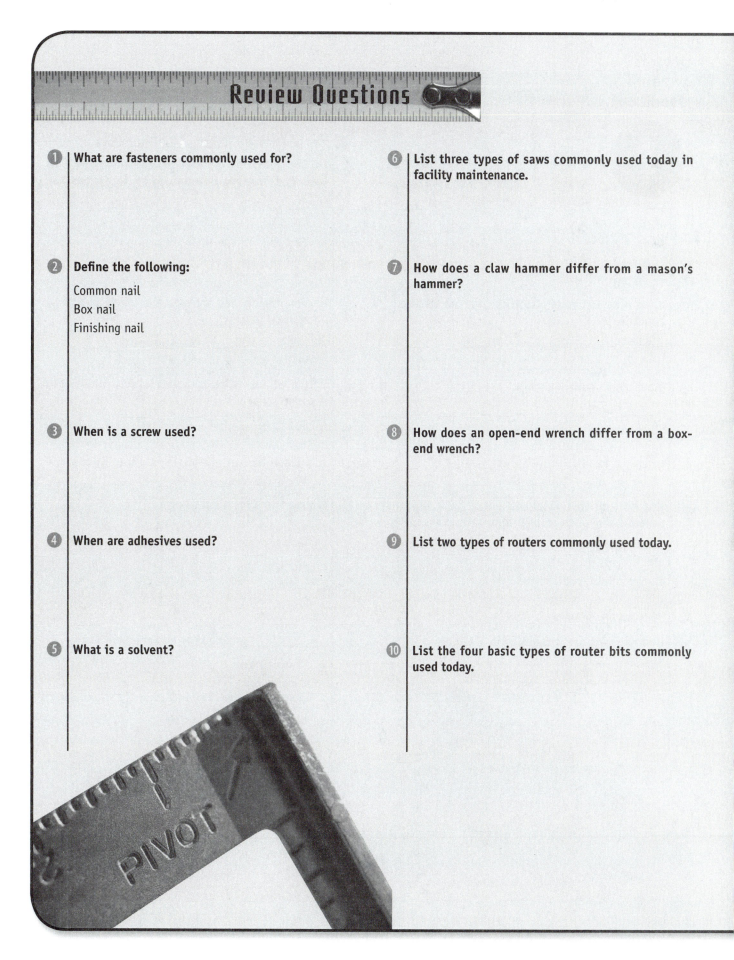

Review Questions

Circle the letter that indicates the correct answer.

11 The term _____ is used to indicate the size of a common nail.

a. Inch
b. Diameter
c. Length
d. Penny

12 Using nuts and bolts also allows for the disassembly of parts

a. True
b. False

13 Wood screws generally have _____.

a. A flat head
b. A round head
c. An oval head
d. Any of the above

14 The hole drilled in sheet metal for a tapping screw should have the approximate size of _____.

a. The thread diameter
b. The thread root diameter
c. The shank diameter
d. None of the above

15 Hollow rivets are assembled in a pin often called a(n) _____.

a. Mandrel   b. Nozzle
c. Anchor   d. Clinch

16 In the machine screw dimension 5/16"-18 UNC-2, the measurement 5/16 indicates the _____.

a. Length
b. Number of threads per inch
c. Outside thread diameter
d. Thread diameter

17 Anchor shields, wall anchors, and toggle bolts may all be used in _____.

a. Wood   b. Masonry walls
c. Hollow walls   d. All of the above

18 Which of the following heads may identify a machine screw?

a. Flat head
b. Round head
c. Fillister head
d. Any of the above

19 Pin rivets are installed using a _____.

a. Drill screw
b. Pin rivet gun
c. Pin rivet drill
d. Hammer

20 _____ are made of hardened steel.

a. Finishing nails
b. Masonry nails
c. Roofing nails
d. None of the above

Name: _____

Date: _____

Tools and Equipment

Fasteners

Upon completion of this job sheet, you should be able to demonstrate your ability to select the correct fastener for the appropriate repair project.

1 Take an inventory and list the different types of fasteners used at your facilities. Write a brief description of where they would be used.

_____

_____

2 Are the correct fasteners used for the appropriate jobs at your facilities?
a. If no, what, if anything, should be done?

_____

_____

3 What are the different types of screws available and what determines which ones are used?

_____

_____

4 Write a brief explanation of what determines whether a nail or a screw should be used.

_____

_____

5 For each numbered item, enter the letter of the correct answer:

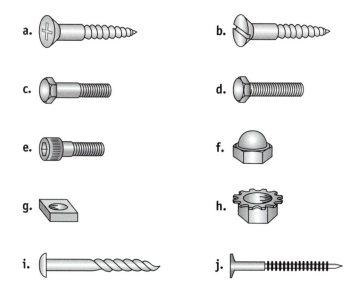

1. Socket head _____   2. K-lock nut _____

3. Annular nail _____   4. Phillips flat head wood screw _____

5. Slotted flat head wood screw _____   6. Cap nut _____

7. Square nut _____   8. Standard bolt _____

9. Full thread tap bolt _____   10. Square twisted nail _____

Instructor's Response:

_____

_____

_____

Name: _____

Date: _____

Tools and Equipment

Adhesives

Upon completion of this job sheet, you should be able to demonstrate your ability to select the correct adhesive for the appropriate repair project.

1 Write a brief explanation of what determines when to use an adhesive.

_____

_____

2 Take an inventory of the types of adhesives used at your facility. Briefly describe where the different adhesives should be used.

_____

_____

Instructor's Response:

_____

_____

_____

Name: _____

Date: _____

Tools and Equipment

Hand Tools

Upon completion of this job sheet, you should be able to demonstrate your ability to safely use hand tools.

1 Compare four different brands of tape measures.

a. How are the symbols different from one tape measure to another and what do the symbols mean?

_____

_____

b. What are the increments of the scales?

_____

_____

2 Check your combination square for accuracy.

a. Tape a sheet of paper to a board that has a perfectly straight edge.

b. Hold the square against the edge and draw a line along the outer edge of the blade.

c. Flip the square over so the opposite side of the blade faces up, align the square on the edge of the stock, and draw a second line about 1/32 in. from the first.

d. If the square is accurate, the lines will be parallel.

3 Investigate the different types of drill bits available at your facility. Write a brief description of the differences and the materials they are used on.

_____

_____

4 Investigate the different types of hand saws available at your local home improvement center and write a brief description of the physical differences of the saws and how each one should be used.

_____

_____

5 Compare and contrast the uses of the portable power tools discussed in this course with their comparable hand tools and/or stationary tools. Discuss how you decide which tool is most appropriate for the project.

Instructor's Response:

_____

_____

_____

Chapter 5 Practical Electrical Theory

OBJECTIVES

By the end of this chapter, you will be able to:

Knowledge-Based

- ⊗ Understand the principle of basic electricity.
- ⊗ Describe the difference between AC and DC currents.
- ⊗ Understand the properties of common electrical wires used by facilities maintenance technicians and understand and correctly measure wire size and load-carrying capacity.
- ⊗ Understand the operation and functions of emergency circuits.
- ⊗ Describe different types of emergency backup electrical power systems.

Skill-Based

- ⊗ Calculate electrical load using Ohm's law.

Glossary of Terms

Atom the smallest particle of an element

Matter a substance that takes up space and has weight

Element any of the known substances (of which ninety-two occur naturally) that cannot be separated into simpler compounds

Law of charges states that like charges repel and unlike charges attract

Law of centrifugal force states that spinning object has a tendency to pull away from its center point and that the faster it spins, the greater the centrifugal force

Valence shell the outermost shell of an atom

Coulomb one coulomb is the amount of electric charge transported in one second by one ampere of current

Ampere (amp) unit of current flow

Electron theory states that current flows from negative to positive in a circuit

Conventional current flow theory states that current flows from positive to negative in a circuit

Voltage the potential electrical difference for electron flow from one line to another in an electrical circuit

Watt a unit of power applied to electron flow. One watt equals 3.414 Btu

Alternating current electron flow that occurs in one direction and then reverses direction at regular intervals

Direct current electron flow that occurs in only one direction; used in the industry only for special applications such as solid-state modules and electronic air filters

Ohm's law relationship between voltage and current and a material's ability to conduct electricity

Continuous load a load in which the maximum current is expected to continue for 3 hours or more

Introduction

Electricity is the driving force that provides most of the power for the industrialized world. It is used to light homes, cook meals, heat and cool buildings, and drive motors, and serves as the ignition for most automobiles. This module provides the technician with a basic understanding of practical electrical theory.

Basic Electrical Theory

Although the practical use of electricity has become common only within the last hundred years, it has been known as a force for much longer. The Greeks were the first to discover electricity, about 2,500 years ago. They noticed that when amber was rubbed with other materials, it got charged with an unknown force that had the power to attract objects such as dried leaves, feathers, bits of cloth, or other lightweight materials. The Greeks called amber *elektron*. The term "electric" was derived from it, meaning "to be like amber" or to have the ability to attract other objects.

> *Do not make any electrical measurements without specific instructions from a qualified person. Use only electrical conductors of proper size to avoid overheating and possibly fire. Electrical circuits must be protected from current overloads.*

Structure of Matter

To understand electricity, starting with the study of atoms is necessary. The atom is the basic building block of the universe. All matter is made by combining atoms into groups (see Figure 5-1). Matter is any substance that has mass and occupies space. An object's weight comes from the earth's gravitational pull. Matter constitutes atoms, which are small parts of a substance and may combine to form molecules. Atoms of one substance may be combined chemically with those of another to form a new substance. When molecules are formed they cannot be broken down. They can exist in any further without changing the chemical nature of the substance. Matter also exists in three states: solids, liquids, and gases. Water, for example, can exist as a solid in the form of ice, as a liquid, or as a gas in the form of steam (see Figure 5-2).

An atom is the smallest part of an element. An **element** is any of the known substances (of which ninety-two occur naturally) that cannot be separated into simpler compounds. It is composed of three principal parts: protons, neutrons, and electrons. Protons and neutrons are located at the center (or nucleus) of the atom. Protons have a positive charge, whereas neutrons have no charge and have little or no effect when

State: Ⓢ Solid
Ⓛ Liquid
Ⓖ Gas
Ⓧ Not found in nature

◻ Metals
◻ Transition metals, lanthanide series, actinide series
◻ Metalloids
◻ Nonmetals, noble gases

Atomic number
Symbol
Atomic weight

92 Ⓧ
U
Uranium
238.03

| 1A | 2A | | 3A | 4A | 5A | 6A | 7A | 8A |

1 Ⓧ
H
Hydrogen
1.01

2 Ⓧ
He
Helium
4.00

3 Ⓢ
Li
Lithium
6.94

4 Ⓢ
Be
Beryllium
9.01

11 Ⓢ
Na
Sodium
22.99

12 Ⓢ
Mg
Magnesium
24.31

3B 4B 5B 6B 7B ┌─── 8B ───┐ 8 9 10 11B 12B

19 Ⓢ
K
Potassium
39.10

20 Ⓢ
Ca
Calcium
40.08

21 Ⓢ
Sc
Scandium
44.96

22 Ⓢ
Ti
Titanium
47.87

23 Ⓢ
V
Vanadium
50.94

24 Ⓢ
Cr
Chromium
52.00

25 Ⓢ
Mn
Manganese
54.94

26 Ⓢ
Fe
Iron
55.85

27 Ⓢ
Co
Cobalt
58.93

28 Ⓢ
Ni
Nickel
58.69

29 Ⓢ
Cu
Copper
63.55

30 Ⓢ
Zn
Zinc
65.41

5 Ⓢ
B
Boron
10.81

6 Ⓢ
C
Carbon
12.01

7 Ⓧ
N
Nitrogen
14.01

8 Ⓧ
O
Oxygen
16.00

9 Ⓧ
F
Fluorine
19.00

10 Ⓧ
Ne
Neon
20.18

13 Ⓢ
Al
Aluminum
26.98

14 Ⓢ
Si
Silicon
28.09

15 Ⓢ
P
Phosphorus
30.97

16 Ⓢ
S
Sulfur
32.07

17 Ⓧ
Cl
Chlorine
35.45

18 Ⓧ
Ar
Argon
39.95

31 Ⓢ
Ga
Gallium
69.72

32 Ⓢ
Ge
Germanium
72.64

33 Ⓢ
As
Arsenic
74.92

34 Ⓢ
Se
Selenium
78.96

35 Ⓛ
Br
Bromine
79.90

36 Ⓧ
Kr
Krypton
83.80

37 Ⓢ
Rb
Rubidium
85.47

38 Ⓢ
Sr
Strontium
87.62

39 Ⓢ
Y
Yttrium
88.91

40 Ⓢ
Zr
Zirconium
91.22

41 Ⓢ
Nb
Niobium
92.91

42 Ⓢ
Mo
Molybdenum
95.94

43 Ⓢ
Tc
Technetium
(98)

44 Ⓢ
Ru
Ruthenium
101.07

45 Ⓢ
Rh
Rhodium
102.91

46 Ⓢ
Pd
Palladium
106.42

47 Ⓢ
Ag
Silver
107.87

48 Ⓢ
Cd
Cadmium
112.41

49 Ⓢ
In
Indium
114.82

50 Ⓢ
Sn
Tin
118.71

51 Ⓢ
Sb
Antimony
121.76

52 Ⓢ
Te
Tellurium
127.60

53 Ⓢ
I
Iodine
126.90

54 Ⓧ
Xe
Xenon
131.29

55 Ⓢ
Cs
Cesium
132.91

56 Ⓢ
Ba
Barium
137.32

57 Ⓢ
La
Lanthanum
138.91

72 Ⓢ
Hf
Hafnium
178.49

73 Ⓢ
Ta
Tantalum
180.95

74 Ⓢ
W
Tungsten
183.84

75 Ⓢ
Re
Rhenium
186.21

76 Ⓢ
Os
Osmium
190.23

77 Ⓢ
Ir
Iridium
192.22

78 Ⓢ
Pt
Platinum
195.08

79 Ⓢ
Au
Gold
196.97

80 Ⓛ
Hg
Mercury
200.59

81 Ⓢ
Tl
Thallium
204.38

82 Ⓢ
Pb
Lead
207.2

83 Ⓢ
Bi
Bismuth
208.98

84 Ⓢ
Po
Polonium
(209)

85 Ⓧ
At
Astatine
(210)

86 Ⓧ
Rn
Radon
(222)

87 Ⓧ
Fr
Francium
(223)

88 Ⓢ
Ra
Radium
(226)

89 Ⓢ
Ac
Actinium
(227)

104 Ⓧ
Rf
Rutherfordium
(261)

105 Ⓧ
Db
Dubium
(262)

106 Ⓧ
Sg
Seaborgium
(266)

107 Ⓧ
Bh
Bohrium
(264)

108 Ⓧ
Hs
Hassium
(277)

109 Ⓧ
Mt
Meitnerium
(268)

110 Ⓧ
Ds
Darmstadtium
(271)

111 Ⓧ
Rg
Roentgenium
(272)

112 Ⓧ
Uub
Ununbium
(285)

113 Ⓧ
Uut
Ununtrium
(284)

114 Ⓧ
Uuq
Ununquadium
(289)

115 Ⓧ
Uup
Ununpentium
(288)

116 Ⓧ
Uuh
Ununhexium
(292)

1 2 3 4 5 6 7

58 Ⓢ
Ce
Cerium
140.12

59 Ⓢ
Pr
Praseodymium
140.91

60 Ⓢ
Nd
Neodymium
144.24

61 Ⓧ
Pm
Promethium
(145)

62 Ⓢ
Sm
Samarium
150.36

63 Ⓢ
Eu
Europium
151.96

64 Ⓢ
Gd
Gadolinium
157.25

65 Ⓢ
Tb
Terbium
158.93

66 Ⓢ
Dy
Dysprosium
162.50

67 Ⓢ
Ho
Holmium
164.93

68 Ⓢ
Er
Erbium
167.26

69 Ⓢ
Tm
Thulium
168.93

70 Ⓢ
Yb
Ytterbium
173.04

71 Ⓢ
Lu
Lutetium
174.97

90 Ⓢ
Th
Thorium
232.04

91 Ⓢ
Pa
Protactinium
231.04

92 Ⓢ
U
Uranium
238.03

93 Ⓧ
Np
Neptunium
(237)

94 Ⓧ
Pu
Plutonium
(244)

95 Ⓧ
Am
Americium
(243)

96 Ⓧ
Cm
Curium
(247)

97 Ⓧ
Bk
Berkelium
(247)

98 Ⓧ
Cf
Californium
(251)

99 Ⓧ
Es
Einsteinium
(252)

100 Ⓧ
Fm
Fermium
(257)

101 Ⓧ
Md
Mendelevium
(258)

102 Ⓧ
No
Nobelium
(259)

103 Ⓧ
Lr
Lawrencium
(262)

Figure 5-1: Periodic table of the elements

Figure 5-2: Three states of water

SOLID

ICEBERG

GAS

LIQUID

considering electrical characteristics. Because the neutron has no charge, the nucleus will have a net positive charge. Electrons have a negative charge and travel around the nucleus in orbits. The number of electrons in an atom is the same as that of protons. Electrons in the same orbit travel at the same distance from the nucleus but do not follow the same orbital paths (see Figure 5-3). The hydrogen atom is a simple atom to illustrate because it has only one proton and one electron (see Figure 5-4). Not all atoms are as simple as the hydrogen atom. Most wiring used to conduct an electrical current is made of copper. Figure 5-5 shows a copper atom, which has twenty-nine protons and twenty-nine electrons. Some electron orbits are farther away from the

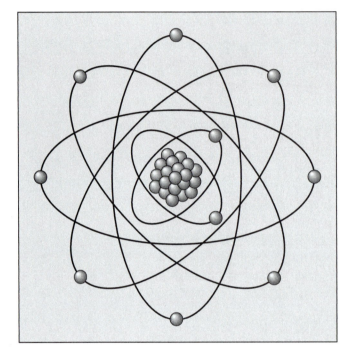

Figure 5-3: **The orbital paths of electrons**

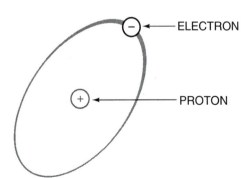

Figure 5-4: **A hydrogen atom with one electron and one proton**

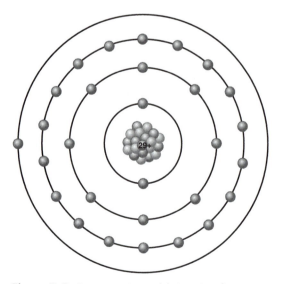

Figure 5-5: **A copper atom with twenty-nine protons and twenty-nine electrons**

nucleus than others. As can be seen, two travel in an inner orbit, eight in the next, eighteen in the next, and one in the outer orbit. It is this single electron in the outer orbit that makes copper a good conductor.

The Law of Charges

To understand atoms, first understand two basic laws of physics. One of these laws is the **law of charges**, which states that opposite charges attract and like charges repel (see Figure 5-6). If two objects with unlike charges come close to each other, the lines of force attract (Figure 5-7). Likewise, if two objects with like charges come close to each other, the lines of force repel (Figure 5-8). For example, consider what happens if you try to connect the north poles of two magnets. It is this principle that helps the electrons to maintain their orbit around the nucleus.

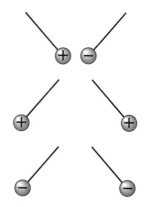

Figure 5-6: **Unlike charges attract and like charges repel**

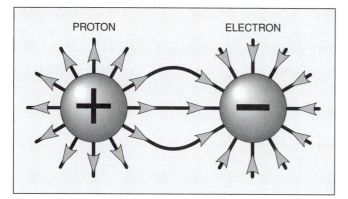

Figure 5-7: **Unlike charges attract each other**

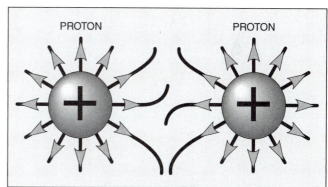

Figure 5-8: **Like charges repel each other**

The Law of Centrifugal Force

Another law that is important to understanding the behavior of an atom is the **law of centrifugal force**, which states that a spinning object will pull away from its center point and that the faster it spins, the greater the centrifugal force will be (see Figure 5-9). If you tie an object to a string and spin it around, it will try to pull away from you. The faster the object spins, the greater will be the force that tries to pull the object away.

Although atoms are often drawn flat, electrons orbit the nucleus in a spherical fashion (see Figure 5-10). Electrons travel at such a high rate of speed that they form a shell around the nucleus. For this reason, electron orbits are often referred to as shells.

The number of electrons that any one orbit, or shell, can contain is found by using the formula $(2N^2)$. The letter N represents the number of the orbit or shell (see Figure 5-11).

Figure 5-9: **Centrifugal force causes an object to pull away from its axis point**

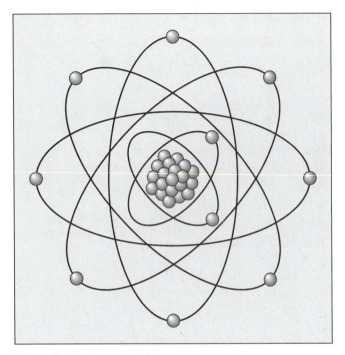

Figure 5-10: **Electron orbits**

The outer shell of an atom is known as the **valence shell**, and electrons located there are known as valence electrons (Figure 5-12). The valence shell of an atom cannot hold more than eight electrons. The valence electrons are of primary concern in the study of electricity, because they explain much of electrical theory. A conductor, for instance, is generally made from a material that contains one or two valence electrons. Atoms with one or two valence electrons are unstable and can be made to give up these electrons with little effort. Conductors are materials that permit electrons to flow through them easily. Examples of good conductors are gold, silver, copper, and aluminum.

Conductors

Good conductors contain atoms with few electrons in the outer orbit. Three common metals—copper, silver, and gold—are good conductors, and the atom of each has only one electron in the outer orbit. These electrons are considered to be free electrons because they move easily from one atom to another.

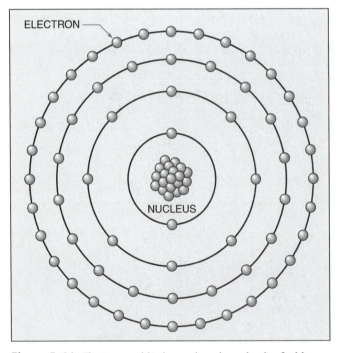

Figure 5-11: **Electrons orbit the nucleus in a circular fashion**

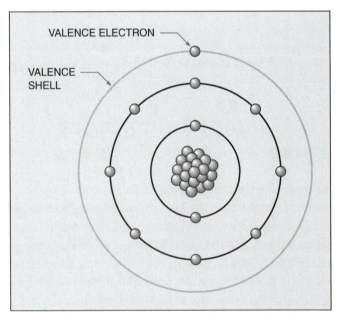

Figure 5-12: **The electrons located in the outer orbit of an atom are valence electrons**

Insulators

Atoms with several electrons in their outer orbits are poor conductors. It is difficult to free these electrons, and materials comprising such atoms are considered to be insulators. Glass, rubber, and plastic are examples of good insulators.

Quantity Measurement for Electrons

A **coulomb** is a quantity measurement for electrons. One coulomb contains 6.25×10^{18}, or 6,250,000,000,000,000,000, electrons. Coulomb's law of electrostatic charges states that the force of electrostatic attraction or repulsion is directly proportional to the product of the two charges and inversely proportional to the square of the distance between them.

The **amp**, or **ampere**, is named for André Ampére, a scientist who lived from the late 1700s to the early 1800s. The amp (A) is defined as one coulomb per second. One amp of current flows through a wire when one coulomb flows past a point in one second (see Figure 5-13).

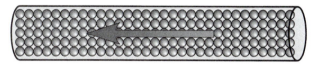

Figure 5-13: **One ampere equals one coulomb per second**

The ampere is a measure of the amount of electricity that flows through a circuit.

There are actually two theories concerning current flow: the electron theory and the conventional current flow theory. The **electron theory** states that since electrons are negative particles, current flows from the most negative point in the circuit to the most positive. The **conventional current flow theory** is older than the electron theory and states that current flows from the most positive point to the most negative (see Figure 5-14).

A complete path must exist before current flows through a circuit. A complete circuit is often referred to as a closed circuit because the power source, conductors, and load form a closed loop (see Figure 5-15). A short circuit, which has very little or no

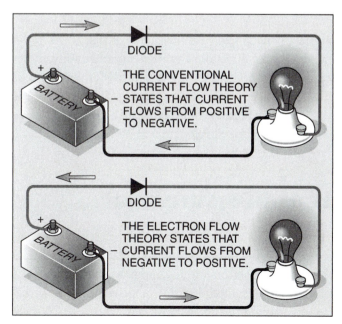

Figure 5-14: **Conventional current flow and electron flow theories**

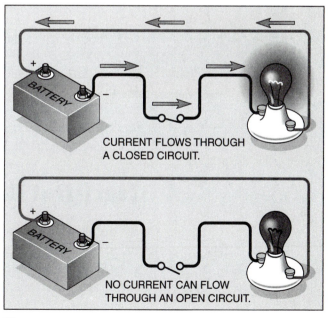

Figure 5-15: **Current flows only through a closed circuit**

resistance, generally occurs when the conductors leading from and back to the power source become connected together (see Figure 5-16). Another type of circuit, one that is often confused with a short circuit, is a grounded circuit (see Figure 5-17). Grounded circuits can also cause an excessive amount of current flow. They occur when a path other than the one intended is established to ground. Many circuits contain an extra conductor called the grounding conductor. The grounding conductor helps prevent a shock hazard when the ungrounded, or hot, conductor comes in contact with the case or frame of the appliance.

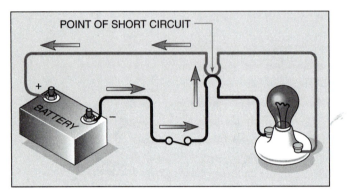

Figure 5-16: **A short circuit bypasses the load and permits too much current to flow**

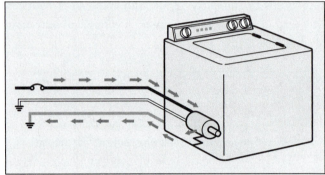

Figure 5-17: **The grounding conductor provides a low resistance path to ground**

Direct Current

Direct current (DC) travels in one direction. Because electrons have a negative charge and travel to atoms with a positive charge, DC is considered to flow from negative to positive.

Alternating Current

Alternating current (AC) is continually and rapidly reversing. The charge at the power source (generator) is continually changing direction; thus the current continually reverses itself. For several reasons, most electrical energy generated for public use is AC. It is much more economical to transmit electrical energy long distances in the form of AC. The voltage of this type can be readily changed so that it has many more uses. DC still has many applications, but it is usually obtained by changing AC to DC or by producing the DC locally where it is to be used.

Electrical Units of Measurement

Electromotive force (emf) or **voltage (V)** is used to indicate the difference of potential in two charges. Voltage is defined as the force that causes electrons to move from atom to atom in a conductor. When an electron surplus builds up on one side of a circuit and a shortage of electrons exists on the other side, a difference of potential or emf is created. The unit used to measure this force is the volt.

All materials oppose or resist the flow of electrical current to some extent. In good conductors this opposition or resistance is very low, whereas in poor conductors it is high. The unit used to measure resistance is ohm. A conductor has a resistance of 1 ohm when a force of 1 volt causes a current of 1 amp to flow.

Volt = electrical force or pressure (V)
Ampere = quantity of electron flow rate (A)
Ohm = resistance to electron flow (Ω)

The Electrical Circuit

An electrical circuit must have a power source, a conductor to carry the current, and a load or device to use the current. Generally, there is also a means for turning the electrical current flow on and off. Figure 5-18 shows an electrical generator for the source, a wire for the conductor, a light bulb for the load, and a switch for opening and closing the circuit.

The generator produces the current by passing many turns of wire through a magnetic field. If it is a DC generator, the current will flow in one direction. If it is an AC generator, the current will continually reverse itself. However, the effect on this circuit will generally be the same for both AC and DC. The wire or conductor provides the path for the electricity to flow to the bulb and complete the circuit. The electrical energy is converted to heat and light energy at the bulb element.

The switch is used to open and close the circuit. When the switch is open, no current will flow. When it is closed, the bulb element will produce heat and light because current is flowing through it.

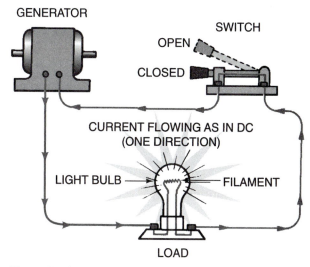

Figure 5-18: **An electric circuit**

Calculate Electrical Load Using Ohm's Law

Ohm's law deals with the relationship between voltage and current and a material's ability to conduct electricity. Typically, this relationship is written as voltage = current × resistance. In its simplest form, Ohm's law states that it takes one volt to push one amp through one ohm.

In a DC circuit, the current is directly proportional to the voltage and inversely proportional to the resistance.

The resistance, voltage, and/or current of a device (load) can be calculated by rearranging Ohm's law (Figure 5-19). The three basic forms of Ohm's law formulas are as follows:

$$E = I \times R$$

Voltage can be found if the current and resistance are known.

$$I = E/R$$

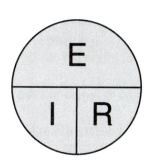

Figure 5-19: **Chart for finding values of voltage, current, and resistance**

Current can be found if the voltage and resistance are known.

$$R = E/I$$

Resistance can be found if the voltage and current are known.
Where

> E = emf or voltage
> I = intensity of current or amperage
> R = resistance

The first formula states that the voltage can be found if the current and resistance are known. The second formula states that the current can be found if the voltage and resistance are known. The third formula states that if the voltage and current are known, the resistance can be found.

For example, determine the voltage for a load that draws 30 amps and has a resistance of 500 ohms (see Figures 5-20 and 5-21).

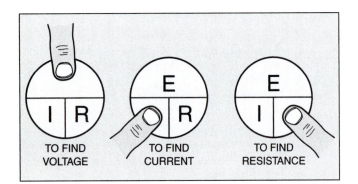

Figure 5-20: **Using the Ohm's law chart**

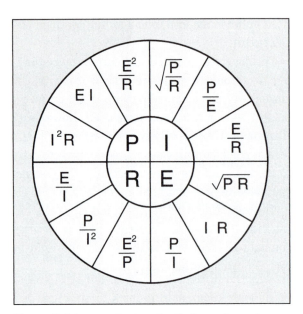

Figure 5-21: **Formula chart for finding values of voltage, current, resistance, and power**

> E = I × R
> E = 30 amps × 500 ohms
> E = 15,000 volts

Determine the current for a 120-volt load that has a resistance of 0.5 ohms.

> I = E/R
> I = 120 volts/0.5 ohms
> I = 60 amps

Determine the resistance for a 120-volt load that draws 15 amps.

> R = E/I
> R = 120 volts/15 amps
> R = 8 ohms

As stated earlier, wattage is proportional to the amounts of voltage and current flow. It can be calculated using the formula: **watt** = voltage × current. For example: determine the wattage of a 120-volt load that draws 15 amps.

R = E/I
R = 120 volts/15 amps
R = 8 ohms
Watts = E × I
Watts = 120 volts × 15 amps
Watts = 1,800 watts

Characteristics of Circuits

There are two kinds of circuits: series and parallel. Three characteristics make a series circuit different from a parallel circuit:

1. The voltage is divided across the electrical loads in a series circuit. For example, if there are three resistances of equal value in a circuit, and the voltage applied to the circuit is 120 volts, the voltage is equally divided across each resistance (see Figure 5-22). If the resistances are not equal, the voltage is divided across each according to its resistance (Figure 5-23).

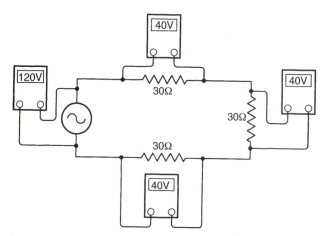

Figure 5-22: **Three resistors of equal value divide the voltage equally in a series circuit**

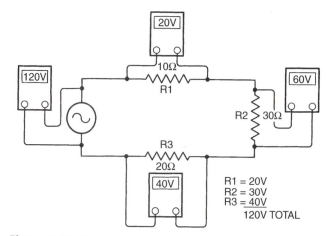

Figure 5-23: **Three resistors of unequal value divide the voltage according to their values**

2. The total current for the circuit flows through each electrical load in the circuit. With one power supply and three resistances, the current must flow through each to reach the others.

3. The total resistance in the circuit is equal to the sum of the resistances in the circuit. For example, when three resistances are in a circuit, the total resistance of the circuit is the sum of the three resistances (see Figure 5-24).

Three characteristics of a parallel circuit make it different from a series circuit:

1. The total voltage for the circuit is applied across each circuit resistance. The power supply feeds each power consuming device (load) directly (see Figure 5-25).

2. The current is divided among the different loads, or the total current is equal to the sum of the currents in each branch (see Figure 5-26).

3. The total resistance is less than the value of the smallest resistance in the circuit. Calculating the resistances in a parallel circuit requires a procedure that is different from simply adding them. A parallel circuit allows current along two or more paths at the same time. This type of circuit applies equal voltage to all

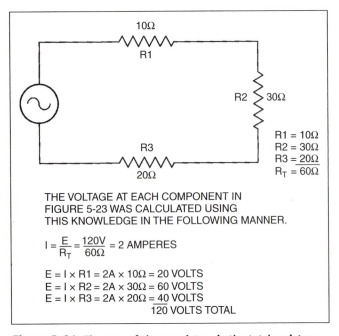

THE VOLTAGE AT EACH COMPONENT IN
FIGURE 5-23 WAS CALCULATED USING
THIS KNOWLEDGE IN THE FOLLOWING MANNER.

$$I = \frac{E}{R_T} = \frac{120V}{60\Omega} = 2 \text{ AMPERES}$$

E = I × R1 = 2A × 10Ω = 20 VOLTS
E = I × R2 = 2A × 30Ω = 60 VOLTS
E = I × R3 = 2A × 20Ω = 40 VOLTS
                      120 VOLTS TOTAL

Figure 5-24: The sum of three resistors is the total resistance of the circuit

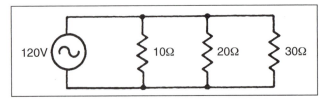

Figure 5-25: The total voltage for a parallel circuit is applied across each component

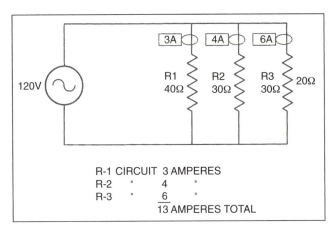

R-1 CIRCUIT 3 AMPERES
R-2   "   4   "
R-3   "   6   "
             13 AMPERES TOTAL

Figure 5-26: The current is divided between the different loads in a parallel circuit

loads. The general formula used to determine total resistance in a parallel circuit is as follows:

$$R_{total} = \frac{1}{\frac{1}{R_1} + \frac{1}{R_2} + \frac{1}{R_3} + \cdots}$$

For example, the total resistance of the circuit in Figure 5-27 is determined as follows:

$$R_{total} = \frac{1}{\frac{1}{R_1} + \frac{1}{R_2} + \frac{1}{R_3}}$$

$$= \frac{1}{\frac{1}{40} + \frac{1}{30} + \frac{1}{20}}$$

$$= \frac{1}{0.025 + 0.0333 + 0.05}$$

$$= \frac{1}{0.1083}$$

$$= 9.234 \text{ ohms}$$

To determine the total current draw, Ohm's law can be used. For example, in the previous problem, the voltage is known and the current flow for each component can be calculated from the total resistance, as shown in Figure 5-28. Again, notice that the total resistance is less than that of the smallest resistor.

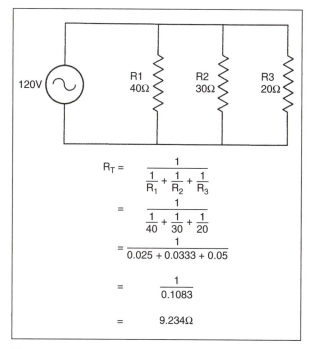

Figure 5-27: **The total resistance for a parallel circuit must be calculated**

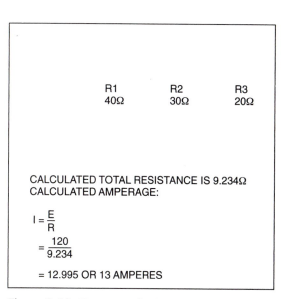

Figure 5-28: **The current in Figure 5-27 may be calculated from the total resistance when the applied voltage is known**

Electrical Power

Electrical power (P) is measured in watts. A watt (W) is the power used when 1 amp flows with a potential difference of 1 volt. Therefore, power can be determined by multiplying the voltage and the current flowing in a circuit.

Watts = volts × amperes

or

P = E × I

The consumer of electrical power pays the electrical utility company according to the number of kilowatts (kW) used for a certain time span usually billed as kilowatt hours (kW h). A kilowatt is equal to 1000 W. To determine the power being consumed, divide the number of watts by 1000:

$$P \text{ (in kW)} = \frac{E \times I}{1000}$$

Wire Sizes and Load-Carrying Capacity

The National Electrical Code (NEC) establishes important fundamentals that weave their way through the decision-making process for an electrical installation. The NEC defines **continuous load** as "a load where the maximum current is expected to continue for three hours or more." General lighting outlets and receptacle outlets in residences are not considered to be continuous loads.

When wiring a house, it is all but impossible to know which appliances, lighting, heating, and other loads will be turned on at the same time. Different families have different lifestyles. The rules for doing the calculations are found in Article 220 of the NEC. For lighting and receptacles, the computations are based on volt-amperes per square foot. For small-appliance circuits such as those in kitchens and dining rooms, the basis is 1,500 volt-amperes per circuit. For large appliances such as dryers, electric ranges, ovens, cooktops, water heaters, air conditioners, heat pumps, and so on, all of which are not used continuously or at the same time, demand factors are used in the calculations. Following the requirements in the NEC, the various calculations roll together in steps that result in the proper sizing of branch-circuits, feeders, and service equipment.

Emergency Circuits

An emergency circuit is intended to supply illumination and power automatically to designated areas and equipment when the normal source of power fails. For the emergency circuit, a portable or temporary alternate source must be available whenever the emergency generator is out of service for major maintenance or repair.

Ground fault circuit interrupts (GFCIs) are used to prevent people from being electrocuted. They work by sensing the amount of current flow on both the ungrounded (hot) and grounded (neutral) conductors supplying power to a device. A ground fault occurs when a path to ground, other than the intended path, is established (see Figure 5-29).

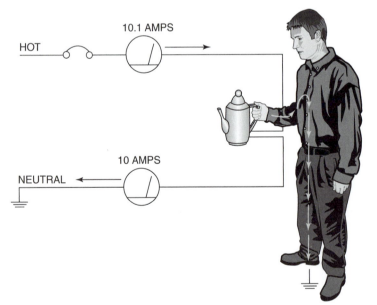

Figure 5-29: **A ground fault**

Ground-fault indication is required for emergency systems operating at more than 150 volts to ground and over-current devices rated at 1,000 amps or more. Wiring for emergency circuits must be kept entirely independent of all other wiring unless required to be associated with normal source wiring.

Emergency Backup Electrical Power Systems

Emergency backup electrical power systems are used to provide continued access to electrical services during power outages. Systems currently available on the market make it possible to achieve this. These systems can be based on fossil fuel–powered generators, battery-based storage systems, or propane or natural gas supply, or can be directly wired into the household circuit.

Permanent generators can be set up to power the whole building structure during an outage or just the essential loads, such as the furnace, security systems, and various appliances.

Review Questions

1. What are the three states of matter?

2. What are the three principal parts of an atom?

3. State the law of charges.

4. What is a coulomb?

5. What is an amp?

6. What is electricity?

7. What is a watt?

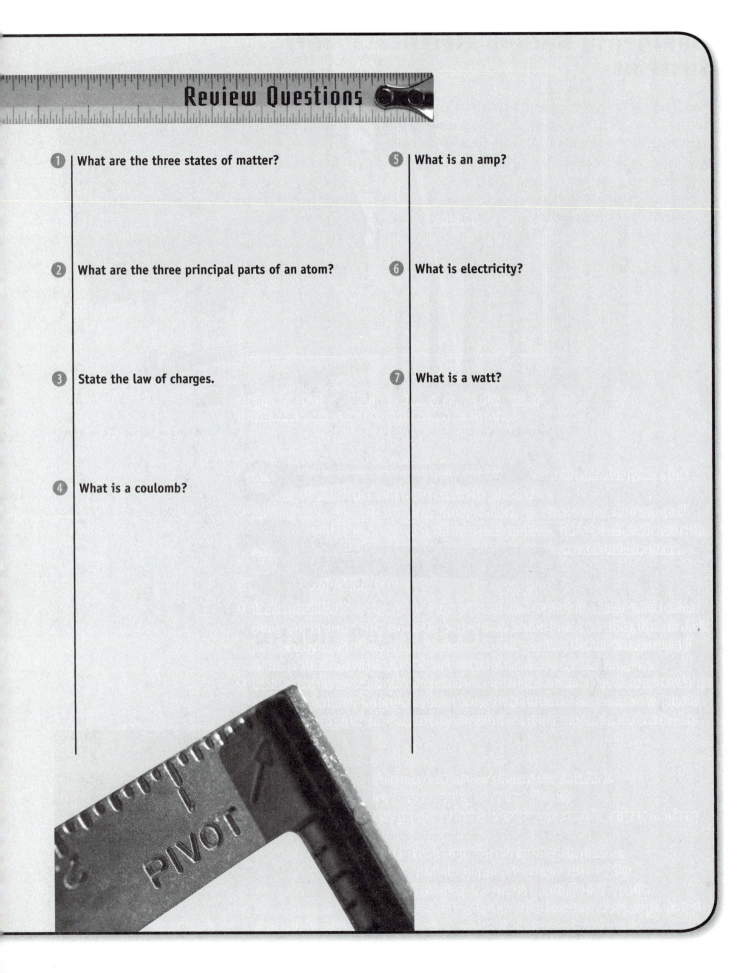

Name: _____

Date: _____

Practical Electrical Theory

Understanding Ohm's Law

Upon completion of this job sheet, you should have a basic understanding of Ohm's law.

1. Determine the voltage of a load drawing 20 amps and having a resistance of 25 ohms.
2. Determine the wattage of a device drawing 10 amps and having a resistance of 75 ohms.

Instructor's Response:

_____
_____
_____

Chapter 6 Electrical Facilities Maintenance

OBJECTIVES

By the end of this chapter, you should be able to:

Knowledge-Based

- ✪ Understand and apply OSHA regulations that cover electrical installations.
- ✪ Describe the difference between AC and DC.
- ✪ Correctly identify single-phase and three-phase electrical systems.
- ✪ Correctly identify and select the boxes most commonly used in electrical installations.
- ✪ Correctly identify and select different types of electrical devices and fixtures.
- ✪ Describe the different types of emergency backup systems.

Skill-Based

- ✪ Follow systematic, diagnostic, and troubleshooting practices.
- ✪ Perform tests on smoke alarms, fire alarms, medical alert systems, and emergency exit lighting.
- ✪ Perform tests on GFCI receptacles.
- ✪ Repair and/or replace common electrical devices such as receptacles and switches.
- ✪ Repair and/or replace lighting fixtures and/or bulbs, and ballasts.

Glossary of Terms

Voltage is the amount of electrical pressure in a given circuit and is measured in *volts*

Current (or amperage) the flow of electrons through a given circuit, which is measured in *amps*

Resistance the opposition to current flow in a given electrical circuit, which is measured in *ohms*

Power the electrical work that is being done in a given circuit, which is measured in wattage (watts) for a purely resistive circuit and volt-amps (VA) for an inductive/capacitive circuit

Introduction

Once the facilities maintenance technician has a good understanding of basic electrical theory and electrical safety, the technician can attempt to troubleshoot, repair, and install basic electrical circuits and appliances. The previous chapter introduced the technician to basic electrical theory. This chapter introduces the concepts of safety and the basic procedures for troubleshooting and repairing basic electrical circuits and devices (switches, receptacles, and so on).

Practical Electrical Theory

Before a technician can safely work with an electrical system, he or she should have a good understanding of the basic principles of electrical theory. A direct relationship exists among voltage, current, and resistance. Power is the product of voltage and current.

- **Voltage** is the amount of electrical pressure in a given circuit and is measured in *volts*.
- **Current** (or amperage) is the flow of electrons through a given circuit and is measured in *amps*.
- **Resistance** is the opposition to current flow in a given electrical circuit and is measured in *ohms*.
- **Power** is the electrical work that is done in a given circuit and is measured in *wattage* (*watts*) for a purely resistive circuit and *volt-amps* (*VA*) for an inductive/ capacitive circuit.

AC vs. DC

Two types of electrical current are used to power electrical circuits today: alternating current (AC) and direct current (DC). AC is electrical current whose magnitude and direction vary cyclically, whereas that of DC does not modulate and therefore remains constant. AC is primarily used in residential, commercial, and industrial applications as the primary source of power. DC is used primarily in electronic, low-voltage applications, batteries, and so on.

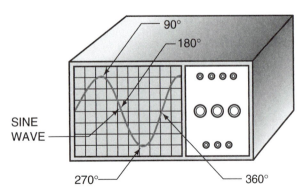

Figure 6-1: **A sine wave displayed on an oscilloscope**

Sine Waves

AC continually reverses direction. An oscilloscope is an instrument that measures the amount of voltage over a period of time and can display this voltage, called a waveform, on a screen. Many types of waveforms exist, but we will discuss the one most refrigeration and air-conditioning technicians are involved with, the sine wave (Figure 6-1). The sine wave displays the voltage of one cycle through 360°. In the United States and Canada, the standard voltage in most locations is produced with a frequency of 60 cycles per second. Frequency is measured in hertz (Hz). Therefore, the standard frequency in the United States and

Canada is 60 Hz. Figure 6-1 is a sine wave as it would be displayed on an oscilloscope. At the 90° point the voltage reaches its peak (positive); at 180° it is back to 0; at 270° it reaches its negative peak; and at 360° it is back to 0. If the frequency is 60 Hz, this cycle would be repeated 60 times every second. The sine wave is a representation of a trigonometric function of an AC cycle.

Figure 6-2 shows the peak and peak-to-peak values. As the sine wave indicates, the voltage is at its peak value briefly during the cycle. Therefore, the peaks of the peak-to-peak values are not the effective voltage values. The effective voltage is the root-mean-square (RMS) voltage (Figure 6-3). This is the AC value measured by most voltmeters and ammeters. The RMS voltage is 0.707 × the peak voltage. If the peak voltage were 170 volt, the effective voltage measured by a voltmeter would be 120 V (170 V × 0.707 = 120.19 V).

Sine waves can illustrate a cycle of an AC electrical circuit that contains only a pure resistance, such as a circuit with electrical heaters. In such a pure resistive circuit, the voltage and current will be in phase. This is illustrated with sine waves such as those in Figure 6-4. Notice that the voltage and current reach their negative and positive peaks at the same time.

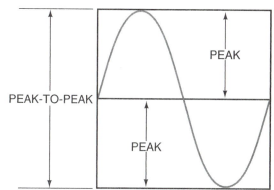

Figure 6-2: **Peak and peak-to-peak AC voltage values**

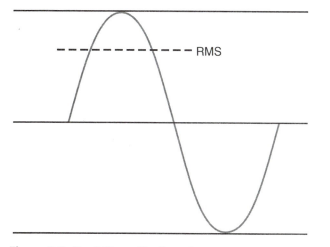

Figure 6-3: **The RMS or effective voltage**

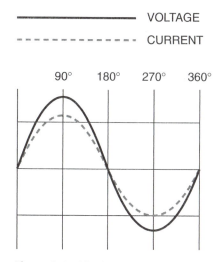

Figure 6-4: **This sine wave represents both the voltage and the current in phase in a resistive circuit**

Sine waves can also illustrate a cycle of an AC electrical circuit that contains a fan relay coil, which will produce an inductive reactance. The sine wave will show the current lagging the voltage in this circuit (Figure 6-5).

Figure 6-6 shows a sine wave illustrating a cycle of an AC electrical circuit that has a capacitor producing a pure capacitive circuit. In this case the current leads the voltage.

Single-Phase AC vs. Three-Phase AC

Single-phase AC is the most commonly used electrical supply for single-family and multifamily dwellings. Single phase consists of two ungrounded conductors (hot wires) and one grounded conductor (neutral wire). When measuring voltage between the two hot wires of this type of system, you will read approximately 240 volts.

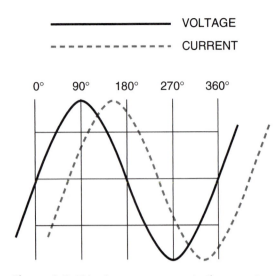

Figure 6-5: **This sine wave represents the current lagging the voltage (out of phase) in an inductive circuit**

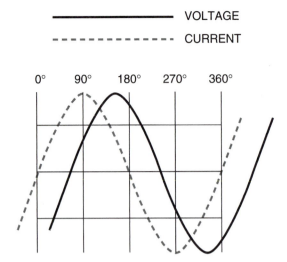

Figure 6-6: **This sine wave represents a capacitive AC cycle with the voltage lagging the current. The voltage and current are out of phase**

If you read from any one hot wire to the neutral, you should read approximately 120 volts. Make sure that the meter that you are using to test voltage is rated appropriately.

Three-phase AC is most often used for commercial buildings, healthcare facilities, and industrial facilities. There are always three ungrounded conductors (hot wires) in this type of system. The three-phase system may or may not have a neutral wire. The common voltage levels for the three-phase system will be 208, 230, 240, 480, 575, or 600 when measured between the two hot wires. This is referred to as the line voltage. If a neutral is present, then voltage read from any one hot wire to the neutral wire will equal the line voltage divided by the square root of 3 (1.732).

Safety, Tools, and Test Equipment

One of the most import aspects of any assigned task is to perform the assignment in a safe manner. This includes using tools and equipment that are properly maintained and designed for that particular task.

Safety

Many safety considerations must be adhered to while working on electrical systems, equipment, and devices. Many of the safety procedures are outlined in the OSHA standard (29 CFR Part 1926) and must be followed while working on electrical systems, equipment, and devices.

In any installation, repair, or removal of electrical equipment or devices, use the proper wiring methods and practices stipulated in the National Electrical Code, better known as the NEC (NFPA 70). The NEC is written to ensure the protection of people and property from the hazards that arise from installation and/or repair of electrical systems or equipment. One must also follow all local codes and regulations that might exceed the minimum standards required by the NEC.

Prior to beginning any work on electrical systems, equipment, or devices, always consult with the local authority having jurisdiction (building/electrical inspector) on the local codes and regulations that might affect the proper and safe installation of electrical systems, equipment, and devices in your area.

The OSHA and NEC safety articles mandate that only qualified personnel should work on electrical systems, equipment, or devices. This is to protect not only the property owner but also the person performing the installation. A qualified person is one who has the skills and knowledge related to the construction and operation of the electrical equipment and installations and has received safety training on the hazards involved.

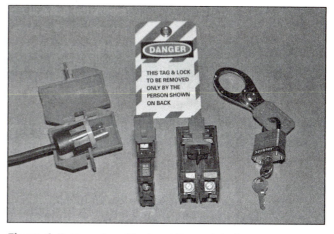

Figure 6-7: **Examples of lock-out/tag-out devices**

Lock-out/tag-out procedures, as stated in the OSHA standard 1926.417 and the NFPA 70E standard, should be followed at all times to prevent electrical shock or even death by electrocution. Various forms of lockouts exist, which can be placed on safety disconnects, switches, and breakers. You must use the lockout that will prevent current from flowing through the circuit that you are working on. Figure 6-7 shows some examples of the different lockouts that are available.

The NFPA 70E mandates that qualified personnel should never work on any part of an energized electrical system that is over 50 volts AC without the proper arc flash/ arc blast personnel protective clothing (PPE) and without using 1,000-volt-rated tools. This mandate is in effect for your protection only. Also, safety glasses should be worn at all times while working on electrical systems, equipment, or devices.

Working Space around Electrical Equipment

Adequate lighting and space is required around electrical equipment so that maintenance on the electrical equipment can be performed safely. The NEC 110.26 covers the minimum working space, access, headroom, and lighting requirements for electrical equipment such as switchboards, panelboards, and motor control centers operating at 600 volts or less.

Tools and Test Equipment

Always ensure that all hand tools and power tools are inspected before and after use. Look for any defects or damage that may cause injury while the hand tool or power tool is in use. If you notice any damage, the tool should not be used until it has been repaired by a qualified person (someone trained to work on the damaged tool). If a qualified person is not available to repair the damaged tool, the tool should be either discarded or put away until such a time when a qualified person can repair it. Regularly maintain and clean all hand tools and power tools to ensure their proper operation.

Always use calibrated test equipment when testing electrical systems, equipment, or devices to ensure accurate measurements. Never use test equipment on any energized circuit above what it is rated for. This could cause the test equipment to explode, possibly leading to serious injury or even death.

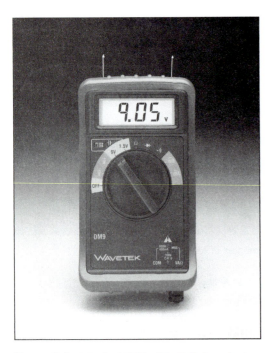

Figure 6-8: A typical VOM with digital readout. (Reproduced with Permission from Fluke Corporation)

Volt-Ohm-Milliammeter

A Volt-Ohm-Milliammeter (VOM), often referred to as a multimeter, is an electrical instrument that measures voltage (volts), resistance (ohms), and current (milliamperes). This instrument has several ranges in each mode. It is available in many types, ranges, and quality, either with a regular dial (analog) or with a digital readout. Figure 6-8 illustrates a typical multimeter with a digital readout. If you purchase one, be sure to select one with the features and ranges used by technicians in this field. The voltmeter that you use is very important. Your life may depend on knowing what voltages you are working with. Be sure to get the best one you can afford. Inexpensive ones are tempting. See page 128–129 for the procedure for "Using a Voltage Tester" and "Using a Noncontact Voltage Tester."

AC Clamp-On Ammeter

An AC clamp-on ammeter is a versatile instrument, which is also called clip-on, tang-type, snap-on, or other names. Some can also measure voltage or resistance or both. Unless you have an ammeter like this, you must interrupt the circuit to place the ammeter in the circuit. With this instrument you simply clamp the jaws around a single conductor, as shown in Figure 6-9. See page 129 for the procedure for "Using a Clamp-On Ammeter."

Figure 6-9: Measuring amperage by clamping the meter around the conductor. (Courtesy Bill Johnson)

Megohmmeter

A megohmmeter is used for measuring very high resistances. This particular device can measure up to 4,000 megohms (Figure 6-10).

Wiring and Crimping Tools

Wiring and crimping tools are available in many designs. Figure 6-11 illustrates a combination tool for crimping solderless connectors, stripping wire, cutting wire, and cutting small bolts. This figure also illustrates an automatic wire stripper. To use this tool, insert the wire into the proper strip-die hole. The length of the strip is determined by the amount of wire extending beyond the die away from the tool. Hold the wire in one hand and squeeze the handles with the other. Release the handles and remove the stripped wire.

Electrical Conductors

When working with electricity, the facilities maintenance technician will be exposed to various conductors (wire, cables, etc.). Having a good understanding of different types of conductors (wires) used in an electrical circuit is essential when working with electricity. Using the wrong size or type of conductor can result in a loss of property and/or even a fatality.

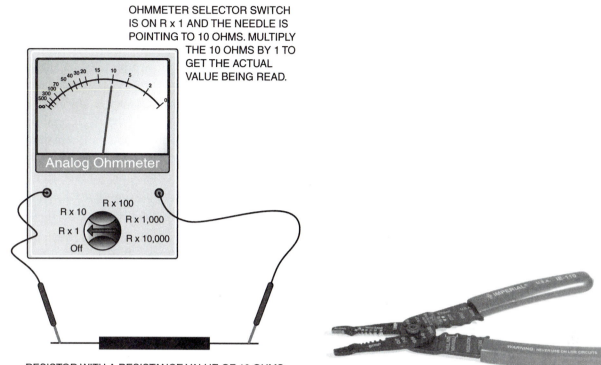

OHMMETER SELECTOR SWITCH IS ON R x 1 AND THE NEEDLE IS POINTING TO 10 OHMS. MULTIPLY THE 10 OHMS BY 1 TO GET THE ACTUAL VALUE BEING READ.

RESISTOR WITH A RESISTANCE VALUE OF 10 OHMS

Figure 6-10: **Analog ohmmeter**

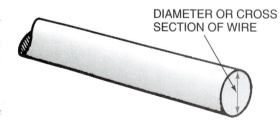

Figure 6-11: **Wire stripping and crimping tool**

Wire Sizes

All conductors have some resistance. The resistance depends on the material, the cross-sectional area, and the length of the conductor. A conductor with low resistance carries a current more easily than that with high resistance.

The proper wire (conductor) size must always be used. The size of a wire is determined by its diameter or cross section (Figure 6-12). A larger-diameter wire has more current-carrying capacity than a smaller-diameter wire.

Standard copper wire sizes are identified by American Standard Wire Gauge numbers and measured in circular mils. A circular mil is the area of a circle 1/1,000 in. in diameter. Temperature is also considered because resistance increases as temperature increases. Increasing wire size numbers indicate smaller wire diameters and greater resistance. For example, number 12 wire size is smaller than number 10 wire size and has less current-carrying capacity. The technician should not determine and install a conductor of a particular size unless licensed to do so. The technician should, however, be able to recognize an undersized conductor and bring it to the attention of a qualified person. As mentioned previously, an undersized wire may cause voltage to drop, breakers or fuses to trip, and conductors to overheat.

The conductors are sized by their amperage-carrying capacity, which is also called ampacity. Figure 6-13 contains a small part of a chart from the NEC and a

> *If a wire is too small for the current passing through, it will overheat and possibly burn the insulation and could cause a fire.*

DIAMETER OR CROSS SECTION OF WIRE

Figure 6-12: **The cross section of a wire**

106 RESIDENTIAL CONSTRUCTION ACADEMY FACILITIES MAINTENANCE

WIRE SIZE	60°C (140°F)
	Types TW, UF
AWG or kcmil	
COPPER	
18	—
16	—
14*	20
12*	25
10*	30
8	40
6	55
4	70
3	85
2	95
1	110

*Small Conductors. Unless specifically permitted in (e) through (g), the overcurrent protection shall not exceed 15 amperes for No. 14, 20 amperes for No. 12, and 30 amperes for No. 10 copper.

Figure 6-13: This section of the NEC shows an example of how a wire is sized for only one type of conductor. It is not the complete and official position but is a representative example

partial footnote for one type of conductor. This chart as shown and the footnote should not be used when determining a wire size. It is shown here only to familiarize you with the way in which it is presented in the NEC. The footnote actually reduces the amount of amperes for number 12 and number 14 wires listed in the NEC table in Figure 6-13. The reason is that number 12 and number 14 wires are used in residential houses where circuits are often overloaded unintentionally by the homeowner. The footnote exception simply adds more protection. An example of a procedure that might be used for calculating the wire size for the outdoor unit for a heat pump follows (Figure 6-14).

This outdoor unit is a 3½-ton unit (42 amp means 42,000 Btu/hour or 3½ tons). The electrical data show that the unit compressor uses 19.2 full-load amperage (FLA), and the fan motor uses 1.4 FLA for a total of 20.6 FLA. The specifications round this up to an ampacity of 25. In other words, the conductor must be sized for 25 amperes.

According to the NEC, the wire size would be 10. The specifications for the unit indicate that a maximum fuse or breaker size would be 40 amp. This will give the compressor some extra amperage for the locked-rotor amperage (LRA) at start-up. Some manufacturers specify the ampacity and the recommended wire size in the directions printed on the unit nameplate. When you find a unit that has low voltage while operating, you should first check the voltage at the entrance panel to the building. If the voltage is correct there and low at the unit, either there is a loose connection, there is undersized wire, or the wire run is too long.

NM Cable

Using improper wire types and wire sizes can result in a fire! It is important to know what type of wire or cable is used for a given application. The most common type of cable used in single-family and multifamily dwellings is NM cable (commonly referred to as Romex) (Figure 6-14). NM cable is covered in Article 334 of the NEC, where you will find its proper use and installation methods.

NM cable is available in many sizes and conductor pairs. This means that an NM cable comes with two, three, or four current-carrying conductors and a ground. The ground that is present in NM cable is typically bare (without insulation); all of the current-carrying conductors will have an insulation covering that is identified by the following colors:

- Black
- Red
- Blue
- White

General Data

4TWX4042A1000A

OUTDOOR UNIT ①②	4TWX4042A1000A	
SOUND RATING (DECIBELS) ②⑨	77/75	
POWER CONNS. — V/PH/HZ ③	208/230/1/60	
MIN. BRCH. CIR. AMPACITY	25	25 AMPACITY
BR. CIR. } MAX. (AMPS)	40	
PROT. RTG. } MIN. (AMPS)	40	
COMPRESSOR	CLIMATUFF® -SCROLL	
NO. USED - NO. SPEEDS	1 - 1	
VOLTS/PH/HZ	208/230/1/60	
R.L. AMPS ⑦ - L.R. AMPS	19.2-104	19.2 A
FACTORY INSTALLED		
START COMPONENTS ⑧	NO	
INSULATION/SOUND BLANKET	YES	
COMPRESSOR HEAT	YES	
OUTDOOR FAN — TYPE	PROPELLER	
DIA. (IN.) - NO. USED	27.6 - 1	
TYPE DRIVE - NO. SPEEDS	DIRECT - 2	
CFM @ 0.0 IN. W.G. ④	4200	
NO. MOTORS - HP	1 - 1/6	
MOTOR SPEED R.P.M.	825	
VOLTS/PH/HZ	200/230	
F.L. AMPS	1.4	1.4 A
OUTDOOR COIL — TYPE	SPINE FIN™	20.6 A TOTAL
ROWS - F.P.I.	1 - 24	
FACE AREA (SQ. FT.)	27.81	
TUBE SIZE (IN.)	5/16	
REFRIGERANT CONTROL	EXPANSION VALVE	
REFRIGERANT		
LBS. — R-410A (O.D. UNIT) ⑤	7/04 -LB/OZ	
FACTORY SUPPLIED	YES	
LINE SIZE - IN. O.D. GAS ⑥	3/4	
LINE SIZE - IN. O.D. LIQ. ⑥	3/8	
FCCV		
RESTRICTOR ORIFICE SIZE	0.071	
DIMENSIONS	H X W X D	
OUTDOOR UNIT - CRATED (IN.)	53.4 X 35.1 X 38.7	
UNCRATED	SEE OUTLINE DWG.	
WEIGHT		
SHIPPING (LBS.)	315	
NET (LBS.)	267	

OUTDOOR UNIT WITH HEAT PUMP COILS

① CERTIFIED IN ACCORDANCE WITH THE AIR-SOURCE UNITARY HEAT PUMP EQUIPMENT CERTIFICATION PROGRAM WHICH IS BASED ON A.R.I. STANDARD 210/240.
② RATED IN ACCORDANCE WITH A.R.I. STANDARD 270.
③ CALCULATED IN ACCORDANCE WITH NATIONAL ELECTRIC CODE. ONLY USE HACR CIRCUIT BREAKERS OR FUSES.
④ STANDARD AIR - DRY COIL - OUTDOOR
⑤ THIS VALUE APPROXIMATE. FOR MORE PRECISE VALUE SEE UNIT NAMEPLATE AND SERVICE INSTRUCTION.
⑥ MAX. LINEAR LENGTH: 80 FT WITH RECIPROCATING COMPRESSOR - 60 FT WITH SCROLL. MAX. LIFT - SUCTION 60 FT; MAX LIFT - LIQUID 60 FT. FOR GREATER LENGTH REFER TO REFRIGERANT PIPING SOFTWARE PUB. NO. 32-3312-01.
⑦ THE VALUE SHOWN FOR COMPRESSOR RLA ON THE UNIT NAMEPLATE AND ON THIS SPECIFICATION SHEET IS USED TO COMPUTE MINIMUM BRANCH CIRCUIT AMPACITY AND MAXIMUM FUSE SIZE. THE VALUE SHOWN IS THE BRANCH CIRCUIT SELECTION CURRENT.
⑧ NO MEANS NO START COMPONENTS
YES MEANS QUICK START KIT COMPONENTS
PTC MEANS POSITIVE TEMPERATURE COEFFICIENT STARTER.
⑨ RATED IN ACCORDANCE WITH ARI STANDARD 270/SECTION 5.3.6.

SPLIT SYSTEM

Figure 6-14: This specification sheet for one unit shows the electrical data that are used to size the wire to the unit. (Courtesy Trane)

The blue insulated conductor is found only in four-conductor NM cable. Admittedly, a four-conductor NM cable with ground is rarely used or seen, but it does exist. Most often, NM cable is sold as a two conductor with ground or a three conductor with ground.

Electrical Devices, Fixtures, and Equipment

Typically the electrical repair that a facilities maintenance technician does will consist of troubleshooting, repairing, and/or replacing electrical devices, fixtures, and equipment as opposed to replacing conductors. When troubleshooting, repairing, and/or replacing electrical devices, fixtures, and equipment, always follow manufacturer's recommendations, warning, and specifications. In most cases manufacturers of electrical devices (outlet, switch, a light fixture, etc.) supply diagrams on how to wire the device with the device.

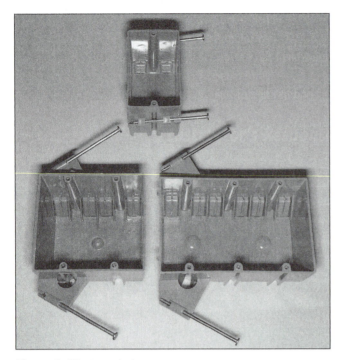

Figure 6-15: **Gang devices**

Figure 6-16: **Cut-in box**

Boxes

A multitude of boxes are used today for electrical installations. Only the most common of these boxes will be discussed here. One of the most common boxes in use is the single gang nail-up box (Figure 6-15). It is made of plastic or fiberglass and is typically used during new construction before the drywall is installed.

This box will have enough room for only one yoked device, such as a switch, a receptacle, or a dimmer switch. A two-gang nail-up box will have enough room for two yoked devices and a three-gang nail-up box for three yoked devices.

The "cut-in" box (see Figure 6-16) is made of plastic or fiberglass and is used when a device must be installed after the drywall is already in place. See page 130 for the procedure for "Installing Old-Work Electrical Boxes in a Sheetrock Wall or Ceiling."

Round ceiling boxes (see Figure 6-17) are available in various diameters, depths, and shapes. Four common types of ceiling boxes are:

- Pancake (the shallow box)
- 4" round nail-up box
- 4" round cut-in box
- Fan-rated ceiling box

Ceiling fan boxes (see Figure 6-18) are used to support a lighting fixture in the ceiling. It is important to know that a ceiling fan should not be installed on a ceiling box that is not fan-rated. If a ceiling box is rated to support a fan, it will be clearly and legibly stamped into the interior of the box. If a fan is mounted to a box that is not rated for a ceiling fan, the fan could fall during operation and cause injury to someone underneath it.

Figure 6-17: **Round ceiling box**

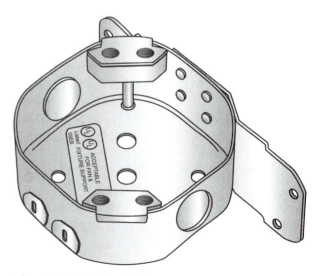

Figure 6-18: **Ceiling fan box**

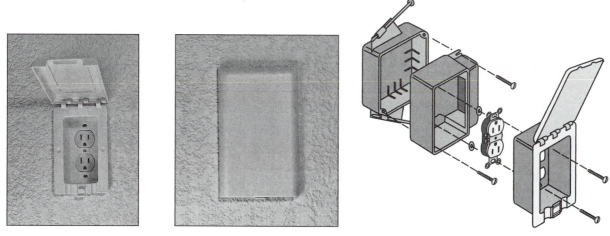

Figure 6-19: **Weatherproof box and box with cover**

Weatherproof boxes (see Figure 6-19) are used whenever a device (receptacle or switch) is installed in a wet location. This box is designed to keep the elements of the weather from affecting the electrical circuit and device that is contained within the box. If this box is installed in the direct weather and it houses a receptacle, it is to have an "in-use" cover. An in-use cover looks like a giant clear bubble.

Feeder Circuits

A feeder is the circuit conductor between the service equipment and the final branch-circuit overcurrent protection device. Some commercial wiring situations may call for another load center (panel), which is called a subpanel, to be located in another part of the structure. The reason for installing a subpanel is usually to locate a load center closer to an area of the house where several circuits are required. It is the wiring from the main panel to the subpanel that is referred to as the feeder. The feeder typically constitutes wiring in a cable or individual wires in an electrical conduit that are large enough to feed the electrical requirements of the subpanel they are servicing.

Surface Metal Raceways

When it is impossible to conceal conductors (for example, around a desk, counter, cabinets) a raceway is often used. Surface raceways, which are governed by Articles 386 and 388 of the NEC, are typically made of either metal or nonmetallic materials.

It should be noted that the number of conductors in a raceway is limited to the design of the raceway. Also that the combined size of the conductors, splices, and tape should not exceed 75 percent of the raceway. Also all conductors that are to be installed in a raceway should be spliced either in a junction box or within the raceway.

Multioutlet Assemblies

The NEC defines a multioutlet assembly as a surface, flush, or freestanding raceway designed to hold conductors and receptacles, assembled in the field or at the factory. They offer a high degree of flexibility to an installation by allowing for the likelihood that the installation and use requirements could change. For example, in an office building where cubicles are utilized, multioutlet assemblies offer a suitable solution. They are also utilized in heavy-use situations, for example, computer rooms and laboratories.

Floor Outlets

Floor outlets are often used in large office complexes where desks will be positioned too far away from a wall to effectively run a power cord. Floor outlets can be contained within a under floor raceway or by installing floor boxes. It should be noted that the installation requirements for an under floor raceway are set forth in the NEC Article 390.

Panelboards

A panelboard is defined by the NEC as a single panel or group of panel units designed for assembly in the form of a single panel, including buses and automatic overcurrent devices, and equipped with or without switches for the control of light, heat, or power circuits; designed to be placed in a cabinet or cutout box placed on or against a wall, partition, or other support; and accessible only from the front.

When separate feeders are to be run from the main service equipment to each of the areas of the commercial building, each feeder will terminate in a panelboard, which is to be installed in the area to be served.

Circuit Protection Devices

Circuit protection is essential to prevent the conductors in the circuit from being overloaded. If one of the power-consuming devices were to cause an overload due to a short circuit within its coil, the circuit protector would stop the current flow before the conductor becomes hot and overloaded. A circuit consists of a power supply, the conductor, and the power-consuming device. The conductor must be sized large enough that it does not operate beyond its rated temperature, typically 140°F (60°C) while in an ambient temperature of 86°F (30°C). For example, a circuit may be designed to carry a load of 20 amp. As long as the circuit is carrying up to its amperage, overheating is not a potential hazard. If the amperage in the circuit is gradually increased, the conductor will become hot (Figure 6-20). Proper understanding of circuit protection is a lengthy process. More details can be obtained from the NEC and further study of electricity.

Fuses

A fuse is a simple device used to protect circuits from overloading and overheating. Most fuses contain a strip of metal that has a higher resistance than the conductors in the circuit. This strip also has a relatively low melting point. Because of its higher resistance, it will heat up faster than the conductor. When the current exceeds the rating on the fuse, the strip melts and opens the circuit.

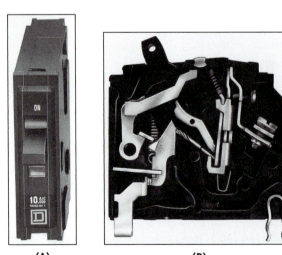

(A) **(B)**

Figure 6-20: **(A) A circuit breaker. (B) A cutaway. (Courtesy Schneider Electric)**

Plug Fuses

Plug fuses have either an Edison base or a type S base (Figure 6-21(A)). Edison-base fuses are used in older installations and can be used only for replacement. Type S fuses can be used only in a type S fuse holder specifically designed for the fuse; otherwise an adapter must be used (Figure 6-21(B)). Each adapter is designed for a specific ampere rating, and these fuses cannot be interchanged. The amperage rating determines the size of the adapter. Plug fuses are rated up to 125 volt and 30 amp.

Figure 6-21: **(A) A type S base plug fuse. (B) A type S fuse adapter**

Electrical circuits must be protected from current over- loads. If too much current flows through the circuit, the wires and components will overheat, resulting in damage and possible fire. Circuits are normally protected with fuses or circuit breakers.

Dual-Element Plug Fuses

Many circuits have electric motors as the load or part of the load. Motors draw more cur- rent when starting and can cause a plain (single element) fuse to burn out or open the circuit. Dual-element fuses are frequently used in this situation (Figure 6-22). One ele- ment in the fuse will melt when there is a large overload such as a short circuit. The other element will melt and open the circuit when there is a smaller current overload lasting more than a few seconds. This allows for the larger starting current of an electric motor.

Cartridge Fuses

For 230-volt to 600-volt service up to 60 amp, the ferrule cartridge fuse is used (Figure 6-23(A)). From 60 amp to 600 amp, knife-blade cartridge fuses can be used (Figure 6-23(B)). A cartridge fuse is sized according to its ampere rating to prevent a fuse with an inadequate rating from being used. Many cartridge fuses have an arc-quenching material around the element to prevent damage from arcing in severe short-circuit situations (Figure 6-24).

Figure 6-22: **A dual-element plug fuse. (Courtesy Cooper Bussmann)**

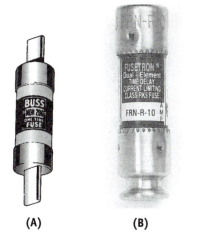

Figure 6-23: **(A) A ferrule-type cartridge fuse. (B) A knife-blade cartridge**

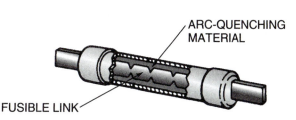

Figure 6-24: **A knife-blade cartridge fuse with arc quenching**

Circuit Breakers

A circuit breaker can function as a switch as well as a means for opening a circuit when a current overload occurs. Most modern installations in houses and many commercial and industrial installations use circuit breakers rather than fuses for circuit protection. Circuit breakers use two methods to protect the circuit. One is a bimetal strip that heats up with a current overload and trips the breaker, opening the circuit. The other is a magnetic coil that causes the breaker to trip and open the circuit when there is a short circuit or other excessive current overload in a short time (Figure 6-25).

Figure 6-25: **Single-pole circuit breaker (left). Double-pole circuit breaker (right)**

Emergency Backup Systems

Emergency backup systems are designed to keep critical parts of an electrical system energized in the event of power loss. Hospitals and assisted living facilities have emergency backup systems in order that life support systems will not be interrupted during the loss of power. These types of emergency backup systems will typically use a diesel engine that drives a generator. The facility will switch over to the generator only in the event of power loss.

Another form of an emergency backup system is the uninterruptible power supply (UPS) (see Figure 6-26). This system uses a battery or batteries to supply constant power to the circuits connected to it in the event of power loss. This system is available in various sizes. A UPS can be small enough to fit on a desk, or it can be so large that it may require its own room or even its own building on the facility grounds. The larger UPS systems require a lot of maintenance to maintain the reliability of the batteries. Many safety issues have to be considered when working on or around these large battery backup systems, and it is for this reason that only qualified personnel work on these systems.

Figure 6-26: **Uninterruptible power supply (UPS)**

Electrical Switches and Receptacles

Switches and receptacles are electrical devices. There are many variations of each, so we will discuss only the most common of each.

Switches

Four of the most commonly used switches are the single-pole, double-pole, three-way, and four-way switches. These switches are available in 15-amp and 20-amp ratings. Be sure to use the correct amp rating when installing a switch in a lighting branch circuit.

- *Single-pole switch* (see Figure 6-27): This switch is used when a light or fan is turned on or off from only one location.
- *Double-pole switch* (see Figure 6-28): This switch is used when two separate circuits must be controlled with one switch. They are used to control 240-volt loads, such as electric heat, motors, and electric clothes dryers.
- *Three-way switch* (see Figure 6-29): This switch is used when a light or fan can be turned on or off from two different locations.
- *Four-way switch* (see Figure 6-30): This switch is used when a light or fan can be turned on or off from three or more different locations. This switch must be used with two three-way switches.

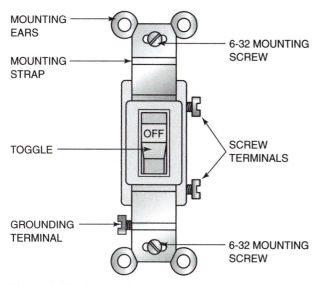

Figure 6-27: **Single-pole switch**

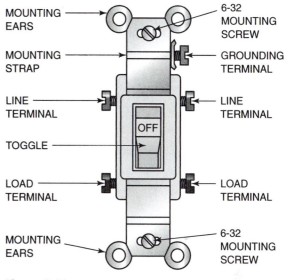

Figure 6-28: **Double-pole switch**

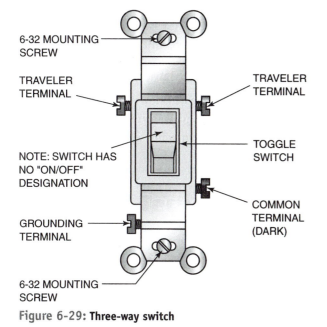

Figure 6-29: **Three-way switch**

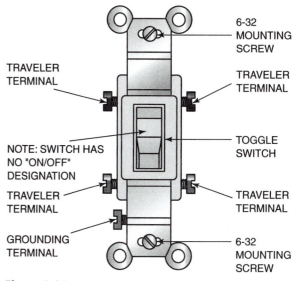

Figure 6-30: **Four-way switch**

Receptacles

Many types of receptacles are available. The most common receptacles are:

- 240-volt 30-amp or 50-amp single receptacle
- 240-volt 20-amp single receptacle
- 120-volt 15-amp or 20-amp single receptacle
- 120-volt 15-amp or 20-amp duplex receptacle
- 120-volt 15-amp or 20-amp ground fault circuit interrupter (GFCI) receptacle

The *240-volt 30-amp single receptacle* is generally used as a clothes dryer receptacle. It has four wires that are connected to it: two ungrounded conductors (hot wires), one grounded conductor (neutral), and one grounding conductor (bare ground wire).

The *240-volt 50-amp single receptacle* is generally used as an electric range (stove) receptacle. It also has four wires that are connected to it, as mentioned above.

The *240-volt 20-amp single receptacle* is generally used as an air conditioner receptacle. As the previous receptacles, it also has four wires that are connected to it. Notice the slots on the front of the receptacle.

The *120-volt 15/20-amp single receptacle* (see Figure 6-31) is used most often for electrical utilization equipment that requires a 20-amp branch circuit. This is so a 15-amp- or a 20-amp-cord can be plugged into it. Any receptacle devices that are 120 volt, 20 amp rated will have this feature.

The *120-volt 15-amp duplex receptacle* (see Figure 6-31) is the most common receptacle in use today.

The *120-volt 15/20-amp duplex receptacle* is similar to the previous receptacle except that it has the "T" slot to accommodate a 20-amp load. See pages 132–134 for the procedures for "Installing Duplex Receptacles in a Nonmetallic Electrical Outlet Box" and "Installing Duplex Receptacles in a Metal Electrical Outlet Box."

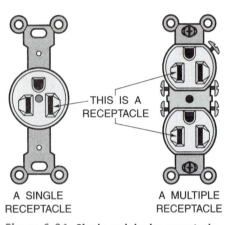

THIS IS A RECEPTACLE

A SINGLE RECEPTACLE

A MULTIPLE RECEPTACLE

Figure 6-31: Single and duplex receptacle

The *GFCI duplex receptacle* (see Figure 6-32) is designed to trip when there is a difference between the current going to the load and the current returning from the load. This device will trip if there is a difference of 4 mA between the two. This device is easily recognizable because of the Trip and Reset buttons that are on the face of the receptacle. Once this receptacle is properly installed on a branch circuit, every device and fixture connected to the branch circuit past the GFCI receptacle is protected. This receptacle is to be used above countertops in bathrooms and kitchens, in wet or damp locations (such as basements), and outside. This receptacle is also available in 15 amp and 20 amp.

See page 134 for the procedure for "Installing Feed-Through GFCI and AFCI Duplex Receptacles in Nonmetallic Electrical Outlet Boxes."

Fixtures

Various types of lighting fixtures are available today. Only the most common types will be discussed in this section.

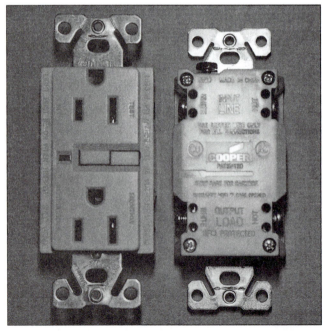

Figure 6-32: Ground fault circuit interrupter receptacles (GFCI)

Fixtures are described by the way they mount and by the type of bulb that is used within them. For example, a surface-mount, incandescent, ceiling fixture is one that mounts against the ceiling surface and has an incandescent bulb, the most commonly used light bulb. It is the standard frosted or clear light bulb that we use in our homes. It tends to give off a yellowish light when compared to a fluorescent light bulb. Fluorescent bulbs, most commonly referred to as "tubes," tend to give off a white light. See page 135 for the procedure for "Installing a Light Fixture Directly to an Outlet Box."

Here is a list of the most common light fixtures:

- Surface-mount incandescent fixture: mounts against ceiling or wall (see Figure 6-33). See page 136 for the procedure for "Installing a Cable-Connected Fluorescent Lighting Fixture Directly to the Ceiling"
- Surface-mount fluorescent fixture: mounts against ceiling or wall (see Figure 6-34). See page 139–140 for the procedure for "Installing a Fluorescent Fixture (Troffer) in a Dropped Ceiling"
- Recessed-can incandescent lighting fixture: mounts in the ceiling (see Figure 6-35). See page 137 for the procedure for "Installing a Strap on a Lighting Outlet Box Lighting Fixture"
- Pendant-type incandescent lighting fixture: hangs from the ceiling on a chain or cable (see Figure 6-36)
- Chandelier-type lighting fixture: multiple lamp fixtures hang from the ceiling on a chain or cable (see Figure 6-37). See page 138 for the procedure for "Installing a Chandelier-Type Light Fixture Using the Stud and Strap Connection to a Lighting Outlet Box"

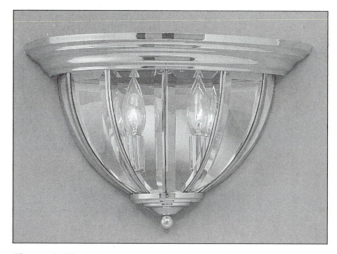

Figure 6-33: **Surface-mount incandescent fixture**

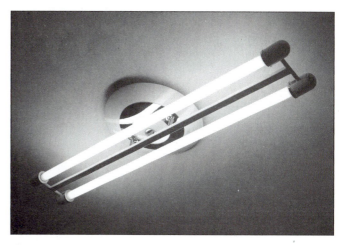

Figure 6-34: **Surface-mount fluorescent fixture (Image copyright JoLin, 2009. Used under license from Shutterstock.com)**

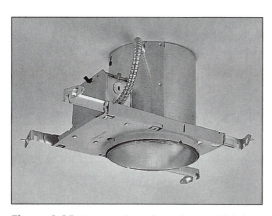

Figure 6-35: **Recessed-can incandescent lighting fixture**

Figure 6-36: **Pendant-type incandescent lighting fixture**

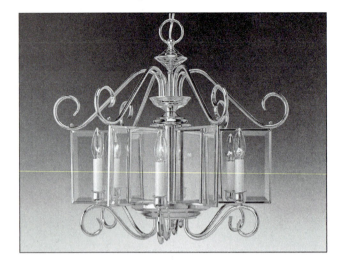

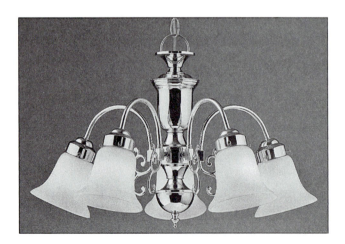

Figure 6-37: Chandelier-type lighting fixture

Electrical Maintenance Procedures

As mentioned earlier, facilities maintenance technicians are more likely to maintain an existing electrical system as opposed to installing a new system. Therefore the technician should have a good understanding of the troubleshooting process as well as the proper technique for maintaining electrical systems.

Troubleshooting

Troubleshooting is a process in which a person gathers information and forms a logical conclusion as to the problems that may be present within the system. All problems should be looked at logically. Take time and consider the most logical problems that would cause the symptoms that are present in the faulty circuit. If you do this, you have a higher chance of success in finding and fixing the problem. All problems that occur in a system will give tell-tale signs that will help you find the problem. Training is always beneficial. This gives you a knowledge base that will help you form the logical conclusions needed to solve problems with the system.

Troubleshooting occurs while you are gathering the information about the faulty system. Diagnosing begins while the data are being collected and is completed when a decision as to what the problem may be is formulated. Once you have diagnosed a problem, you must prove the diagnosis and repair the problem.

A simplified, step-by-step guideline that could be used while troubleshooting follows. *Remember: Do not perform any of the following steps unless you are a qualified individual.*

1. You are notified of a problem.
2. Ask the person who has reported the problem as many questions as possible as to what was witnessed during the failure. This may include something that was seen, smelling a distinct or peculiar odor, or feeling heat in the general vicinity of the problem.
3. Begin troubleshooting while making sure that all safety standards are adhered to. It is a good idea to start troubleshooting at the most logical area that would cause the described symptoms. You should start diagnosing the problem as soon as you receive the descriptions given to you and as you start receiving data from troubleshooting.
4. Safely remove any covers or panels that will give you access to the part of the electrical system that you are troubleshooting.
5. Visually inspect the equipment and devices for any signs of overheating or disintegration. If signs are visible, go to step 6a. If signs are not visible, go to step 6b.
6a. *Be sure that you are qualified to work on the device and equipment before working on any part of an electrical system.* De-energize the circuit or system, attach your lock-out/tag-out device to the disconnecting means, and place the key in *your* pocket. Go to step 7.
6b. Take the appropriate step to safely acquire voltage and/or current readings at the suspected device or equipment. If your readings indicate that voltage is present and current is not following as it should be, the suspected device or equipment may be faulty and it may need to be replaced. If you decide to re- place the faulty device or equipment, de-energize the circuit or system, attach

your lock-out/tag-out device to the disconnecting means, and place the key in *your* pocket; then go to step 7.

7. Go back to the device or equipment and verify that the circuit is in fact de-energized.

8. Once it has been verified that the circuit is de-energized, begin working on your fault.

   **In the event that you may need to repair and/or replace common electrical devices such as receptacles, switches, interior and exterior lighting fixtures, bulbs, or ballasts, follow the simple procedures listed in the last section of this chapter.*

9. Once the fault is repaired, and all covers are back in place, remove the lock-out/tag-out from the source of energy and re-energize the circuit.

10. Go back to the device or equipment that was replaced and verify that it is working properly. If it is, inform the person who called the job in. If it is not working, you may want to consider calling a qualified electrician to troubleshoot and diagnose the problem.

Perform Tests

Regularly perform tests on the following to ensure that they are operating properly before an emergency arises:

- Smoke alarms
- Fire alarms
- Medical alert systems
- Emergency exit lighting
- GFCI receptacles

Test Smoke Alarms and Fire Alarms

Individual smoke and fire alarms typically have a test button that can be pushed. Be aware that pushing the test button on any one alarm may set off all alarms that are on the system as the fire alarm code requires all of them to be tied together.

Some fire alarm systems may require you to put the system in test mode before testing. If this is not done and the fire alarm is activated during the test, the sprinkler systems may activate. Placing the fire alarm system in test mode would allow the electrical portion of the system to be tested without the sprinklers activating during the test. Be sure to repair or replace any defective alarm or smoke detector according to manufacturer's specifications and wiring diagrams, and report any malfunctions to your supervisor.

Many of the smoke detectors have a 9-volt battery in them so that the alarm can continue to work in the event that there is a loss of power. The batteries should be checked on a regular basis. If a detector is found to have a dead or weak battery, the battery should be changed immediately. If the smoke detector does not operate properly, it should be changed immediately (see Figure 6-38).

Figure 6-38: **Smoke detector**

To test a smoke detector:

1. Press the battery-test button on the unit to make sure the battery is properly connected.
2. If the unit has a battery that's more than a year old, replace the battery.
3. Light a candle and hold it approximately 6 inches below the detector so that heated air will rise into the unit.
4. If the alarm doesn't sound within 20 seconds, blow out the candle and let the smoke rise into the unit.
5. If the alarm still doesn't sound, open the unit up and make sure it is clean and that all electrical connections are solid.
6. If, again, the alarm doesn't sound, replace the smoke detector.

To replace a smoke detector battery:

1. Remove the smoke detector cover, typically by carefully pulling down on the case's perimeter or twisting the case counterclockwise.
2. Locate and remove the battery.
3. Replace it with a new one.
4. Close the case and test the smoke detector.

Read the owner's manual for additional troubleshooting tips and possible adjustments.

Test GFCI Receptacles

1. Go to the receptacle and locate the "Test" button.
2. Press the test button and listen for a very light "pop" in the receptacle. You may also notice a small indicator that will light up after you have tripped the GFCI. If the GFCI tripped, then it is working correctly.
3. Press the "Reset" button, and the GFCI should reset and the indicator (if preset) will turn off.
4. If it did not trip, press the test button again.

If the GFCI still does not trip, then it is faulty and needs to be replaced, or it could have been wired incorrectly and therefore would need to be rewired correctly. For the correct installation procedures of a GFCI receptacle, see page 000.

Test Medical Alert Systems

Assisted living facilities may have and hospitals will have medical alert systems. If your facility has medical alert systems, all tests that are conducted on these systems must be performed according to the manufacturer's specifications and should be performed only by qualified personnel who have been properly trained on the proper test procedures for that system. If any defects are found in the system, they should be reported to your supervisor immediately.

Replace Detectors, Devices, Fixtures, and Bulbs

Be sure that as you replace or repair any defective equipment or devices, you do so according to the manufacturer's specifications. Also, don't forget to report any malfunctions to your supervisor.

Figure 6-39: **Lock-out/tag-out devices**

Replace Smoke Detectors

1. Acquire a fiberglass or nonconductive ladder that will be tall enough for you to reach the detector once you are on it. *Do not use an aluminum ladder.* Set the ladder up under the faulty detector, making sure that all four legs of the ladder are solidly in place.
2. Go to the panel and de-energize the branch circuit that supplies the smoke detectors.
3. Lock out/tag out the breaker (see Figure 6-39).
4. Go back to the faulty detector and remove it from its mounting base by twisting the detector body in a counterclockwise direction. The detector should release from the base (see Figure 6-40).

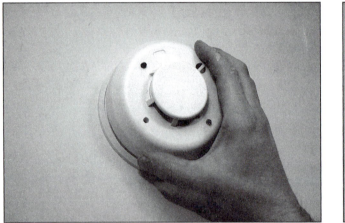

Figure 6-40: **Removing faulty detector**

5. Make a note of how the smoke detector is wired. If necessary, make a sketch on a piece of paper to follow when you are reconnecting the power leads. You may find that the smoke detector you are replacing has a quick connector plug on the back of the smoke detector. If you are replacing the faulty detector with a new one that is the same model, simply unplug the quick connect from the smoke detector. If a quick connect is not present, you may have to disconnect the power leads from the smoke detector by removing the wirenuts. There should be three leads on a newer type of smoke detector (see Figure 6-41). One will be black, one white, and one orange. The black wire is the hot wire, the white wire is the neutral, and the orange wire is for the repeating circuit. (This wire triggers all of the other smoke alarms.)
6. If you are replacing the faulty detector with a new one that is the same model, you may be able to leave the mounting base mounted to the ceiling. If you are replacing the faulty smoke detector with a different brand or model, you may find that the base of the old one will not work for the new one, and that the old mounting base must therefore be removed. This can be accomplished by loosening the two screws that are holding the base to the box. Once the base is loose, simply twist it in a counterclockwise direction as you did with the detector. The

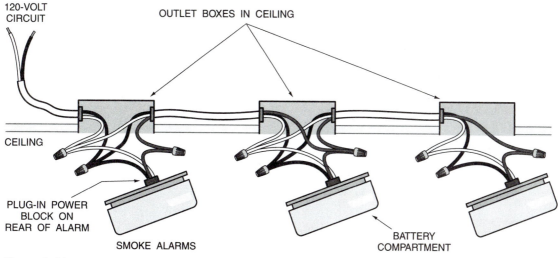

Figure 6-41: **Interconnected smoke detectors**

mounting base should release from the box. Now replace the old base with the new base and tighten the screws (see Figure 6-42).

7. Reconnect the power leads. If you are replacing an older two-wire model with a newer three-wire model, you will not need to use the orange lead. Simply cap the orange lead and stuff it into the box.

8. Set the new detector up against the mounting base, and twist it in a clockwise direction until it is securely mounted.

9. Go to the panel, remove your lockout from the breaker, and turn on the breaker. If the breaker does not trip, go to the next step. If the breaker trips, set the breaker to the off position and lock it out again. Go back to the detector, remove it from the base, and visually inspect the connections. Be sure that there is no exposed copper wire on the black lead that could be touching the bare copper ground wire. Once you think everything is correct, put the detector back on its base, go back to the panel, remove your lockout from the breaker, and turn the breaker back on. If the breaker trips again, call a qualified electrician to troubleshoot the problem. If the breaker does not trip, go to the next step.

10. Go back to the detector and verify that it is working correctly.

11. Gather all tools and ladders, and clean up the area that you were working in.

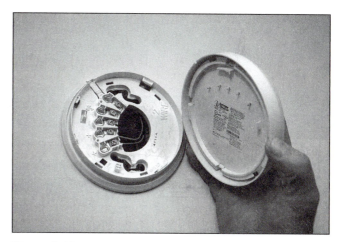

Figure 6-42: **Mounting base still attached to box**

Replace a Switch

To remove the old switch:

1. Shut off the power at the circuit breaker box.
2. Remove the cover plate (see Figures 6-43 and 6-44) and test the terminals with the circuit tester. If the tester does not light up, there is no power going to the switch.

Figure 6-43: **Cover plate (Image copyright Dani Simmonds, 2009. Used under license from Shutterstock.com)**

Figure 6-44: **Cover plate removed (Image copyright Tootles, 2009. Used under license from Shutterstock.com)**

Figure 6-45: **Replacement switch (Image copyright Bob Hosea, 2009. Used under license from Shutterstock.com)**

Figure 6-46: **New switch (Image copyright Ervstock, 2009. Used under license from Shutterstock.com)**

3. Remove the screws that hold the switch in place, and pull the switch from the wall.
4. Loosen the screws holding the wires to the switch, remove the wires, and remove the switch. In some newer switches, the wires may go directly into the switch, where they are held in place by clamps inside the switch. These switches usually have a slot into which you can insert a small screwdriver to loosen the clamps.

To install the new switch:

1. Begin by bending the end of the ground wire into a small hook, placing the hook over the ground terminal of the switch, and tightening it into place (see Figure 6-45).
2. Attach the remaining wires to the terminals and tighten them in the same way. Your new switch may have holes that allow you to insert the wires without using the terminal screws. If so, straighten the wires, press them into the holes as far as they will go, and then tug on them to make sure they are held securely in place.
3. Gently press the switch back into position and secure it in place with screws. Then replace the switch cover (see Figure 6-46), and turn on the power at the circuit breaker box.

Replace a Receptacle

The following steps should be used to replace an existing receptacle:

1. Turn the power off to the receptacle.
2. Test the receptacle to ensure that the power is off.
3. Remove the cover plate and receptacle from the box.
4. Mark the common wire with a piece of tape. The common wire terminal of the receptacle will have a different color from that of the other terminals.
5. Detach the common wire from the old receptacle, followed by the neutral and ground wires.

6. Locate the common terminal on the new receptacle and attach the wire. Remember that the common terminal will have a different color.
7. Attach the neutral and ground wires to the new receptacle.
8. Push the receptacle back into the box.
9. Install the receptacle cover onto the new receptacle.
10. Turn the power back on.

Replace a GFCI Receptacle

A GFCI receptacle has a line side and a load side (see Figure 6-47).

If the GFCI receptacle has only one black and one white wire terminated to it (not counting the ground wire), then both wires are to be terminated to the line side of the GFCI receptacle while making sure to connect the black wire to the brass-colored screw, the white wire to the silver screw, and the bare wire (if present) to the green-colored screw.

If other receptacles on the branch circuit are fed through the GFCI receptacle, there will be five wires on the receptacle: two black wires, two white wires, and one ground wire. If this is the case, terminate the wires that are feeding the line voltage into the GFCI receptacle to its line side. The wires that are feeding the remaining receptacles on the branch circuit should be connected to the load side of the GFCI receptacle (see the procedure on page 134).

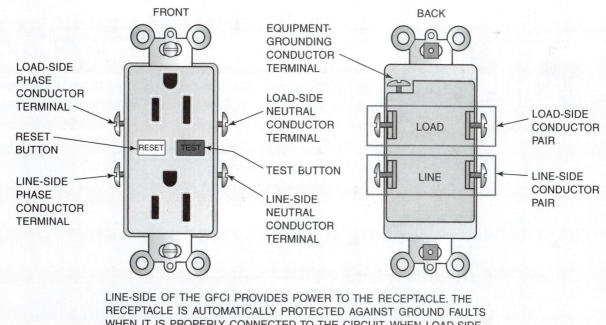

LINE-SIDE OF THE GFCI PROVIDES POWER TO THE RECEPTACLE. THE RECEPTACLE IS AUTOMATICALLY PROTECTED AGAINST GROUND FAULTS WHEN IT IS PROPERLY CONNECTED TO THE CIRCUIT. WHEN LOAD-SIDE CONDUCTORS ARE CONNECTED TO THE TERMINALS MARKED LOAD, EVERY DEVICE, APPLIANCE, OR OTHER EQUIPMENT ON THE LOAD-SIDE OF THE GFCI WILL ALSO HAVE GFCI PROTECTION. THE PROTECTION OF LOAD-SIDE COMPONENTS IS SOMETIMES REFERRED TO AS A FEED-THROUGH. A GFCI RECEPTACLE TAKES UP A LOT OF ROOM IN THE BOX.

Figure 6-47: Ground fault circuit interrupter (GFCI)

Figure 6-48: **Face plate**

Replace or Repair a Light Fixture (120-Volt Outlet)

1. Turn off both the branch circuit and the main power at the service panel. Work in the daytime so that you can use natural light to light the work area. Also use a flashlight if needed.
2. Remove the plate and the outlet mounting screws (see Figure 6-48).
3. Pull the outlet with wires still attached about 4–6 inches out of the junction box.
4. Note the color of the wires and identify the hot ground, neutral ground, and device ground.
5. Unscrew the terminal screws that attach the wire to the outlet and remove the wire. Start with the hot, then the neutral, and finally the ground.
6. Examine the new outlet. Identify which wire connects to which terminal. It does not matter which set of vertical screws you attach the wire to. If the outlet does not have markings indicating the polarity, then remember that the bright brass screw connects to your hot wire.
7. Using needle-nose pliers, connect the ground to the green terminal on the bottom of the outlet. Then connect the white wire to the neutral or silver terminal, and finally the hot wire to the hot or brass terminal. The wire should be wrapped completely around the terminal screws.
8. Finally, tuck the wires back in the junction box, and mount the outlet and the outlet plate.

Note: Make sure the outlet is rated for the amperage of the circuit. Do not use an outlet rated at 15 amp on a 20-amp circuit.

Replace a Light Bulb

1. Turn off the lamp or light fixture.
2. Allow a hot bulb to cool before touching it.
3. Grasp the bulb lightly but firmly and turn counterclockwise until it is released from the socket (see Figure 6-49).
4. Insert a replacement bulb lightly but firmly into the socket, and turn it clockwise until it is snug.
5. Turn the lamp or fixture back on.
6. Dispose of the used bulb.

Figure 6-49: **Removing light bulb**

Replace a Fluorescent Bulb

1. Make sure the light switch is turned off.
2. Remove the lens or diffuser to access the bulb (see Figure 6-50). On most fluorescent lights, the lens or diffuser is a plastic panel below the bulb (see Figure 6-51). Push the panel up and tilt to remove.
3. Check to make sure that the problem is not something as simple as a poor contact. This can usually be corrected by giving the bulb a gentle turn a few degrees and then back to the lock position.

Figure 6-50: **Fluorescent light bulb (Image copyright JoLin, 2009. Used under license from Shutterstock.com)**

Figure 6.51: **Lens/diffuser on a fluorescent fixture**

4. Hold the old bulb firmly at one end and rotate it one quarter-turn clockwise. This should put the end prongs in line with the loading slot.
5. Slide the bulb free.
6. Lower the end of the bulb carefully out of the socket. When one end is free, pull slightly and the other end should come out also.
7. Set the old bulb aside and lift a new bulb into the fixture.
8. Hold the bulb horizontally and rotate the new bulb until the prongs on each end are lined up with the grooves in the socket.
9. Insert the prongs in the socket, and rotate the bulb a quarter-turn in a counter-clockwise direction. The bulb should click into place on each end.
10. Test the light at the switch. If the light still doesn't come on, you may need to replace the ballast.

Replacing a Fluorescent Ballast

1. Find the ballast, which is usually seated near one end of the bulb. It is a silver, cylinder-shaped item with a diameter similar to that of a quarter. The ballast provides the starting voltage and then stabilizes the current for the fluorescent bulb (see Figure 6-52).
2. Loosen the ballast by turning it ¼- to ½-turn counterclockwise.
3. Pull the ballast out of the fixture. Take it to a hardware or electrical supply store to match it with a new one.
4. Insert a new ballast into the light fixture; twist clockwise to lock into position. Always check manufacturer's instruction before installing any electrical parts and/or devices.
5. Replace the lens cover.

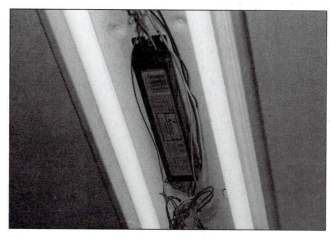

Figure 6-52: **Ballast on a fluorescent fixture**

Figure 6-53: **Screw in fuse (Image copyright Tootles, 2009. Used under license from Shutterstock.com)**

Test and Replace Fuses

Identify and Replace Blown Fuses

1. Open the door to your service panel, and examine it to locate the blown fuse. For light and receptacle circuits, look for a break or blackened area visible through the glass of a screw-in plug fuse (see Figure 6-53). If all the fuses look good, identify the fuse according to the circuit map printed on the door of the service panel or next to each fuse.
2. If the circuits are not mapped, locate the fuse by trial and error. Remove the fuses one at a time, and insert a new fuse to test the circuit. You can also test the fuse with a continuity tester. Touch the pointed probe of a continuity tester to the fuse's tip and the clip to its threaded shaft. If the tester does not glow, the fuse is bad.
3. For fuse blocks (see Figure 6-54), which protect an electric stove and the main circuit, pull straight out on the handle and remove the individual cartridge fuses from the block using a cartridge-fuse puller. Test the fuses with a continuity tester by probing the two ends.

Install a New Fuse

1. Screw in a new plug fuse, or install a new cartridge fuse in the fuse block and press the block back in. The replacement should always have the same rating as the original.
2. If all of the circuits have stopped working, remove and test the cartridge fuses in the main fuse block, usually located at the top left. (Occasionally it is reversed with the stove circuit on the top right.) Replace any faulty ones.

Reset a Circuit Breaker

Figure 6-54: **Fuse block**

1. Push the breaker switch to the "off" position and then back fully to the "on" position.
2. There will be a click as it snaps into the "on" position.

If the breaker trips again, you need to determine the reason for the overcurrent condition and correct the root cause of the problem. The breaker may be tripping due to excessive amperage in the circuit or may be shorting out.

Procedures

Using a Voltage Tester

- In this example, we will determine which conductor of a circuit is grounded using a voltage tester.

- Put on safety glasses, and observe regular safety procedures.

A Connect the tester between one circuit conductor and a well-established ground.

- If the tester indicates a voltage, the conductor being tested is not grounded.

B Continue this procedure with each conductor until zero voltage is indicated between the tested conductor and the known ground. Zero voltage indicates that you have found a grounded circuit conductor.

- In this example, we will determine the approximate voltage between two conductors using a voltage tester.

- Put on safety glasses and observe regular safety procedures.

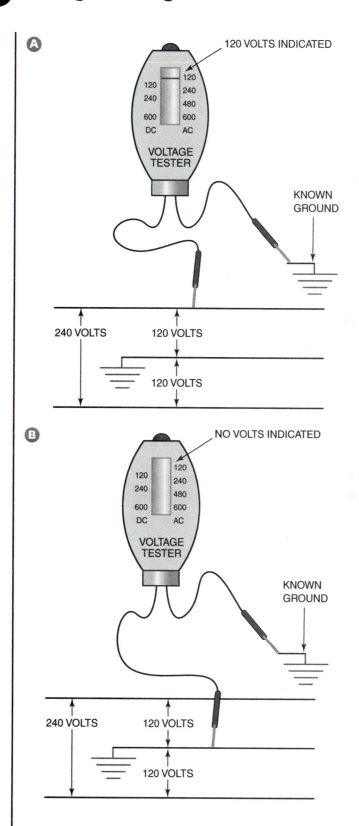

Procedures

Using a Voltage Tester (Continued)

C Connect the tester between the two conductors.

- Read the indicated voltage value on the meter. *Note:* With a solenoid type tester, you should also feel a vibration that is another indication of voltage being present.

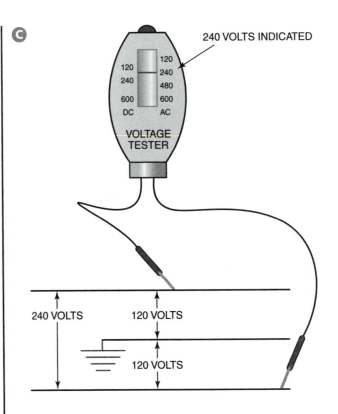

Procedures

Using a Noncontact Voltage Tester

- In this example, we will determine if an electrical conductor is energized using a noncontact voltage tester.

- Put on safety glasses and observe regular safety procedures.

- Identify the conductor to be tested.

(A) Bring the noncontact voltage tester close to the conductor. Note that some noncontact voltage testers may have to be turned on before using.

- Listen for the audible alarm, observe a light coming on, or feel a vibration to indicate that the conductor is energized.

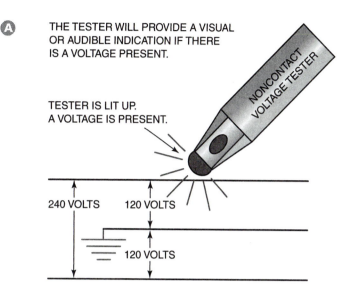

A THE TESTER WILL PROVIDE A VISUAL OR AUDIBLE INDICATION IF THERE IS A VOLTAGE PRESENT.

TESTER IS LIT UP. A VOLTAGE IS PRESENT.

NONCONTACT VOLTAGE TESTER

240 VOLTS   120 VOLTS

120 VOLTS

Procedures

Using a Clamp-On Ammeter

- In this example, we will be measuring current flow through a conductor with a clamp-on ammeter. Note that you can take a current reading with a clamp-on ammeter clamped around only one conductor. For example, a clamp-on meter will not give a reading when clamped around a two-wire Romex cable.

- Put on safety glasses and observe regular safety procedures.

- If the meter is analog and has a scale selector switch, set it to the highest scale. Skip this step if the meter is digital and has an autoranging feature.

(A) Open the clamping mechanism and clamp it around the conductor.

- Read the displayed value.

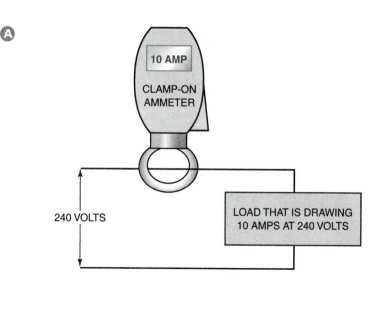

A

10 AMP

CLAMP-ON AMMETER

240 VOLTS

LOAD THAT IS DRAWING 10 AMPS AT 240 VOLTS

Procedures

Installing Old-Work Electrical Boxes in a Sheetrock Wall or Ceiling

- Put on safety glasses and follow all applicable safety rules.

- Determine the location where you want to mount the box and make a mark. Make sure there are no studs or joists directly behind where you want to install the box.

- Turn the old-work box you are installing backward, and place it at the mounting location so the center of the box is centered on the box location mark. Trace around the box with a pencil. Do not trace around the plaster ears.

- Using a keyhole saw, carefully cut out the outline of the box. There are two ways to get the cut started. One way is to use a drill and a flat blade bit (say a ½-inch size), drill out the corners, and start cutting in one of the corners. The other way used often is to simply put the tip of the keyhole saw at a good starting location and, with the heel of your hand, hit the keyhole saw handle with enough force to cause the blade to go through the sheetrock. It usually does not require much force to start the cut this way.

Ⓐ Assuming that a cable has been run to the box opening, secure the cable to the box and insert it into the hole. Secure the box to the wall or ceiling surface with Madison hold-its, or use a metal or nonmetallic box with built-in drywall grips.

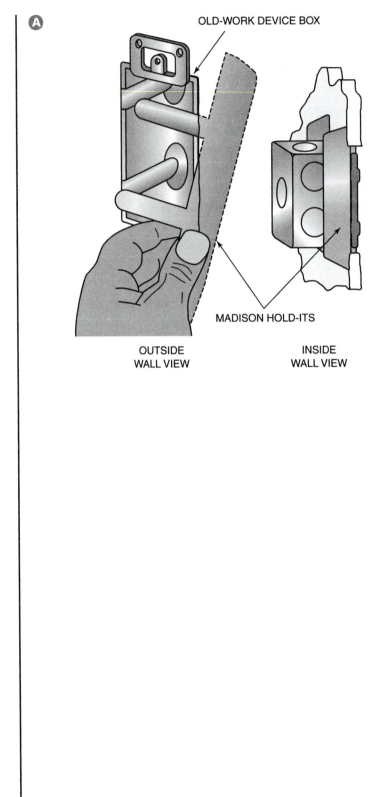

Ⓐ

OLD-WORK DEVICE BOX

MADISON HOLD-ITS

OUTSIDE WALL VIEW

INSIDE WALL VIEW

Procedures

Installing Duplex Receptacles in a Nonmetallic Electrical Outlet Box

- Wear safety glasses and observe all applicable safety rules.

A Using a wire stripper, remove approximately ¾ inch of insulation from the end of the insulated wires.

B Using long-nose pliers or wire strippers, make a loop at the end of each of the wires.

- Place the loop on the black wire around a brass terminal screw so that the loop is going in the clockwise direction. While pulling the loop snug around the screw terminal, tighten the screw to the proper amount with a screwdriver.

- Place the loop on the white wire around a silver terminal screw so that the loop is going in the clockwise direction. While pulling the loop snug around the screw terminal, tighten the screw to the proper amount with a screwdriver.

C Complete the installation by placing the loop on the bare grounding wire around the green terminal screw so that the loop is going in the clockwise direction. While pulling the loop snug around the screw terminal, tighten the screw to the proper amount with a screwdriver.

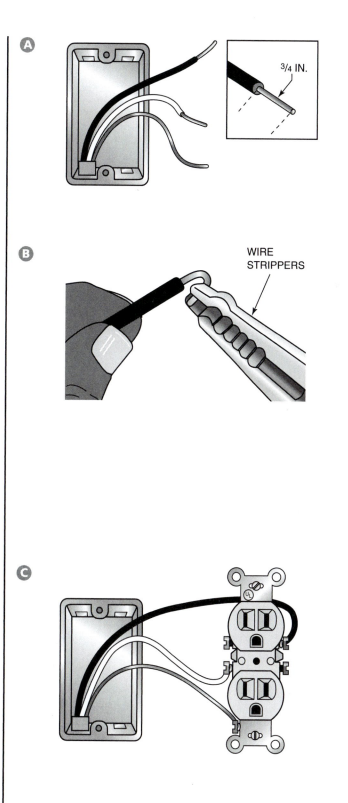

A

¾ IN.

B

WIRE STRIPPERS

C

Procedures

Installing Duplex Receptacles in a Nonmetallic Electrical Outlet Box (Continued)

D Place the receptacle into the outlet box by carefully folding the conductors back into the device box.

E Secure the receptacle to the device box using the 6-32 screws. Mount the receptacle so that it is vertically aligned.

F Attach the receptacle cover plate to the receptacle. Be careful not to tighten the mounting screw(s) too much. Plastic faceplates tend to crack very easily.

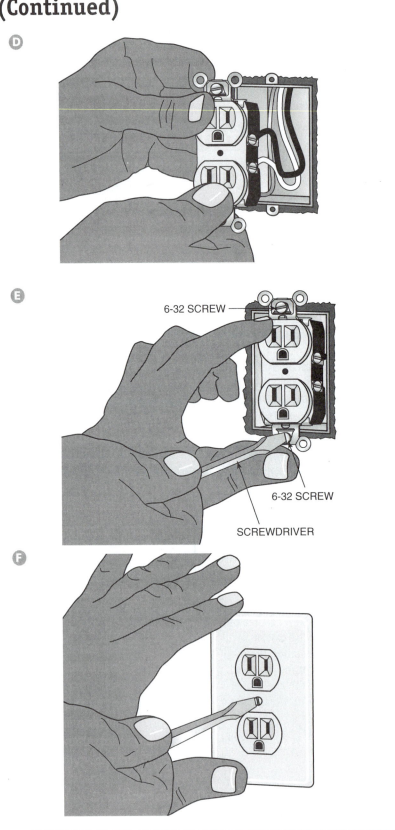

D

E

6-32 SCREW

6-32 SCREW

SCREWDRIVER

F

Procedures

Installing Duplex Receptacles in a Metal Electrical Outlet Box

- Wear safety glasses and observe all applicable safety rules.

A Attach a 6–8-inch-long grounding pigtail to the metal electrical outlet box with a 10–32 green grounding screw. The pigtail can be a bare or green insulated copper conductor.

B Attach another 6–8-inch-long grounding pigtail to the green screw on the receptacle.

C Using a wirenut, connect the branch-circuit grounding conductor(s), the grounding pigtail attached to the box, and the grounding pigtail attached to the receptacle together.

- Using a wire stripper, remove approximately ¾ inch of insulation from the end of the insulated wires.

- Using long-nose pliers or wire strippers, make a loop at the end of each of the wires.

- Place the loop on the black wire around a brass terminal screw and the loop on the white wire around a silver terminal screw so that the loops are going in the clockwise direction. Tighten the screws to the proper amount with a screwdriver.

- Place the receptacle into the outlet box by carefully folding the conductors back into the device box.

- Secure the receptacle to the device box using the 6-32 screws. Mount the receptacle so that it is vertically aligned.

- Attach the receptacle cover plate to the receptacle. Be careful not to tighten the mounting screw(s) too much. Plastic faceplates tend to crack very easily.

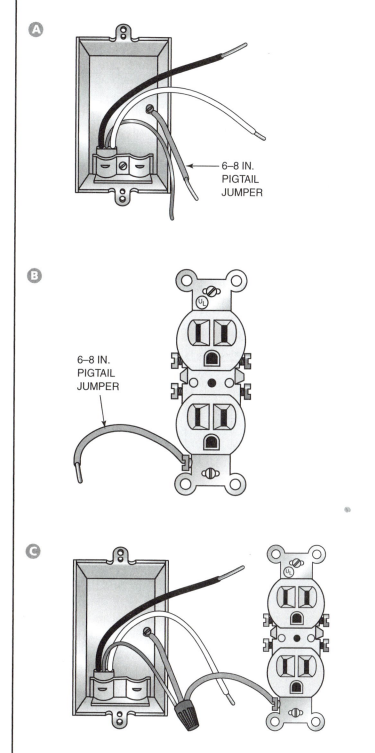

A

6–8 IN. PIGTAIL JUMPER

B

6–8 IN. PIGTAIL JUMPER

C

Procedures

Installing Feed-Through GFCI and AFCI Duplex Receptacles in Nonmetallic Electrical Outlet Boxes

- Wear safety glasses and observe all applicable safety rules.

Ⓐ At the electrical outlet box containing the GFCI or AFCI feed-through receptacle, use a wirenut to connect the branch-circuit grounding conductors and the grounding pigtail together. Connect the grounding pigtail to the receptacle's green grounding screw.

- At the electrical outlet box containing the GFCI or AFCI feed-through receptacle, identify the incoming power conductors, and connect the white grounded wire to the line-side silver screw and the incoming black ungrounded wire to the line-side brass screw.

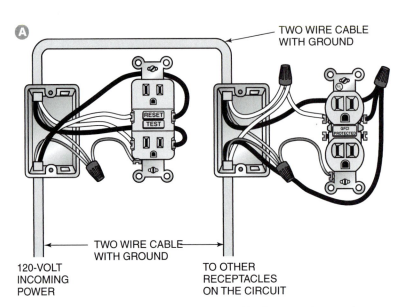

Ⓐ

TWO WIRE CABLE WITH GROUND

RESET TEST

GFCI PROTECTED

TWO WIRE CABLE WITH GROUND

120-VOLT INCOMING POWER

TO OTHER RECEPTACLES ON THE CIRCUIT

- At the electrical outlet box containing the GFCI or AFCI feed-through receptacle, identify the outgoing conductors and connect the white grounded wire to the load-side silver screw and the outgoing black ungrounded wire to the load-side brass screw.

- Secure the GFCI or AFCI receptacle to the electrical box with the 6-32 screws provided by the manufacturer.

- A proper GFCI or AFCI cover is provided by the device manufacturer; attach it to the receptacle with the short 6-32 screws provided.

- At the next "downstream" electrical outlet box containing a regular duplex receptacle, connect the white grounded wire(s) to the silver screw(s) and the black ungrounded wire(s) to the brass screw(s) in the usual way. Place a label on the receptacle that states "GFCI Protected." These labels are provided by the manufacturer.

- Continue to connect and label any other "downstream" duplex receptacles as outlined in the previous step.

Procedures — Installing a Light Fixture Directly to an Outlet Box

- Put on safety glasses and observe all applicable safety rules.

- Using a voltage tester, verify that there is no electrical power at the lighting outlet where the fixture will be installed. If electrical power is present, turn off the power and lock out the circuit.

- Locate and identify the ungrounded, grounded, and grounding conductors in the lighting outlet box.

- **A** The grounding conductor will not be connected to this fixture. If there is a grounding conductor in a nonmetallic box, simply coil it up and push it to the back or bottom of the electrical box. Do not cut it off, as it may be needed if another type of light fixture is installed at that location. If there are two or more grounding conductors in a nonmetallic box, connect them together with a wirenut and push them to the back or bottom of the lighting outlet box. If there is a grounding conductor in a metal outlet box, it must be connected to the outlet box by means of a listed grounding screw or clip. If there are two or more grounding conductors in a metal box, use a wirenut to connect them together along with a grounding pigtail. Attach the grounding pigtail to the metal box with a listed grounding screw.

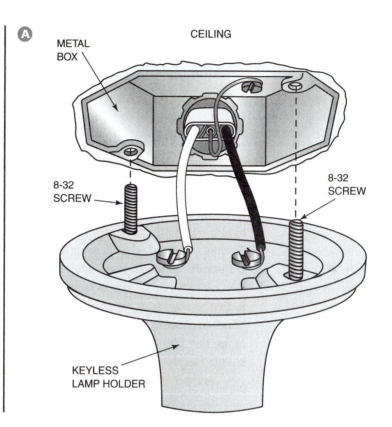

- Connect the white grounded conductor(s) to the silver screw or white fixture pigtail. If there is one grounded conductor in the box, strip approximately ¾ inch (19 mm) of insulation from the end of the conductor, and form a loop at the end of the conductor using an approved tool such as a T-stripper. Once the loop is made, slide it around the silver terminal screw on the fixture, so the end is pointing in a clockwise direction. Hold the conductor in place and tighten the screw. If there are two or more grounded conductors in the box, strip the ends as described previously, and use a wirenut to connect them and a white pigtail together. Attach the pigtail to the silver grounded screw as described previously.

- The black ungrounded conductor(s) is connected to the brass-colored terminal screw on the fixture. The connection procedure is the same as for the grounded conductor(s).

- Now the fixture is ready to be attached to the lighting outlet box. Make sure the grounding conductors are positioned so they will not come in contact with the grounded or ungrounded screw terminals. Align the mounting holes on the fixture with the mounting holes on the lighting outlet box. Insert the 8-32 screws that are usually equipped with the fixture through the fixture holes. Then thread them into the mounting holes in the outlet box.

- Tighten the screws until the fixture makes contact with the ceiling or wall. Be careful not to overtighten the screws, as you may damage the fixture.

- Install the proper lamp, remembering not to exceed the recommended wattage.

- Turn on the power and test the light fixture.

Procedures

Installing a Cable-Connected Fluorescent Lighting Fixture Directly to the Ceiling

- Put on safety glasses and observe all applicable safety rules.

- Using a voltage tester, verify that there is no electrical power at the lighting outlet where the fixture will be installed. If electrical power is present, turn off the power and lock out the circuit.

- Place the fixture on the ceiling in the correct position, making sure it is aligned and the electrical conductors have a clear path into the fixture.

- Mark on the ceiling the location of the mounting holes.

- Use a stud finder to determine if the mounting holes line up with the ceiling trusses. If they do, screws will be used to mount the fixture. If they do not, toggle bolts will be necessary. For some installations, a combination of screws and toggle bolts will be required.

- If screws are used, drill holes into the ceiling using a drill bit that has a smaller diameter than the screws to be used. This will make installing the screws easier. If toggle bolts are to be used, use a flat-bladed screwdriver to punch a hole in the sheetrock only large enough for the toggle to fit through.

- Remove a knockout from the fixture where you wish the conductors to come through. Install a cable connector in the knockout hole.

- Place the fixture in its correct position, and pull the cable through the connector and into the fixture.

Tighten the cable connector to secure the cable to the fixture. This part of the process may require the assistance of a coworker. Using toggle bolts, put the bolt through the mounting hole and start the toggle on the end of the bolt.

- With a coworker holding the fixture, install the mounting screws or push the toggle through the hole until the wings spring open. This will hold the fixture in place until the fixture is secured to the ceiling.

- Make the necessary electrical connections. The grounding conductor should be properly wrapped around the fixture grounding screw and the screw tightened. The white grounded conductor is connected to the white conductor lead; then the black ungrounded conductor is connected to the black fixture conductor.

- Install the wiring cover by placing one side in the mounting clips, squeezing it, and then snapping the other side into its mounting clips.

- Install the recommended lamps. Usually, they have two contact pins on each end of the lamp. Align the pins vertically, slide them up into the lamp holders at each end of the fixture, and rotate the lamp until it snaps into place.

- Test the fixture and lamps for proper operation.

- Install the fixture lens cover.

Procedures   Installing a Strap on a Lighting Outlet Box Lighting Fixture

- Put on safety glasses and observe all applicable safety rules.

- Using a voltage tester, verify that there is no electrical power at the lighting outlet where the fixture will be installed. If electrical power is present, turn off the power and lock out the circuit.

- Before starting the installation process, read and understand the manufacturer's instructions.

Ⓐ Mount the strap to the outlet box using the slots in the strap. With metal boxes, the screws are provided with the box. With nonmetallic boxes, you must provide your own 8-32 mounting screws. Put the 8-32 screws through the slot and thread them into the mounting holes on the outlet box. Tighten the screws to secure the strap to the box.

- Identify the proper threaded holes on the strap, and install the fixture-mounting headless bolts in the holes so the end of the screw will point down.

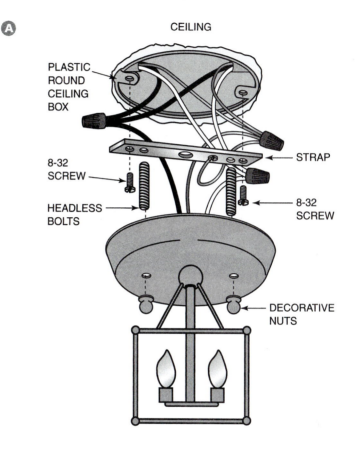

- Make the necessary electrical connections. Make sure that all metal parts (including the outlet box), the strap, and the fixture are properly connected to the grounding conductor in the power feed cable.

- Neatly fold the conductors into the outlet box. Align the headless bolts with the mounting holes on the fixture. Slide the fixture over the headless bolts until the screws stick out through the holes. Do not be alarmed if the mounting screws seem to be too long. Thread the provided decorative nuts onto the headless bolts. Keep turning the nuts until the fixture is secured to the ceiling or wall.

- Install the recommended lamp, and test the fixture operation.

- Install any provided lens or globe. They are usually held in place by three screws that thread into the fixture. Start the screws into the threaded holes, position the lens or globe so it touches the fixture, and tighten the screws until the globe or lens is snug. Do not overtighten the screws. You may return the next day and find the globe or lens cracked or broken.

Procedures

Installing a Chandelier-Type Light Fixture Using the Stud and Strap Connection to a Lighting Outlet Box

- Put on safety glasses and observe all applicable safety rules.

- Using a voltage tester, verify that there is no electrical power at the lighting outlet where the fixture will be installed. If electrical power is present, turn off the power and lock out the circuit.

- Before starting the installation process, read and understand the manufacturer's instructions.

A Install the mounting strap to the outlet box using 8-32 screws.

- Thread the stud into the threaded hole in the center of the mounting strap. Make sure that enough of the stud is screwed into the strap to make a good secure connection.

- Measure the chandelier chain for the proper length, remove any unneeded links, and install one end to the light fixture.

- Thread the light fixture's chain-mounting bracket on to the stud. Remove the holding nut and slide it over the chain.

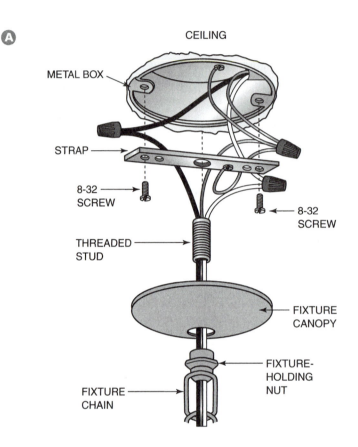

- Slide the canopy over the chain.

- Attach the free end of the chain to the chain-mounting bracket.

- Weave the fixture wires and the grounding conductor up through the chain links, being careful to keep the chain links straight. Section 410.28(F) of the NEC states that the conductors must not bear the weight of the fixture. As long as the chain is straight and the conductors make all the bends, the chain will support the fixture properly.

- Now run the fixture wires up through the fixture stud and into the lighting outlet box.

- Make all necessary electrical connections.

- Slide the canopy up the chain until it is in the proper position. Slide the nut up the chain and thread it on to the chain-mounting bracket until the canopy is secure.

- Install the recommended lamp, and test the fixture for proper operation.

Procedures

Installing a Fluorescent Fixture (Troffer) in a Dropped Ceiling

- Put on safety glasses and observe all applicable safety rules.

- Before starting the installation process, read and understand the manufacturer's instructions.

- During the rough-in stage, mark the location of the fixtures on the ceiling.

- Using standard wiring methods, place lighting outlet boxes on the ceiling near the marked fixture locations, and connect them to the lighting branch circuit.

- Once the dropped ceiling grid has been installed by the ceiling contractor, install the fluorescent light fixtures in the ceiling grid at the proper locations. Some electricians refer to this action as "laying in" the fixture. Once the fixture is installed, some electricians refer to the fixtures as being "laid in."

- Ⓐ Support the fixture according to NEC requirements. Section 410.16(C) requires that all framing members used to support the ceiling grid be securely fastened to each other and to the building itself. The fixtures themselves must be securely fastened to the grid by an approved means, such as bolts, screws, rivets, or clips. This is to prevent the fixture from falling and injuring someone.

- Using a voltage tester, verify that there is no electrical power at the lighting outlet where the fixture will be installed. If electrical power is present, turn off the power and lock out the circuit.

Ⓐ

IMPORTANT: TO PREVENT THE LUMINAIRE (FIXTURE) FROM INADVERTENTLY FALLING, *410.16(C)* OF THE CODE REQUIRES THAT (1) SUSPENDED CEILING FRAMING MEMBERS THAT SUPPORT RECESSED LUMINAIRES (FIXTURES) MUST BE SECURELY FASTENED TO EACH OTHER AND MUST BE SECURELY ATTACHED TO THE BUILDING STRUCTURE AT APPPROPRIATE INTERVALS, AND (2) RECESSED LUMINAIRES (FIXTURES) MUST BE SECURELY FASTENED TO THE SUSPENDED CEILING FRAMING MEMBERS BY BOLTS, SCREWS, RIVETS, OR SPECIAL LISTED CLIPS PROVIDED BY THE MANUFACTURER OF THE LUMINAIRE (FIXTURE) FOR THE PURPOSE OF ATTACHING THE LUMINAIRE (FIXTURE) TO THE FRAMING MEMBER.

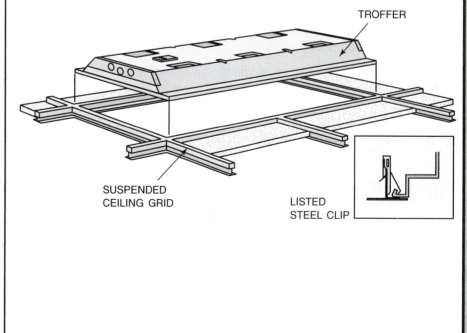

TROFFER

SUSPENDED CEILING GRID

LISTED STEEL CLIP

Procedures

Installing a Fluorescent Fixture (Troffer) in a Dropped Ceiling (Continued)

B Connect the fixture to the electrical system. This is done by means of a "fixture whip." A fixture whip is often a length of Type NM, Type AC, or Type MC cable. It can also be a raceway with approved conductors such as flexible metal conduit or electrical nonmetallic tubing. The fixture whip must be at least 18 inches (450 mm) long and no longer than 6 feet (1.8 m).

- Make all necessary electrical connections. The fixture whip should already be connected to the outlet box mounted in the ceiling. Using an approved connector, connect the cable or raceway to the fixture outlet box and run the conductors into the outlet box. Make sure that all metal parts are properly connected to the grounding system. Connect the white grounded conductors together and then the black ungrounded conductors together. Close the connection box.

- Install the recommended lamps, and test the fixture for proper operation.

- Install the lens on the fixture.

B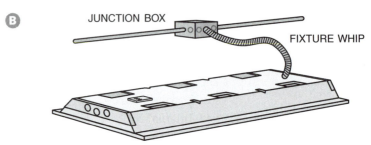

JUNCTION BOX

FIXTURE WHIP

RECESSED LUMINAIRE (FIXTURE)

Review Questions

1. What is an emergency backup system used for?

2. What is NM cable?

3. What is a single-pole switch?

4. What is a double-pole switch?

5. How do three-way switches differ from a single-pole switch?

6. What is a GFCI?

7. What is a continuity tester?

8. How is a smoke detector tested?

9. How are GFCI tested?

10. List the steps for replacing a smoke detector.

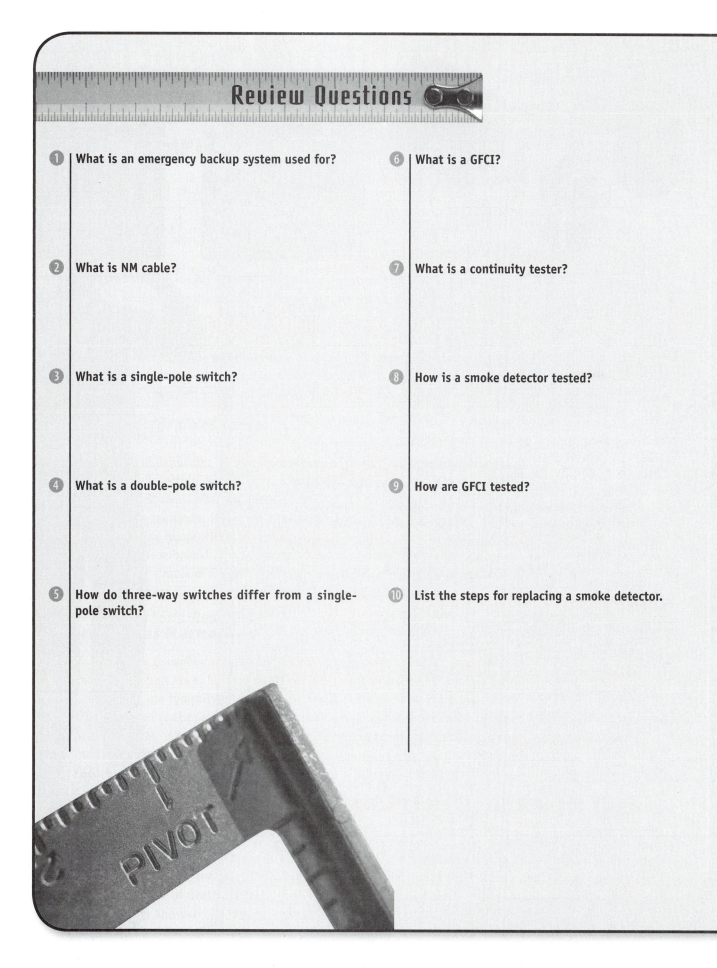

Name: _____

Date: _____

Electrical Facilities Maintenance

Electrical Troubleshooting and Maintenance

Upon completion of this job sheet, you should be able to demonstrate your ability to perform basic maintenance and electrical troubleshooting.

1 Choose an area in your facility, and take an inventory of the switches being used in the area. What types are being used, and what are they made of?

_____

_____

2 List five types of receptacles and their uses.

_____

_____

3 What is the most important thing to remember when working with electricity?

_____

_____

4 Define troubleshooting.

_____

_____

5 What is a GCFI receptacle, and what is its purpose?

_____

_____

6 What is the purpose of a lock-out/tag-out device?

_____

_____

Instructor's Response:

_____

_____

_____

Chapter 7 Carpentry

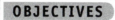

Introduction

Although facility maintenance technicians will not be responsible for new construction, they will be responsible for maintaining the integrity of the facility. This responsibility often includes performing minor construction and/or repairs using carpentry skills and techniques.

Tools

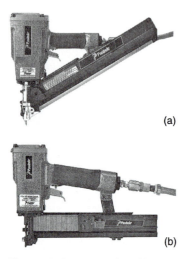

In addition to many of the tools discussed in Chapter 4, portable fastening tools are used extensively in carpentry for both construction and repair to assist in speeding up production. The two most commonly used power fastening tools are pneumatic (compressed air to drive the fastener, see Figures 7-1 through 7-6) and power-actuated (explosive charge to drive the fastener, see Figure 7-7) tools. Before operating any pneumatic tool, read and follow the manufacturer's instruction.

(a)

(b)

Figure 7-1: Pneumatic nailer and staplers are widely used to fasten building parts. (Courtesy of Paslode)

General Properties of Hardwood and Softwood

The carpenter works with wood more than any other material and must understand its characteristics to use it intelligently. Wood is a remarkable substance and is classified as hardwood or softwood. Different methods of classifying these woods exist. The most common method of classifying wood is by its source (see Figure 7-8). **Hardwood** comes from deciduous trees that shed their leaves each year. **Softwood** comes from coniferous, or cone-bearing, trees, commonly known as evergreens.

Common hardwoods include:

- Ash
- Birch
- Cherry
- Hickory
- Maple
- Mahogany
- Oak
- Walnut

Common softwoods include:

- Pine
- Fir
- Hemlock
- Spruce
- Cedar
- Cypress
- Redwood

The best way to learn the different kinds of woods is to work with them and examine them.

- Look at the color and the grain.
- Feel if it is heavy or light.
- Feel if it is hard or soft.
- Smell it for a characteristic odor.

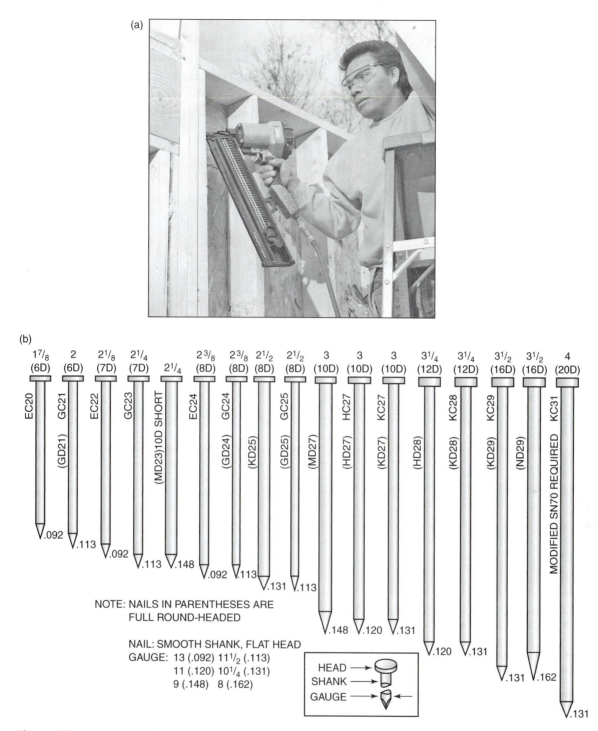

Figure 7-2: **Heavy-duty framing nailers are used for floor, walls, and roof framing. (Courtesy of Senco Products, Inc.)**

Effects of Moisture Content

When a tree is first cut down, it contains a great amount of water. Lumber, when first cut from the log, is called **green lumber** and is very heavy because most of its weight is water (see Figure 7-9). A piece 2 inches thick, 6 inches wide, and 10 feet long may contain as much as 4¼ gallons of water, weighing about 35 pounds.

SMOOTH SHANK   RING SHANK

1½" 1⅞" 2⅛" 2⅜"   1½" 1¾"

EC17 EC20 EC22 EC24   EE17 EE19

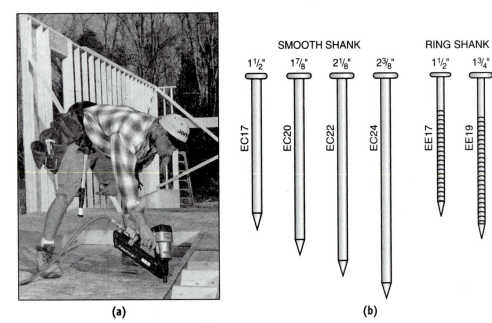

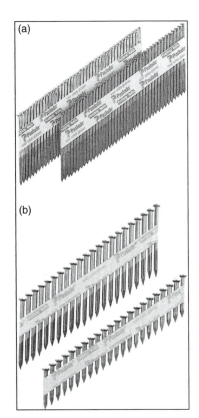

(a)   (b)

Figure 7-3: A light-duty nailer is used to fasten light framing, subfloors, and sheathing. (Courtesy of Senco Products, Inc.)

Figure 7-4: Both headed and finish nails used in nail guns come glued together in strips. (Courtesy of Paslode)

FINISH NAILS

1"(2D) 1¼"(3D) 1½"(4D) 1¾"(5D) 2"(6D)

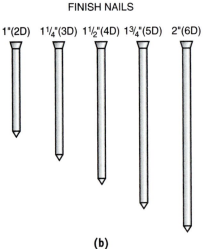

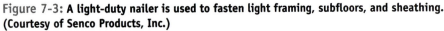

(a)   (b)

Figure 7-5: The finish nailer is used to fasten all kinds of interior trim. (Courtesy of Senco Products, Inc.)

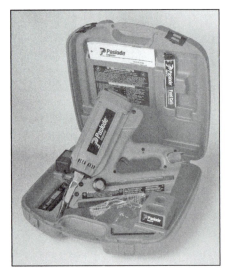

Figure 7-6: A coil roofing nailer is used to fasten asphalt roof shingles. (Courtesy of Paslode)

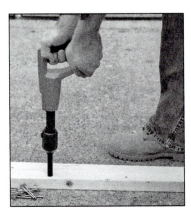
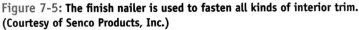

Figure 7-7: Power-actuated drivers are used for fastening into masonry or steel

Figure 7-8: **Softwood and hardwood**

Green lumber should not be used for construction. As green lumber dries, it shrinks considerably unequally as the large amount of water leaves it. When it shrinks, it usually warps, depending on the way it was cut from the log (see Figure 7-10).

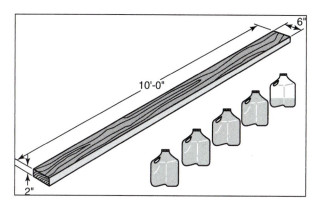

Figure 7-9: **Green lumber contains a large amount of water**

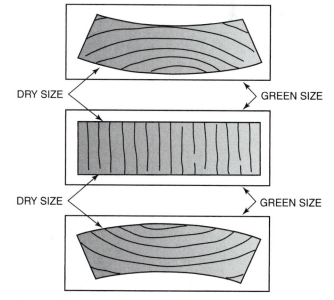

Figure 7-10: **Lumber shrinks in the direction of the annular rings**

Realizing that lumber undergoes certain changes when moisture is absorbed or lost, the experienced carpenter uses techniques to deal with this characteristic of wood (see Figure 7-11).

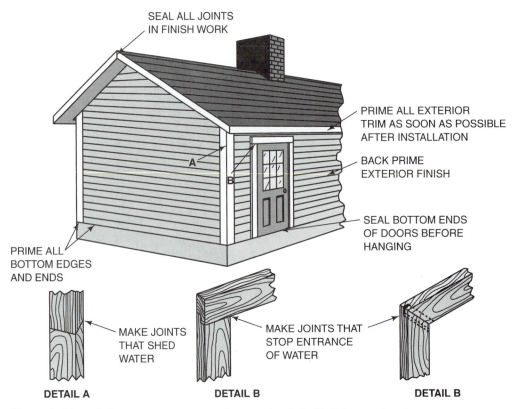

SEAL ALL JOINTS IN FINISH WORK

PRIME ALL EXTERIOR TRIM AS SOON AS POSSIBLE AFTER INSTALLATION

BACK PRIME EXTERIOR FINISH

SEAL BOTTOM ENDS OF DOORS BEFORE HANGING

PRIME ALL BOTTOM EDGES AND ENDS

MAKE JOINTS THAT SHED WATER

MAKE JOINTS THAT STOP ENTRANCE OF WATER

DETAIL A

DETAIL B

DETAIL B

Figure 7-11: **Techniques to prevent water from getting in behind the wood surface**

Figure 7-12: **Laminated strand lumber**

Correctly Identify and Select Engineered Products, Panels, and Sheet Goods

Before a construction project of any scope involving lumber or engineered products can be completed, the technician must correctly select the type of product that will best deliver the desired results. Engineered products are typically stronger than traditional lumber products. In addition engineered products can be used to span greater distances than traditional lumber products (see Figure 7-12).

Engineered Panels

The term **engineered panels** refers to human-made products in the form of large reconstituted wood sheets, sometimes called panels or boards.

- Plywood—one of the most extensively used engineered panels (Figure 7-13). It is a sandwich of wood. Most plywood panels constitute sheets of veneer called **plies**.
- Oriented strand board (OSB)—a nonveneered, performance-rated structural panel composed of small oriented (lined up) strand-like wood pieces arranged in three to five layers with each layer at right angles to the other (Figure 7-14).

Figure 7-13: **APA performance-rated panels**

Figure 7-14: **Oriented strand board (OSB)**

Nonstructural Panels

- Hardwood plywood—available with hardwood face veneers, of which the most popular are birch, oak, and lauan.
- Particleboard—reconstituted wood panels made of wood flakes, chips, sawdust, and planer shavings (Figure 7-15). These wood particles are mixed with an adhesive, formed into a mat, and pressed into sheet form.

Figure 7-15: **Particleboard is made from wood flakes, shaving, resins, and waxes**

- Fiberboards—manufactured as high-, medium-, and low-density boards. Medium-density fiberboard (MDF) is manufactured in a manner similar to that used to make hardboard except that the fibers are not pressed as tightly. Low-density fiberboard, which is called softboard, is light and contains many tiny air spaces because the particles are not compressed tightly.

- Hardboards—high-density fiberboards, which are sometimes known by the trademark Masonite.

Figure 7-16: **Drywall, also known as sheetrock**

Others

- Gypsum board—used extensively for construction. It is sometimes called wallboard, plasterboard, drywall, or Sheetrock (the brand name) (Figure 7-16). Gypsum board is readily available; easy to apply, decorate, or repair; and relatively inexpensive.
- Plastic laminates—used for surfacing kitchen cabinets and countertops. They are also used to cover walls or parts of walls in kitchens, bathrooms, and similar areas where a durable, easy-to-clean surface is desired.

Estimating

To estimate the amount of drywall material needed:

1. Determine the area of the walls and ceiling to be covered. For ceiling, multiply the length of the room by its width and for walls, multiply the perimeter of the room by its height.
2. Subtract only the large wall openings, such as double doors.
3. Combine all areas to find the total number of square feet of drywall.
4. Add about 5 percent of the total for waste.
5. Divide the total area to be covered by the area of one panel to get the number of panels.

Framing Components

The majority of residential and light commercial constructions consist of wood framing. This is because in most cases wood framing construction is generally less expensive than other types of construction. Although many of the different framing methods used today may vary, the components used are the same.

Wall Framing Components

The wall frame consists of a number of different parts. An exterior wall frame consists of the following components:

- Plates—top and bottom horizontal members of the wall frame
- Studs—vertical members of the wall frame
- Headers—run at right angles to the studs (Figure 7-17)
- Rough sills—form the bottom of a window opening at right angles to the studs (Figures 7-18 and 7-19)

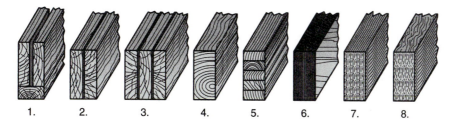

1. A BUILT-UP HEADER WITH A 2 X 4 OR 2 X 6 LAID FLAT ON THE BOTTOM.
2. A BUILT-UP HEADER WITH A 1/2" SPACER SANDWICHED IN BETWEEN.
3. A BUILT-UP HEADER FOR A 6" WALL.
4. A HEADER OF SOLID SAWN LUMBER.
5. GLULAM BEAMS ARE OFTEN USED FOR HEADERS.
6. A BUILT-UP HEADER OF LAMINATED VENEER LUMBER.
7. PARALLEL STRAND LUMBER MAKES EXCELLENT HEADERS.
8. LAMINATED STRAND LUMBER IS USED FOR LIGHT DUTY HEADERS.

Figure 7-17: **Types of solid and built-up headers**

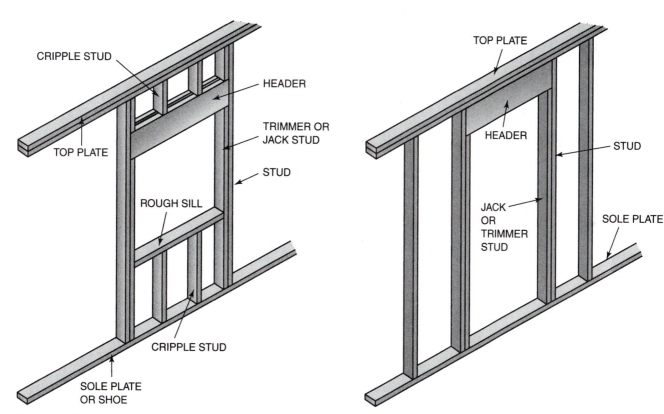

Figure 7-18: **Typical framing for a window opening**

Figure 7-19: **Typical framing for a door opening**

- Trimmers (jacks)—shortened studs that support the headers
- Corner posts—same length as studs (Figure 7-20)
- Partition intersections—framing needed when interior partitions meet an exterior partition (Figure 7-21)

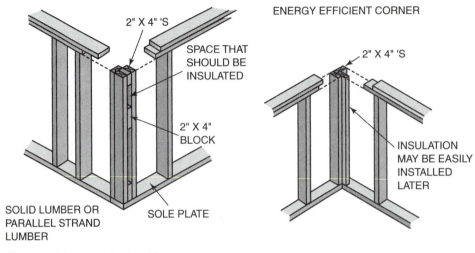

Figure 7-20: **Methods of making corner posts**

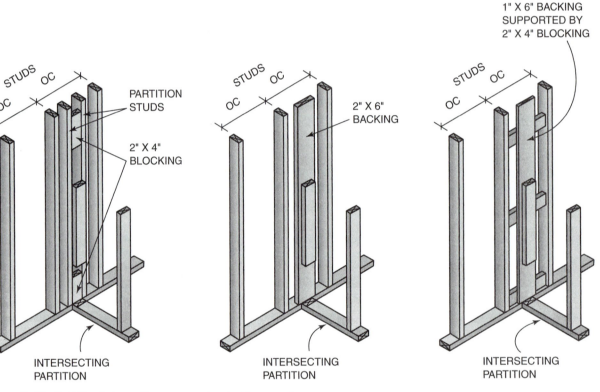

Figure 7-21: **Partition intersections are constructed in several ways**

- Ribbons—horizontal members of the exterior wall frame in balloon construction (Figure 7-22)
- Corner braces—used to brace walls (Figures 7-23 and 7-24)

Interior Doors

Unlike exterior doors that are made of more dense materials, interior doors are typically light weight designed to isolate an interior space. They are not ordinarily exposed to weather and therefore are not usually subjected to the typical problems associated with an exterior door.

Repairing and Replacing Interior Doors

The most common problems associated with interior doors are doors that bind along their edges, doors that don't latch, and loose hinges.

Repairing Doors That Bind along Their Top Edge

The most common cause for interior doors sticking along or near their top edge is from loose hinge plate screws. If the hinge plate screws are loose, try tightening them. This might take care of the problem; however, if the problem should persist, then the screw holes might be stripped. If the screw holes are stripped, then replacing the stripped screw with a longer or wider screw usually works. If a longer screw is used, be sure to get a screw long enough to go through the door jamb, through the shim, and into the stud. Typically a 3-inch screw will be sufficient.

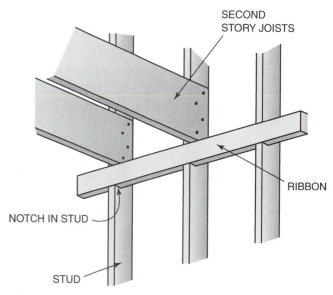

Figure 7-22: **Ribbons are used to support floor joists in balloon frame**

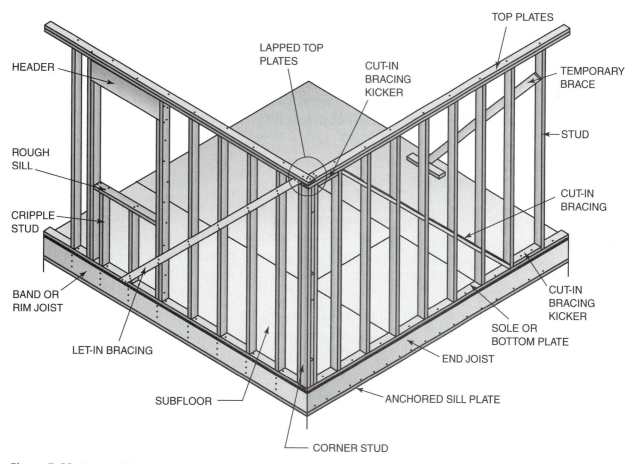

Figure 7-23: **Wood wall bracing may be cut in or let in**

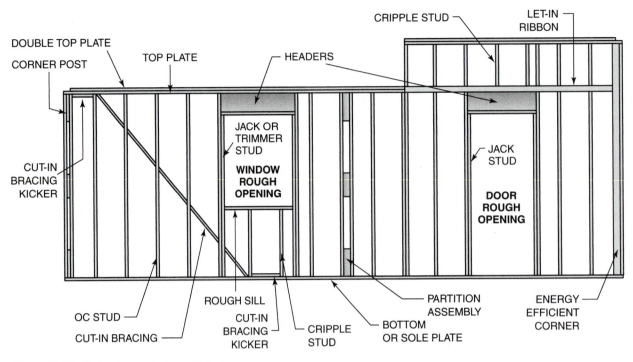

Figure 7-24: Parts of an exterior wall frame

Repairing Doors That Bind along Their Bottom Edge

The most common cause for interior doors sticking along or near the bottom is typically from the jamb shifting or pulling away from its fastener (nail). The problem can be corrected by resetting the door jamb. This is accomplished by prying away the bottom section of the casting and renailing the door jamb back into place followed by repositioning and tacking the casting back into place.

Repairing Doors That Bind along Their Entire Edge

If a door is not sealed properly (painted or varnished along all edges), moisture can enter the door and cause it to swell thus preventing the door from not closing properly. Doors that are not sealed properly swell typically during the rainy season of the year. If the door is not closing properly due to an elevation in moisture, trimming the door will work as long as the door's moisture content remains the same. In other words, as soon as the dry season approaches the door will not be properly gapped. Therefore, waiting for the dry season and then trimming (if needed) and sealing the door's edges is the correct way of repairing the door.

Repairing Doors That Don't Latch

Another common problem a facility maintenance technician will have to resolve is a door that won't latch. This is typically caused by the shifting of either the door or the door's frame, resulting in the misalignment of the strike plate or strike (see the procedure for "Installing the Striker Plate" on page 196). If the misalignment is less than or equal to $1/8$ of an inch, then the simplest approach to resolving the problem is to file the strike plate. If the amount of misalignment is more than $1/8$ inch, then the strike plate should be repositioned instead of filing it.

Any structure built before 1978 may contain lead-based paint. Lead-based paint is considered a hazardous material and must be handled and disposed of properly according to the local, state, and federal regulations.

Replacing Interior Doors

Hollow core interior doors can be easily damaged and may therefore require replacing. To replace a door the following procedure should be followed:

- Remove the old door by removing the hinge pins. When removing the hinge pins always start from the bottom and work your way to the top. Remove the top hinge pin last.
- Using the old door as a pattern, lay it on top of the new door and mark any excess material (from the new door) that should be removed (trimmed). In addition the locations of the hinge mortises and the lockset can be transferred.
- Trim the excess material for the new door. See page 168 for "Hanging Interior Doors."
- Drill and assemble the lockset. See page 194 for "Installing Cylindrical Locksets."

Repairing Interior Door Hardware

Sneaky door hinges, deadbolts that stick, doors that are difficult to open, door knobs that are difficult to turn are often a sign that the hardware needs to be either cleaned or lubricated, or both.

Cleaning Door Hardware

To clean the hardware associated with a door the following steps should be followed:

- Remove the hardware form the door.
- If necessary, dissemble the hardware and soak it in a cleaning solution.
- After soaking it in a cleaning solution dry it thoroughly with either compressed air or a clean rag.
- Lubricate the components using penetrating oil and reassemble (if necessary) and reinstall the hardware on the door.

NOTE: The rest of this chapter outlines various procedures for interior and exterior carpentry maintenance.

Interior Carpentry Maintenance

Procedures

Constructing the Grid Ceiling System

Ⓐ Locate the height of the ceiling, marking elevations of the ceiling at the ends of all wall sections. Snap chalk lines on all walls around the room to the height of the top edge of the wall angle. If a laser is used, the chalk line is not needed since the ceiling is built to the light beam.

• Fasten wall angles around the room with their top edge lined up with the line. Fasten into framing wherever possible, not more than 24 inches apart. If available, power nailers can be used for efficient fastening.

Ⓐ

FROM EXPERIENCE

To fasten wall angles to concrete walls, short masonry nails are sometimes used. However, they are difficult to hold and drive. Use a small strip of cardboard to hold the nail while driving it with the hammer.

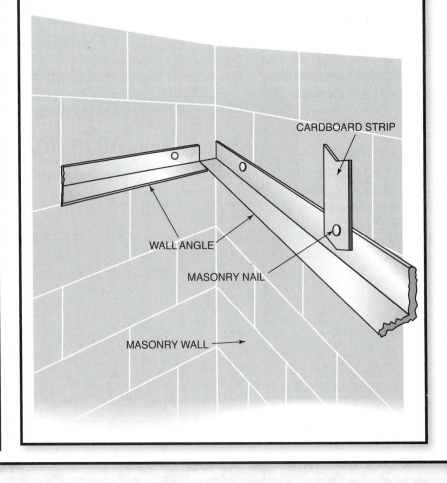

CARDBOARD STRIP

WALL ANGLE

MASONRY NAIL

MASONRY WALL →

B Make miter joints on outside corners and butt joints in interior corners and between straight lengths of wall angle. Use a combination square to lay out and draw the square and angled lines. Cut carefully along the lines with snips.

- From the ceiling sketch, determine the position of the first main runner. Stretch a line at this location across the room from the top edges of the wall angle. The line serves as a guide for installing *hanger lags* or *screw eyes* and *hanger wires* from which main runners are suspended.

- Install the cross tee line by measuring out from the short wall, along the stretched main runner line, a distance equal to the width of the border panel. Mark the line. Stretch the cross tee line through this mark and at right angles to the main runner line.

- Install hanger lags not more than 4 feet apart and directly over the stretched line. Hanger lags should be of the type commonly used for suspended ceilings. They must be long enough to penetrate wood joists a minimum of 1 inch to provide strong support. Hanger wires may also be attached directly around the lower chord of bar joists or trusses.

CAUTION

Use care in handling the cut ends of the metal grid system. The cut ends are sharp and may have jagged edges that can cause serious injury.

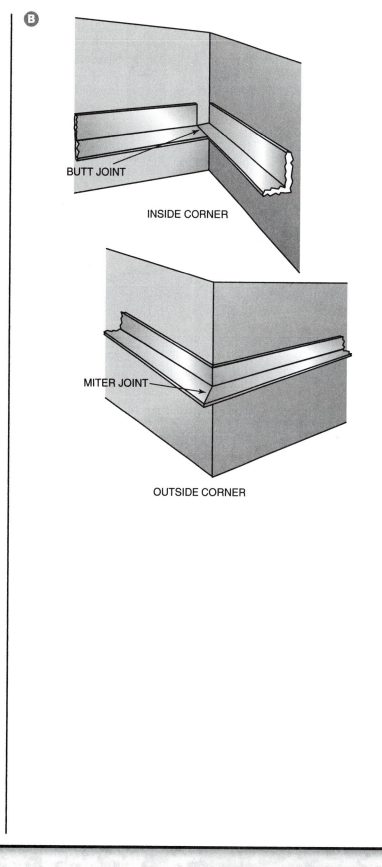

B

BUTT JOINT

INSIDE CORNER

MITER JOINT

OUTSIDE CORNER

Procedures

Constructing the Grid Ceiling System (Continued)

FROM EXPERIENCE

Stretch the line tightly on nails inserted between the wall and the wall angle.

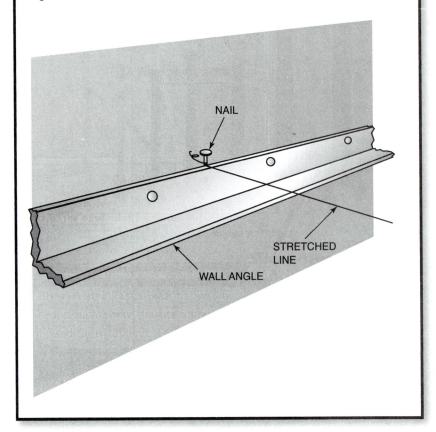

NAIL

STRETCHED LINE

WALL ANGLE

C Cut a number of hanger wires using wire cutters. The wires should be about 12 inches longer than the distance between the overhead construction and the stretched line. Attach the hanger wires to the hanger lags. Insert about 6 inches of the wire through the screw eye. Securely wrap the wire around itself three times. Pull on each wire to remove any kinks. Then make a 90° bend where it crosses the stretched line. If a laser is used, the 90° bend is done later when the main runner is installed.

C

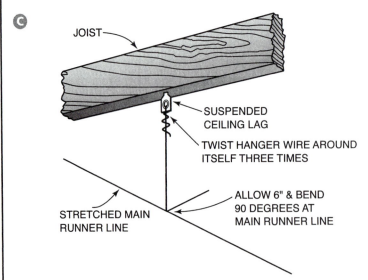

JOIST

SUSPENDED CEILING LAG

TWIST HANGER WIRE AROUND ITSELF THREE TIMES

ALLOW 6" & BEND 90 DEGREES AT MAIN RUNNER LINE

STRETCHED MAIN RUNNER LINE

D Stretch lines, install hanger lags, and attach and bend hanger wires in the same manner at each main runner location. Leave the last line stretched tightly in position. It will be used to align the cross tee slots of the main runner.

- At each main runner location, measure from the wall to the cross tee line. Transfer this measurement to the main runner, measuring from the first cross tee slot beyond the measurement, so as to cut as little as possible from the end of the main runner.

- Example: If the first cross tee will be located 23 inches from the wall, then the main runner will be cut.

- Cut the main runners about ⅛ inch less to allow for the thickness of the wall angle. Backcut the web slightly for easier installation at the wall. Measure and cut main runners individually. Do not use the first one as a pattern to cut the rest. Measure each from the cross tee line.

E Hang the main runners by resting the cut ends on the wall angle and inserting suspension wires in the appropriate holes on the top of the main runner. Bring the runners up to the bend in the wires or to the laser light beam. Twist the wires with at least three turns to hold the main runners securely. More than one length of main runner may be needed to reach the opposite wall. Connect lengths of main runners together by inserting tabs into matching ends. Make sure that end joints come up tight.

D

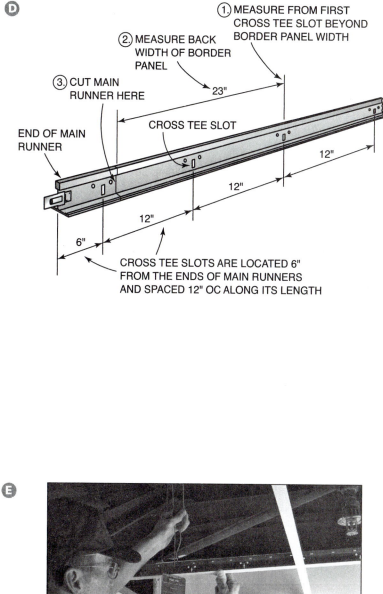

① MEASURE FROM FIRST CROSS TEE SLOT BEYOND BORDER PANEL WIDTH

② MEASURE BACK WIDTH OF BORDER PANEL

③ CUT MAIN RUNNER HERE

END OF MAIN RUNNER

CROSS TEE SLOT

23"

12"

12"

12"

6"

CROSS TEE SLOTS ARE LOCATED 6" FROM THE ENDS OF MAIN RUNNERS AND SPACED 12" OC ALONG ITS LENGTH

E

Courtesy of Trimble.

Procedures

Constructing the Grid Ceiling System (Continued)

F The length of the last section is measured from the end of the last one installed to the opposite wall, allowing about $1/8$ inch less to fit.

- Cross tees are installed by inserting the tabs on the ends into the slots in the main runners. These fit into position easily, although the method of attaching varies from one manufacturer to another. Install all full-length cross tees between main runners first.

- Lay in a few full-size ceiling panels to stabilize the grid while installing the border cross tees.

- Cut and install cross tees along the border. Insert the connecting tab of one end in the main runner and rest the cut end on the wall angle. It may be necessary to measure and cut cross tees for border panels individually, if walls are not straight or square.

- For 2 × 2 panels, install 2-foot cross tees at the midpoints of the 4-foot cross tees. After the grid is complete, straighten and adjust the grid to level and straight where necessary.

F

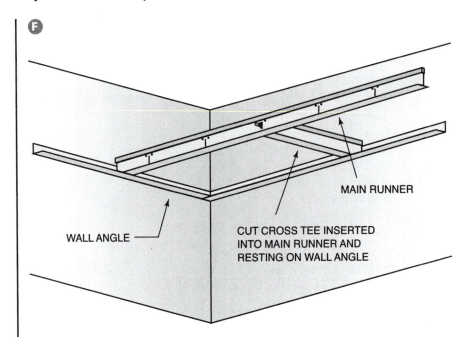

WALL ANGLE

MAIN RUNNER

CUT CROSS TEE INSERTED INTO MAIN RUNNER AND RESTING ON WALL ANGLE

G Ceiling panels are placed in position by tilting them slightly, lifting them above the grid, and letting them fall into place. Be careful when handling panels to avoid marring the finished surface. Cut and install border panels first and install the full-sized panels last. Measure each border panel individually, if necessary. Cut them $1/8$ inch smaller than measured so that they can drop into place easily. Cut the panels with a sharp utility knife using a straightedge as a guide. A scrap piece of cross tee material can be used as a straightedge. Always cut with the finished side of the panel up.

H When a column is near the center of a ceiling panel, cut the panel at the midpoint of the column. Cut semicircles from the cut edge to the size required for the panel pieces to fit snugly around the column. After the two pieces are rejoined around the column, glue scrap pieces of panel material to the back of the installed panel. If the column is close to the edge or end of a panel, cut the panel from the nearest edge or end to fit around the column. The small piece is also fitted around the column and joined to the panel by gluing scrap pieces to its back side.

G

H

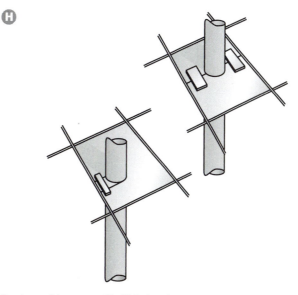

Courtesy of Armstrong World Industries.

Procedures

Replacing Broken Ceiling Tiles

Broken or discolored ceiling tiles can be easily replaced using the following steps:

- After positioning a ladder or scaffolding under the tile or tiles to be replaced, push up on the tile to pop it out of the track.

- Once the tile has been popped out of the track, tilting it an angle will permit you to remove the ceiling tile.

- Use the old ceiling tile as a guide to trim the new ceiling tile.

- The ceiling tile can be trimmed using a straight edge and a utility knife.

- Ceiling panels are placed in position by tilting them slightly, lifting them above the grid, and letting them fall into place.

Procedures   Applying Wall Molding

Ⓐ To snap a line for wall trim, begin by holding a short scrap piece of the molding at the proper angle on the wall. Lightly mark the wall along the bottom edge of the molding. Measure the distance from the ceiling down to the mark.

• Measure and mark this same distance down from the ceiling on each end of each wall to which the molding is to be applied. Snap lines between the marks. Apply the molding so its bottom edge is to the chalk line.

Ⓑ Apply the molding to the first wall with square ends in both corners. If more than one piece is required to go from corner to corner, the butt joints may be squared or mitered. Position the molding in the miter box the same way each time. Mitering the molding with the same side down each time helps make fitting more accurate, faster, and easier.

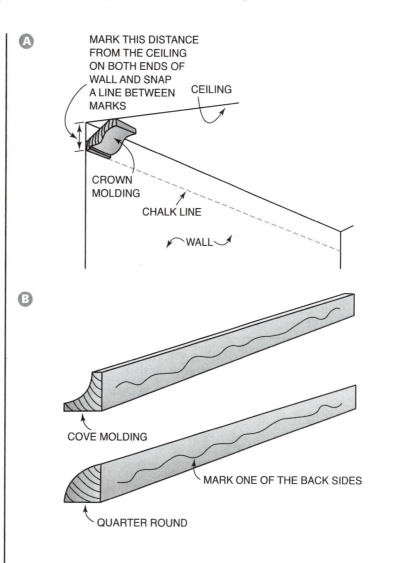

Ⓐ
MARK THIS DISTANCE FROM THE CEILING ON BOTH ENDS OF WALL AND SNAP A LINE BETWEEN MARKS

CEILING

CROWN MOLDING

CHALK LINE

WALL

Ⓑ

COVE MOLDING

MARK ONE OF THE BACK SIDES

QUARTER ROUND

FROM EXPERIENCE

Since the revealed edges of the molding are often not the same, cut the molding with the same orientation. To do this mark one of the back surfaces with a pencil.

Procedures   Applying Wall Molding (Continued)

C If a small-size molding is used, fasten it with finish nails in the center. Use nails of sufficient length to penetrate into solid wood at least 1 inch. If large-size molding is used, fastening is required along both edges. Nail at about 16-inch intervals and in other locations as necessary to bring the molding tight against the surface. End nails should be placed 2–3 inches from the end to keep the molding from splitting. If it is likely that the molding may split, blunt the pointed end of the nail.

• Cope the starting end of the first piece on each succeeding wall against the face of the last piece installed on the previous wall. Work around the room in one direction. The end of the last piece installed must be coped to fit against the face of the first piece.

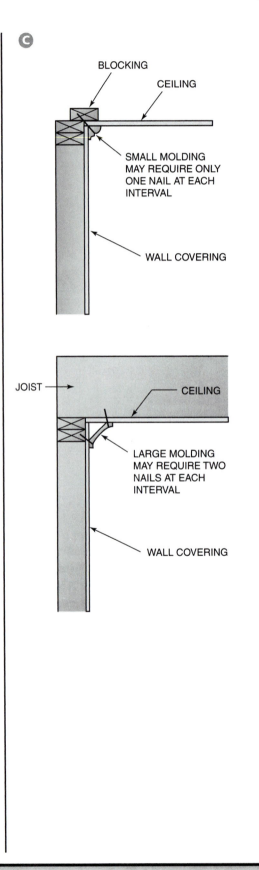

C

BLOCKING

CEILING

SMALL MOLDING MAY REQUIRE ONLY ONE NAIL AT EACH INTERVAL

WALL COVERING

JOIST

CEILING

LARGE MOLDING MAY REQUIRE TWO NAILS AT EACH INTERVAL

WALL COVERING

Procedures   Applying Door Casings

A Set the blade of the combination square so that it extends ⁵/₁₆ inch beyond the body of the square. Gauge lines at intervals along the side and head jamb edges by riding the square against the inside face of the jamb. Let the lines intersect where side and head jambs meet.

- Cut one miter on the ends of the two side casings. Cut them a little long as they will be cut to fit later. Be sure to cut pairs of right and left miters.

- Miter one end of the head casing. Hold it against the head jamb of the door frame so that the miter is on the intersection of the gauged lines. Mark the length of the head casing at the intersection of the gauged lines on the opposite side of the door frame. Miter the casing to length at the mark.

- Fasten the head casing in position with a few tack nails. Move the ends slightly to fit the mitered joint between head and side casings. Keep the casing inside edge aligned to the gauged lines on the head jamb. The mitered ends should be in line with the gauged lines on the side jambs. Use finish nails along the inside edge of the casing into the header jamb. Straighten the casing as necessary as nailing progresses. Drive nails at the proper angle to keep them from coming through the face or back side of the jamb. Fasten the top edge of the casing into the framing.

A

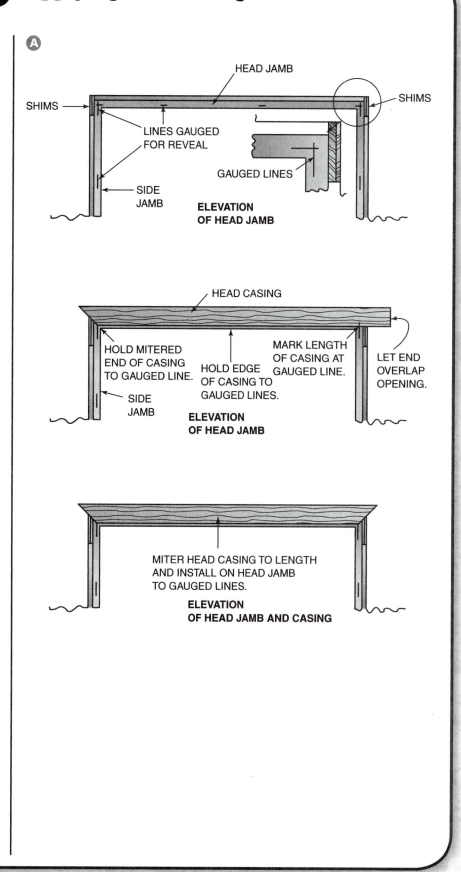

ELEVATION OF HEAD JAMB

ELEVATION OF HEAD JAMB

ELEVATION OF HEAD JAMB AND CASING

Procedures

Applying Door Casings (Continued)

- Cut the previously mitered side casing to length. Mark the bottom end by turning it upside down with the point of the miter touching the floor. Mark the side casing in line with the top edge of the head casing. Make a square cut on the casing at that mark. If the finish floor has not been laid, hold the point of the miter on a scrap block of material that is equal in thickness to the finish floor. Replace the side casing in position and try to fit it at the mitered joint. If the joint needs adjusting, trim with a power miter box or use a sharp block plane.

- When fitted, apply a little glue to the joint. Nail the side casing in the same manner as that of the head casing. Bring the faces flush, if necessary, by shimming between the back of the casing and the wall. Usually, only thin shims are needed. Any small space between the casing and the wall is usually not noticeable or it can be filled later with joint filling compound. Also, the backside of the thicker piece may be planed or chiseled to the desired thinness.

- Drive a 4d finish nail into the edge of the casing and through the mitered joint, and then set all fasteners. Keep nails 2 or 3 inches from the end to avoid splitting the casing.

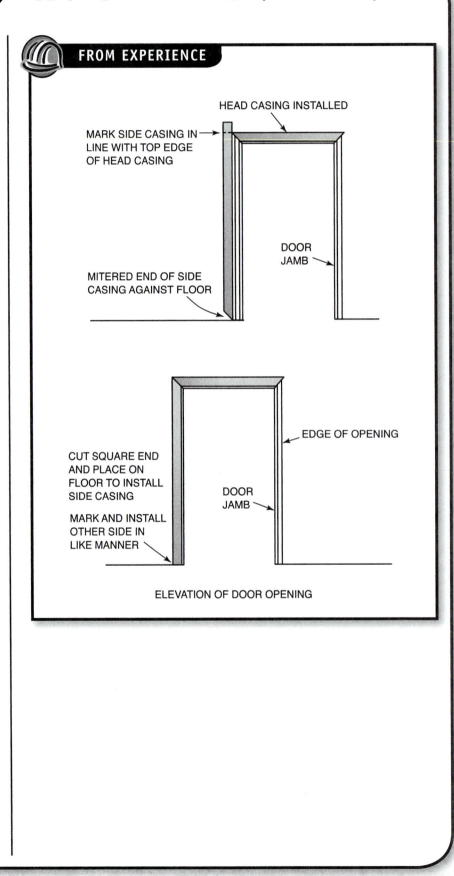

FROM EXPERIENCE

HEAD CASING INSTALLED

MARK SIDE CASING IN LINE WITH TOP EDGE OF HEAD CASING

DOOR JAMB

MITERED END OF SIDE CASING AGAINST FLOOR

EDGE OF OPENING

CUT SQUARE END AND PLACE ON FLOOR TO INSTALL SIDE CASING

MARK AND INSTALL OTHER SIDE IN LIKE MANNER

DOOR JAMB

ELEVATION OF DOOR OPENING

Procedures | Hanging Interior Doors

Setting a Prehung Door Frame

A Remove the protective packing from the unit. Leave the small fiber shims between the door and the jambs to help maintain this space. Cut off the horns if necessary. Remove nail that holds the door closed.

• Center the unit in the opening, so the door will swing in the desired direction. Be sure the door is closed and spacer shims are still in place between the jamb and the door.

B Level the head jamb. Make adjustments by shimming the jamb that is low so that it brings the head jamb level. Adjust a scriber to the thickness of shim and scribe this amount off of the other jamb. Remove frame and cut the jamb. Note that the clearance under the door is being reduced by the amount being cut off.

A

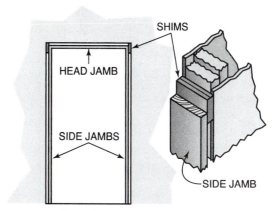

HEAD JAMB

SIDE JAMBS

SHIMS

SIDE JAMB

ELEVATION

SET FRAME IN OPENING. SHIM ON BOTH SIDES OPPOSITE HEAD JAMB. LEVEL HEAD JAMB AND FASTEN AT TOP.

B

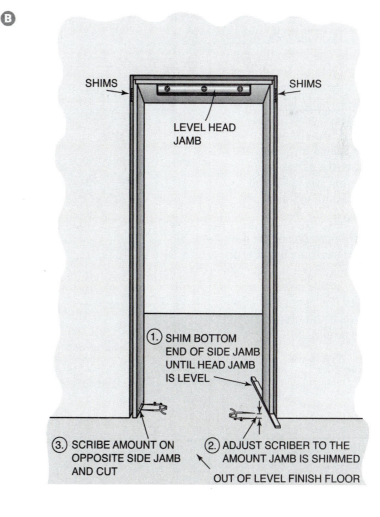

SHIMS

SHIMS

LEVEL HEAD JAMB

1.) SHIM BOTTOM END OF SIDE JAMB UNTIL HEAD JAMB IS LEVEL

3.) SCRIBE AMOUNT ON OPPOSITE SIDE JAMB AND CUT

2.) ADJUST SCRIBER TO THE AMOUNT JAMB IS SHIMMED

OUT OF LEVEL FINISH FLOOR

Procedures | Hanging Interior Doors (Continued)

C Plumb the hinge side jamb of the door unit. A 2-foot carpenter's level may not be accurate when plumbing the sides because of any bow that may be in the jambs. Use a 6-foot level or a plumb bob. Tack the jamb plumb to the wall through the casing with one nail on either side.

D Open the door and move to the other side. Check that the unit is nearly centered. Install shims between the side jambs and the rough opening at intermediate points, keeping side jambs straight. Shims should be located behind the hinges and lockset **strike plates**. Nail through the side jambs and shims. Remove spacers from door edges.

• Check the operation of the door. Make any necessary adjustments. The space between the door and the jamb should be equal on all jambs. The entire door edge should touch the stop or weather strip.

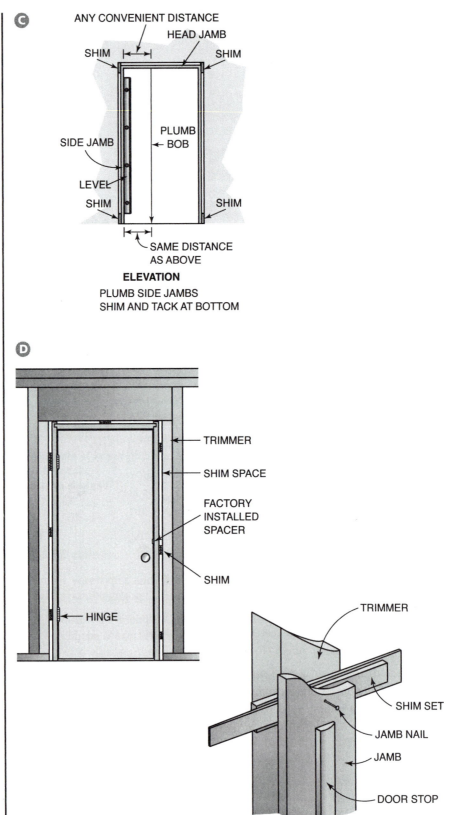

C

ANY CONVENIENT DISTANCE
HEAD JAMB
SHIM
SHIM
SIDE JAMB
PLUMB
BOB
LEVEL
SHIM
SHIM
SAME DISTANCE
AS ABOVE

ELEVATION
PLUMB SIDE JAMBS
SHIM AND TACK AT BOTTOM

D

TRIMMER
SHIM SPACE
FACTORY
INSTALLED
SPACER
SHIM
HINGE

TRIMMER
SHIM SET
JAMB NAIL
JAMB
DOOR STOP

E Finish nailing the casing and install it on the other side of the door. Drive and set all nails. Do not make any hammer marks on the finish.

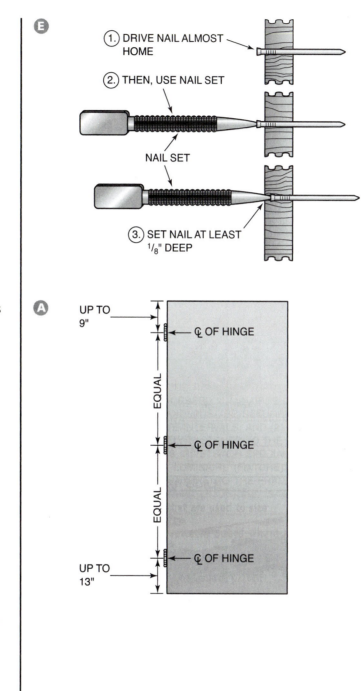

E

1. DRIVE NAIL ALMOST HOME

2. THEN, USE NAIL SET

NAIL SET

3. SET NAIL AT LEAST 1/8" DEEP

Fitting a Door to a Frame

A Begin by checking the door for its beveled edge and the direction of the face of the door. Note the direction of the swing.

- Lightly mark the location of the hinges on the door. On paneled doors, the top hinge is usually placed with its upper end in line with the bottom edge of the top rail. The bottom hinge is placed with its lower end in line with the top edge of the bottom rail. The middle hinge is centered between them. On flush doors, the usual placement of the hinge is approximately 9 inches down from the top and 13 inches up from the bottom, as measured to the center of the hinge. The middle hinge is centered between the two.

- Check the opening frame for level and plumb.

A

UP TO 9"

C OF HINGE

EQUAL

C OF HINGE

EQUAL

C OF HINGE

UP TO 13"

Procedures

Hanging Interior Doors (Continued)

B Plane the door edges so the door fits onto the opening with an even joint of approximately $3/32$ inch between the door and the frame on all sides. A wider joint of approximately $1/8$ inch must be made to allow for the swelling of the door and frame in extremely damp weather. Use a *door jack* to hold the door steady. Do not cut more than $1/2$ inch total from the width of a door. Cut no more than 2 inches from its height. Check the fit frequently by placing the door in the opening, even if this takes a little extra time.

B

DOOR

WEIGHT OF DOOR BOWS PLYWOOD BASE CAUSING BRACKETS TO CLAMP DOOR.

$3/4$" OR THICKER BRACKETS

ADD STRIPS OF CARPET TO PROTECT THE DOOR.

$3/8$" PLYWOOD

SPACE BETWEEN BRACKETS SLIGHTLY MORE THAN DOOR THICKNESS.

A DOOR JACK CAN BE MADE ON THE JOB FROM SCRAP LUMBER.

2" X 4" BLOCKS

C Place the door in the frame. Shim the door so the proper joint is obtained along all sides. Place shims between the lock edge of the door and the side jamb of the frame. Mark across the door and jamb at the desired location for each hinge. Place a small X on both the door and the jamb, to indicate on which side of the mark to cut the gain.

• Remove the door from the frame. Place a hinge leaf on the door edge with its end on the mark previously made. Score a line along edges of the leaf. Score only partway across the door edge.

C

DOOR FITTED & SHIMMED

DOOR FRAME

PENCIL MARK

PLACE X'S ON SIDE OF PENCIL MARK TO BE CUT OUT

D Score the hinge lines, taking care not to split any part of the door. With a chisel, cut small chips from each end of the gain joint. The chips will break off at the scored end marks. Then, with the flat of the chisel down, pare and smooth the excess down to the depth of the gain. Be careful not to slip.

- Press the hinge leaf into the gain joint. It should be flush with the door edge and install screws. Center the screws carefully so the hinge leaf will not shift when the screw head comes in contact with the leaf.

- Place the door in the opening and insert the hinge pins. Check the swing of the door and adjust as needed.

E Apply the *door stops* to the jambs with several tack nails, in case they have to be adjusted when locksets are installed. A **back miter** joint is usually made between molded side and header stops. A butt joint is made between square-edge stops.

D

E

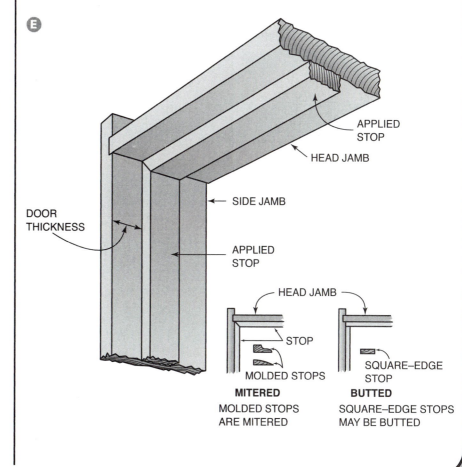

APPLIED STOP

HEAD JAMB

SIDE JAMB

DOOR THICKNESS

APPLIED STOP

HEAD JAMB

STOP

MOLDED STOPS

MITERED
MOLDED STOPS ARE MITERED

SQUARE-EDGE STOP

BUTTED
SQUARE-EDGE STOPS MAY BE BUTTED

Procedures | Hanging Interior Doors (Continued)

Installing Bypass Doors

A Cut the track to length. Install it on the header jamb according to the manufacturer's directions. Bypass doors are installed so that they overlap each other by about 1 inch when closed.

• Install pairs of *roller hangers* on each door. The roller hangers may be offset a different amount for the outside door than the inside door. They are also offset differently for doors of various thicknesses. Make sure that rollers with the same and correct offset are used on each door. The location of the rollers from the edge of the door is usually specified in the manufacturer's instruction sheet.

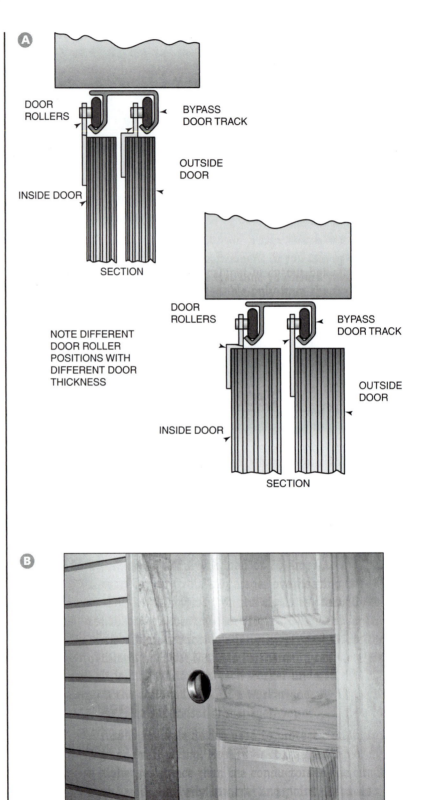

A

DOOR ROLLERS

BYPASS DOOR TRACK

OUTSIDE DOOR

INSIDE DOOR

SECTION

NOTE DIFFERENT DOOR ROLLER POSITIONS WITH DIFFERENT DOOR THICKNESS

DOOR ROLLERS

BYPASS DOOR TRACK

OUTSIDE DOOR

INSIDE DOOR

SECTION

B Mark the location and bore holes for *door pulls*. Use flush pulls so that bypassing is not obstructed. The proper size hole is bored partway into the door. The pull is tapped into place with a hammer and wood block. The press fit holds the pull in place. Rectangular flush pulls, also used on bypass doors, are held in place with small recessed screws.

B

C Hang the doors by holding the bottom outward. Insert the rollers in the overhead track. Then gently let the door come to a vertical position. Install the inside door first, then the outside door.

- Test the door operation and the fit against side jambs. Door edges must fit against side jambs evenly from top to bottom. If the top or bottom portion of the edge strikes the side jamb first, it may cause the door to jump from the track. The door rollers have adjustments for raising and lowering. Adjust one or the other to make the door edges fit against side jambs.

D A *floor guide* is included with bypass door hardware to keep the doors in alignment. The guide is centered on the lap of the two doors to steady them at the bottom. Mark the location and fasten the guide.

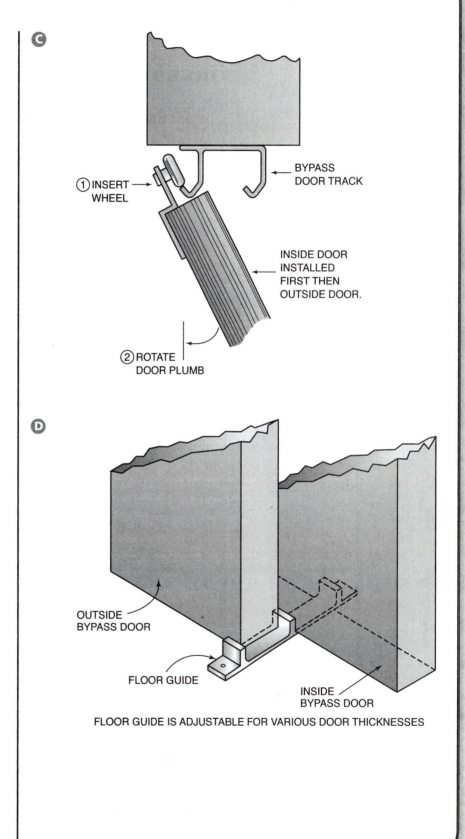

C

① INSERT WHEEL

BYPASS DOOR TRACK

INSIDE DOOR INSTALLED FIRST THEN OUTSIDE DOOR.

② ROTATE DOOR PLUMB

D

OUTSIDE BYPASS DOOR

FLOOR GUIDE

INSIDE BYPASS DOOR

FLOOR GUIDE IS ADJUSTABLE FOR VARIOUS DOOR THICKNESSES

Procedures | Hanging Interior Doors (Continued)

Installing Bifold Doors

A Check that the door and its hardware are all present. The hardware consists of the track, pivot sockets, pivot pins and guides, door aligners, door pulls, and necessary fasteners.

B Cut the track to length. Fasten it to the header jamb with screws provided in the kit. The track contains adjustable *sockets* for the door *pivot pins*. Make sure these are inserted before fastening the track in position. The position of the track on the header jamb is not critical. It may be positioned as desired.

- Locate the bottom pivot sockets. Fasten one on each side, at the bottom of the opening. The pivot socket bracket is L-shaped. It rests on the floor against the side jamb. It is centered on a plumb line from the center of the pivot sockets in the track on the header jamb above.

- Install pivot pins at the top and bottom ends of the door in the prebored holes closest to the jamb. Sometimes the top pivot pin is spring loaded. It can then be depressed for easier installation of the door. The bottom pivot pin is threaded and can be adjusted for height. The guide pin rides in the track. It is installed in the hole provided at the top end of the door farthest away from the jamb.

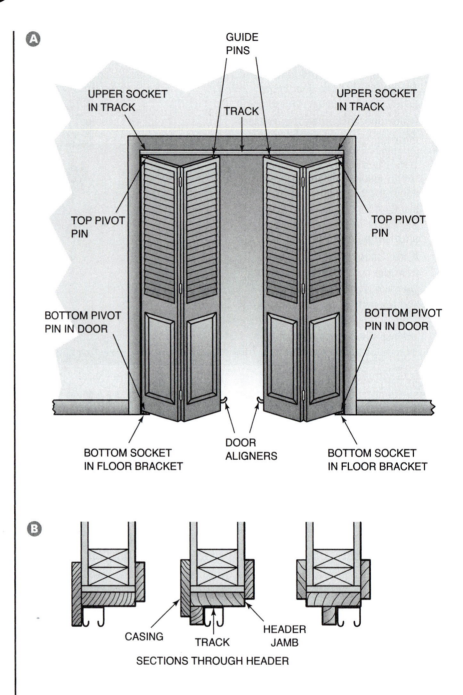

A

GUIDE PINS

UPPER SOCKET IN TRACK

TRACK

UPPER SOCKET IN TRACK

TOP PIVOT PIN

TOP PIVOT PIN

BOTTOM PIVOT PIN IN DOOR

BOTTOM PIVOT PIN IN DOOR

BOTTOM SOCKET IN FLOOR BRACKET

DOOR ALIGNERS

BOTTOM SOCKET IN FLOOR BRACKET

B

CASING   TRACK   HEADER JAMB

SECTIONS THROUGH HEADER

C Loosen the set screw in the top pivot socket. Slide the socket along the track toward the center of the opening about one foot away from the side jamb. Place the bottom door pivot in position by inserting it into the bottom pivot socket. Tilt the door to an upright position, while at the same time inserting the top pivot pin in the top socket. Slide the top pivot socket back toward the jamb where it started from.

• Adjust top and bottom pivot sockets in or out so that the desired joint is obtained between the door and the jamb. Lock top and bottom pivot sockets in position. Adjust the bottom pivot pin to raise or lower the doors, if necessary.

• Install second door in the same manner.

• Install pull knobs and door aligners in the manner and location recommended by the manufacturer. The door aligners keep the faces of the center doors lined up when closed.

C

JOINT BETWEEN SETS OF DOORS

DOOR ALIGNERS

INSIDE OF BIFOLD DOOR

Procedures   Applying Base Moldings

A Cut the first piece with squared ends if it fits between two walls. Miter the butt joint if desired. If one piece fits from corner to corner, determine its length by measuring from corner to corner. Then, transfer the measurement to the baseboard. Cut the piece ½ to 1 inch longer. Place the piece in position with one end tight to the corner and the other end away from the corner. Press the piece tight to the wall near the center. Place small marks on the top of the base trim and onto the wall so that they line up with each other. Reposition the piece with the other end in the corner. Press the base against the wall at the mark. The difference between the mark on the wall and the mark on the base is the amount to cut off.

- After cutting, place one end of the piece in the corner and bow out the center. Place the other end in the opposite corner, and press the center against the wall. Fasten in place. Continue in this manner around the room. Make regular miter joints on outside corners.

- If both ends of a single piece are to have regular miters for outside corners, fasten it in the same position as it was marked. Tack the rough length in position with one finish nail in the center. Mark both ends. Remove and cut the miters. Remember that these marks are to the short side of the miter so that the piece will be longer than these marks indicate. Reinstall the piece by first fastening into the original nail hole.

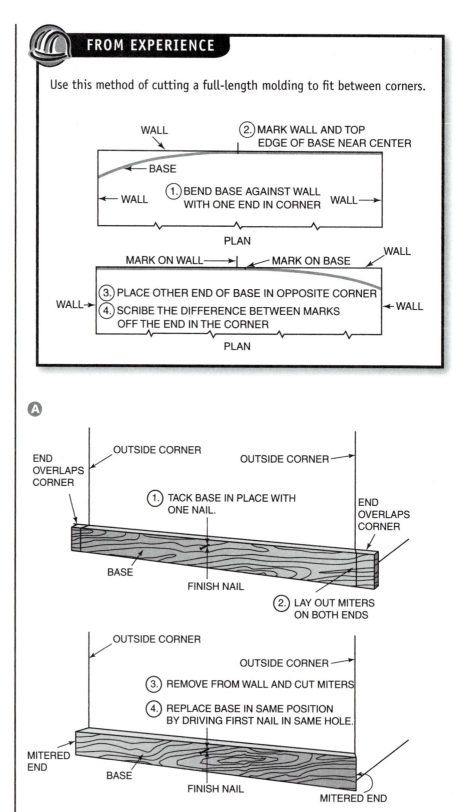

FROM EXPERIENCE

Use this method of cutting a full-length molding to fit between corners.

WALL
②. MARK WALL AND TOP EDGE OF BASE NEAR CENTER
BASE
WALL
①. BEND BASE AGAINST WALL WITH ONE END IN CORNER
WALL
PLAN

MARK ON WALL
MARK ON BASE
WALL
③. PLACE OTHER END OF BASE IN OPPOSITE CORNER
WALL
④. SCRIBE THE DIFFERENCE BETWEEN MARKS OFF THE END IN THE CORNER
WALL
PLAN

A

END OVERLAPS CORNER
OUTSIDE CORNER
OUTSIDE CORNER
END OVERLAPS CORNER
①. TACK BASE IN PLACE WITH ONE NAIL.
②. LAY OUT MITERS ON BOTH ENDS
BASE
FINISH NAIL

OUTSIDE CORNER
OUTSIDE CORNER
③. REMOVE FROM WALL AND CUT MITERS
④. REPLACE BASE IN SAME POSITION BY DRIVING FIRST NAIL IN SAME HOLE.
MITERED END
BASE
FINISH NAIL
MITERED END

B If a *base cap* is applied, it is done so in the same manner as most wall or ceiling molding. Cope interior corners and miter exterior corners. However, it should be nailed into the floor and not into the baseboard. This prevents the joint under the shoe from opening should shrinkage take place in the baseboard.

C When the base shoe must be stopped at a door opening or other location, with nothing to butt against, its exposed end is generally *back-mitered* and sanded smooth. Generally, no base shoe is required if carpeting is to be used as a floor finish.

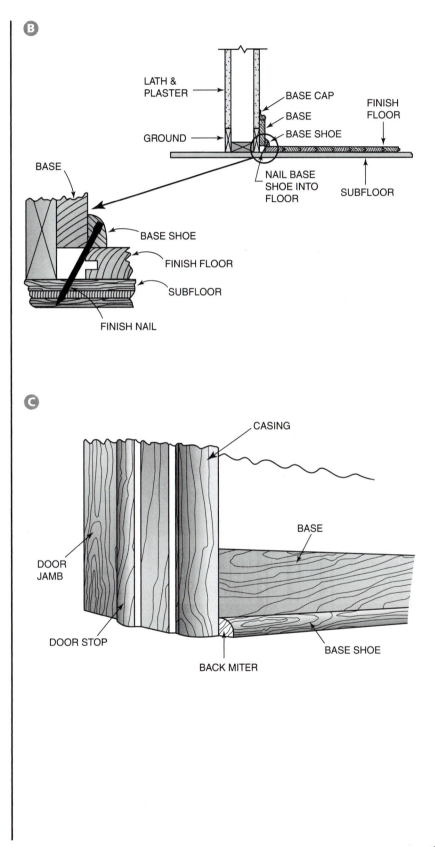

Procedures   Installing Window Trim

Applying the Stool

Ⓐ Hold a piece of side casing in position at the bottom of the window and draw a light line on the wall along the outside edge of the casing stock. Mark a distance outward from these lines equal to the thickness of the window casing. Cut a piece of stool stock to length equal to the distance between the outermost marks.

• Position the stool with its outside edge against the wall. The ends should be in line with the marks previously made on the wall. Lightly square lines, across the face of the stool, even with the inside face of each side jamb of the window frame.

• Set the pencil dividers or scribers to mark the cutout so that, on both sides, an amount equal to twice the casing thickness will be left on the stool. Scribe the stool by riding the dividers along the wall on both sides. Also scribe along the bottom rail of the window sash.

• Cut to the lines using a handsaw. Smooth the sawed edge that will be nearest to the sash. Shape and smooth both ends of the stool the same as the inside edge.

• Apply a small amount of caulking compound to the bottom of the stool. Fasten the stool in position by driving finish nails along its outside edge into the sill. Set the nails.

FROM EXPERIENCE

Raise the lower sash slightly. Place a short, thin strip of wood under it, on each side, which projects inward to support the stool while it is being laid out. Place the stool on the strips. Raise or lower the sash slightly so the top of the stool is level.

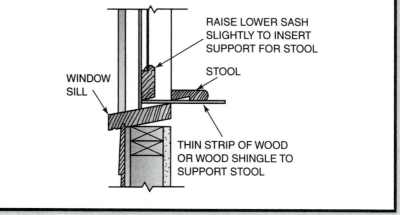

RAISE LOWER SASH SLIGHTLY TO INSERT SUPPORT FOR STOOL

STOOL

WINDOW SILL

THIN STRIP OF WOOD OR WOOD SHINGLE TO SUPPORT STOOL

Ⓐ

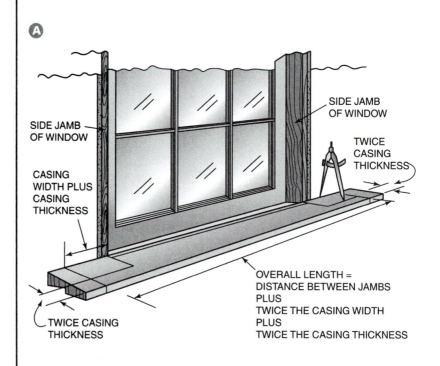

SIDE JAMB OF WINDOW

SIDE JAMB OF WINDOW

CASING WIDTH PLUS CASING THICKNESS

TWICE CASING THICKNESS

TWICE CASING THICKNESS

OVERALL LENGTH = DISTANCE BETWEEN JAMBS PLUS TWICE THE CASING WIDTH PLUS TWICE THE CASING THICKNESS

Applying te Apron

Ⓐ Cut a length of apron stock equal to the distance between the outer edges of the window casings.

- Each end of the apron is then *returned upon itself*. This means that the ends are shaped the same as its face. To return an end upon itself, hold a scrap piece on the apron. Draw its profile flush with the end. Cut to the line with a coping saw. Sand the cut end smooth. Return the other end upon itself in the same manner.

- Place the apron in position with its upper edge against the bottom of the stool. Be careful not to force the stool upward. Keep the top side of the stool level by holding a square between it and the edge of the side jamb. Fasten the apron along its bottom edge into the wall. Then drive nails through the stool into the top edge of the apron.

Ⓐ

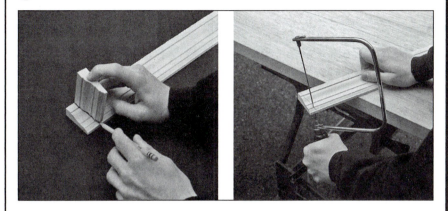

🪖 **FROM EXPERIENCE**

When nailing through the stool, wedge a short length of 1 × 4 stock between the apron and the floor at each nail location. This supports the apron while nails are being driven. Failure to support the apron results in an open joint between it and the stool. Take care not to damage the bottom edge of the apron with the supporting piece.

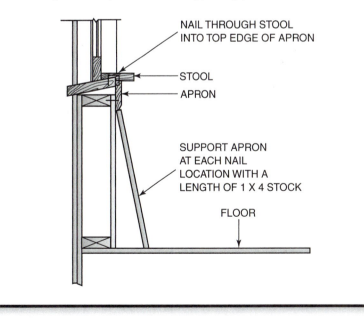

NAIL THROUGH STOOL INTO TOP EDGE OF APRON

STOOL

APRON

SUPPORT APRON AT EACH NAIL LOCATION WITH A LENGTH OF 1 X 4 STOCK

FLOOR

Procedures   Installing Window Trim (Continued)

Installing Jamb Extensions

A Measure the distance from the jamb to the finished wall. Rip the jamb extensions to this width with a slight back bevel on the side toward the jack stud.

- Cut the pieces to length and apply them to the header, side jambs, and stool. Shim them, if necessary, and nail with finish nails that will penetrate the framing at least an inch.

Applying the Casings

- Cut the number of window casings needed to a rough length with a miter on one end. Cut side casings with left- and right-hand miters.

- Install the header casing first and then the side casings in a similar manner as with door casings. Find the length of side casings by turning them upside down with the point of the miter on the stool in the same manner as door casings.

- Fasten casings with their inside edges flush with the inside face of the jamb or with a reveal. Make neat, tight-fitting joints at the stool and at the head.

A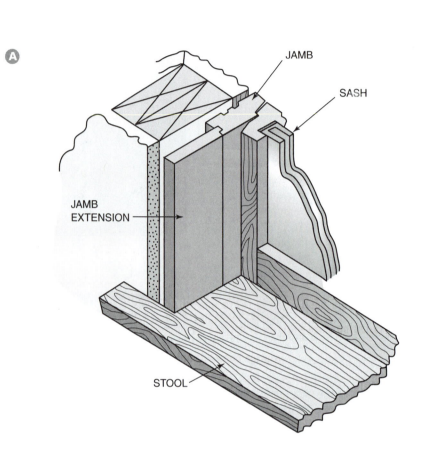

JAMB

SASH

JAMB EXTENSION

STOOL

Procedures   Installing Wood Flooring

Preparation for Installation

- Check the subfloor for any loose areas and add nails where appropriate. Sweep and vacuum the subfloor clean. Scraping may be necessary to remove any unwanted material.

- Cover the subfloor with building paper. Lap it 4 inches at the seams and at right angles to the direction of the finish floor. This paper prevents squeaks in dry seasons and retards moisture from below that could cause warping of the floor.

- Snap chalk lines on the paper showing the centerline of floor joists so that flooring can be nailed into them. For better holding power, fasten flooring through the subfloor and into the floor joists whenever possible. On ½-inch plywood subfloors, all flooring fasteners must penetrate into the joists.

Starting Strip

- The location and straight alignment of the first course is important. Place a strip of flooring on each end of the room, ¾ inch from the starter wall with the groove side toward the wall. The gap between the flooring and the wall is needed for expansion. It will eventually be covered by the base molding.

- Mark along the edge of the flooring tongue. Snap a chalk line between the two points. Hold the strip with its tongue edge to the chalk line.

Procedures | Installing Wood Flooring (Continued)

Ⓐ Face-nail it with 8d finish nails, alternating from one edge to the other, 12–16 inches apart.

Make sure that end joints between strips are driven up tight.

- Cut the last strip to fit loosely against the wall. Use a strip long enough so that the cut-off piece is 8 inches or longer. This scrap piece is used to start the next course back against the other wall.

Ⓐ

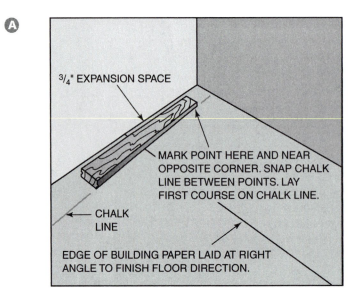

Courtesy of Chickasaw Hardwood Floors.

Ⓑ After the second course of flooring is fastened, lay out seven or eight loose rows of flooring, end to end. This is called *racking the floor*. Racking is done to save time and material.

- Lay out in a staggered end-joint pattern. End joints should be at least 6 inches apart. Find or cut pieces to fit within ½ inch of the end wall. Distribute long and short pieces evenly for the best appearance. Avoid clusters of short strips. Lay out loose flooring.

- Continue across the room. Rack seven or eight courses ahead as work progresses.

Ⓑ

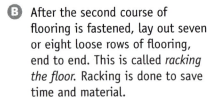

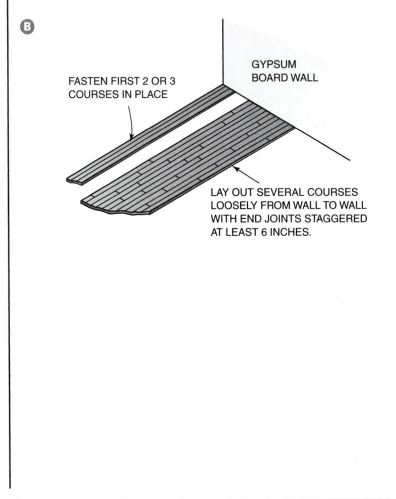

Procedures

Replacing Broken Tiles

- Before a broken tile can be replaced, the grout must be removed from around the tile you wish to replace. The grout is a bonding agent that binds the tiles together and provides a means of sealing the subfloor.

- To remove soft unsanded grout (typically used for wall tiles), use a sharp object such as a utility knife to scratch the grout out.

- To remove sanded grout (typically used for floor tiles), use a cold chisel, which scores the surface of the grout. Once the surface has been scored, remove it using a utility knife.

- If the grout spacing is sufficient (wide enough), then use a grout saw to remove the grout.

- Once you remove the grout, simply lift the tile away from the remaining tiles if the tile is loose. If the tile is not loose, carefully tap the edges of the tile to be removed using a cold chisel and a hammer until the tile is loosened.

- Once the tile has been removed, using a putty knife remove any lumps or bumps in the mortar.

- Ⓐ Once the hole has been clean of debris, lumps, and bumps, test fit the new tile to ensure that it sets properly in the space. The tile should not be higher than the surrounding tiles or have excessive rocking. If the tile is higher than the surrounding tiles or has excessive rocking associated with it, then remove the tile and continue to scrape more mortar from the surface. If the tile is larger than the hole, then trim the tile to fit. For such small jobs, use a hand tile cutter.

Ⓐ

Procedures

Replacing Broken Tiles (Continued)

B Fit the tile to a curved hole using a nibbler. Nibblers are used to make small or irregular cuts.

- Once the tile is properly fit, remove the tile and apply a $\frac{1}{8}$-inch layer of adhesive to the back of the tile. The adhesive should not be any closer than $\frac{1}{2}$" from the edge of the tile.

- Press the tile using a slight back-and-forth motion to evenly spread the adhesive and to ensure a good bond. Allow the tile to set for 24 hours before applying the grout.

- The grout should be mixed per the manufacturer's instructions.

- Using a damp sponge push the grout into the cracks.

B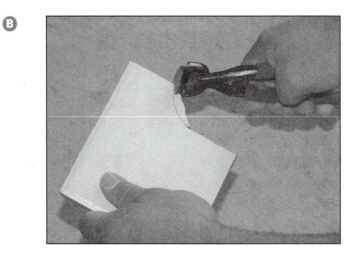

Procedures | Removing an Existing Carpet

- A carpet can be installed over almost any type of smooth substrate (concrete, plywood, particleboard, etc.). However, before the carpet can be installed, the old carpet (if necessary) must first be removed. This is accomplished by removing all the furniture from the area as well as the molding (around the floor).

- Before removing the old carpet, be sure to wear a respirator to keep from breathing in any dust.

- Using a carpet or a utility knife, cut the old carpet into 2-foot wide strips. This will make handling the old carpet much easier.

- Starting at one end of the carpet, pull it up from the tackless strips and rolling it up as you go.

- If the carpet has an underlayment installed, it should be also removed.

- Remove any existing tack strips and ensure that the area is clean of any debris and dry.

- Inspect and repair any loose floorboards.

FROM EXPERIENCE

In most cases when a carpet is being replaced, the underlayment will be worn out. Therefore, even if the underlayment appears to be in good shape, it is recommended to always replace it when installing a new carpet.

Procedures   Installing or Replacing Carpet

- Before installing a new carpet, be sure to remove the existing one using the procedure outlined in "Removing an Existing Carpet."

- Install new tackless strips around the perimeter of the room in which the carpet is to be installed. The tackless strips should be installed about ½ inch from the wall with the pins or tacks facing toward the wall. In addition be sure that tackless strips are butted together. This is especially true in the corners.

A When installing the underlayment, place it down so that the edges overlap the tackless strips. The underlayment should be placed so that the edges are butted together but not overlapped.

- Once the underlayment is in position, it must be either glued or stapled along the inside edge of the tackless strip.

- Trim the excess underlayment. Seal all seams using duct tape.

B The carpet is installed by laying it over the underlayment.

C Bond any seams between sections of carpet using adhesive carpet tape or hot melt tape. This is accomplished by placing the tape between the underlayment and the carpet (adhesive side up).

A
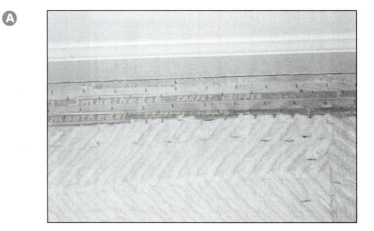

B

C
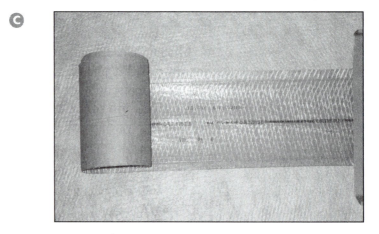

D If hot melt tape is used, then place an electric seaming iron below the carpet and directly onto the adhesive of the tape.

E Press the carpet firmly onto the hot tape butting the edges together.

F Roll the seam using a carpet seam roller.

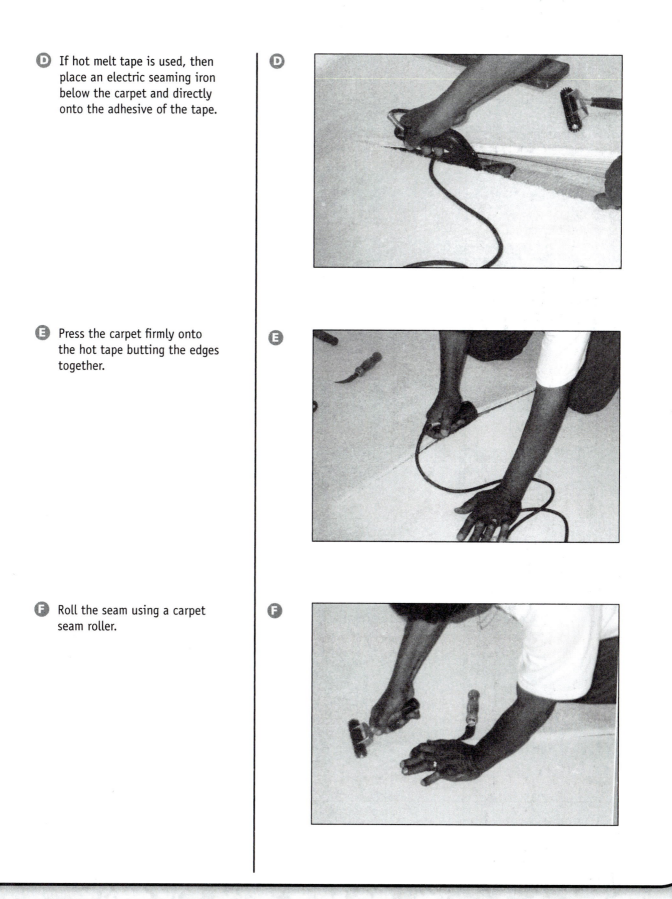

Procedures

Installing or Replacing Carpet (Continued)

Ⓖ Once all the seams have been made (as previously described), stretch the carpet into place using either a knee kicker or a power stretcher.

Ⓗ Press the edge of the carpet into position using a stair tool.

• Trim any excess carpet and replace the base boards.

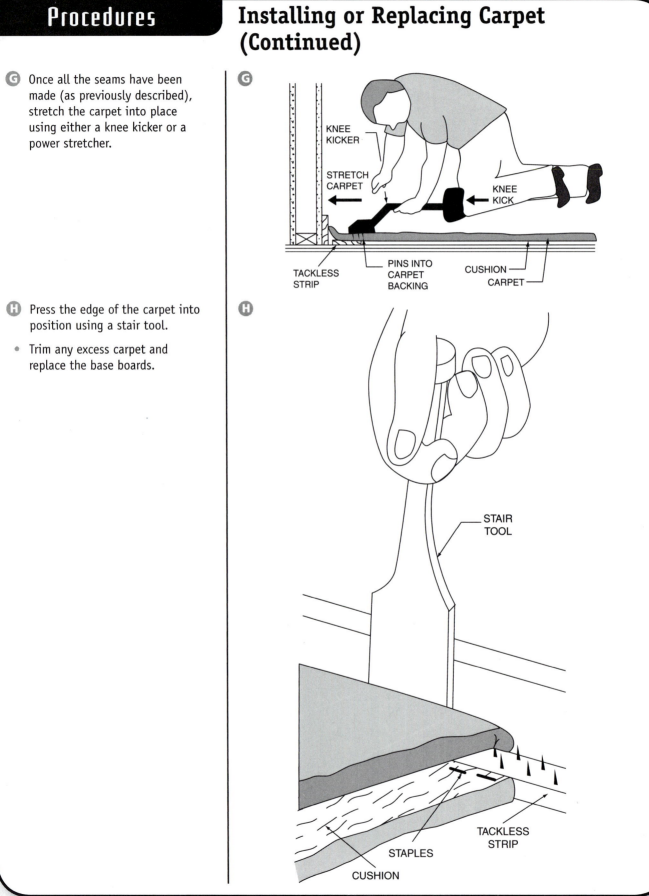

Ⓖ
KNEE KICKER
STRETCH CARPET
KNEE KICK
TACKLESS STRIP
PINS INTO CARPET BACKING
CUSHION
CARPET

Ⓗ
STAIR TOOL
STAPLES
CUSHION
TACKLESS STRIP

Procedures

Installing Manufactured Cabinets

Cabinet Layout Lines

Ⓐ Measure 34½ inches up the wall. Draw a level line to indicate the tops of the base cabinets. Measure and mark another level line on the wall 54 inches from the floor. The bottom of the wall units are installed to this line.

• Then mark the stud locations of the framed wall. Drive cabinet mounting screws into the studs. Lightly tap on and across a short distance of the wall with a hammer. Above the upper line on the wall, drive a finish nail in at the point where a solid sound is heard to accurately locate the stud. Drive nails where the holes will be later covered by a cabinet. Mark the locations of the remaining studs where cabinets will be attached. At each stud location, draw plumb lines on the wall. Mark the outlines of all cabinets on the wall to visualize and check the cabinet locations against the layout.

Ⓐ

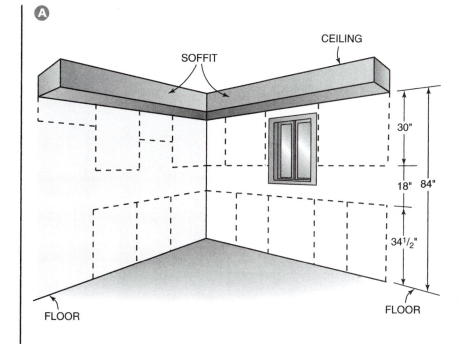

Installing Wall Units

Ⓐ Use a *cabinet lift* to hold the cabinets in position for fastening to the wall. If a lift is not available, remove the doors and shelves to make the cabinet lighter and easier to clamp together. If possible, screw a strip of lumber so that its top edge is on the level line for the bottom of the wall cabinets or strips of wood cut to the proper length. This is used to support the wall units while they are being fastened. If it is not possible to screw to the wall, build a stand on which to support the unit near the line of installation.

Ⓐ

Procedures

Installing Manufactured Cabinets (Continued)

B Start the installation of wall cabinets in a corner. On the wall, measure from the line representing the outside of the cabinet to the stud centers. Transfer the measurements to the cabinets. Drill shank holes for mounting screws through mounting rails usually installed at the top and bottom of the cabinet. Place the cabinet on the supporting strip or stand so that its bottom is on the level layout line. Fasten the cabinet in place with mounting screws of sufficient length to hold the cabinet securely. Do not fully tighten the screws. Install the next cabinet in the same manner.

C Align the adjoining *stiles* so that their faces are flush with each other. Clamp them together with C-clamps. Screw the stiles tightly together. Continue this procedure around the room. After all the stiles are secured to each other, tighten all mounting screws. If a filler needs to be used, it is better to add it at the end of a run. It may be necessary to scribe the filler to the wall.

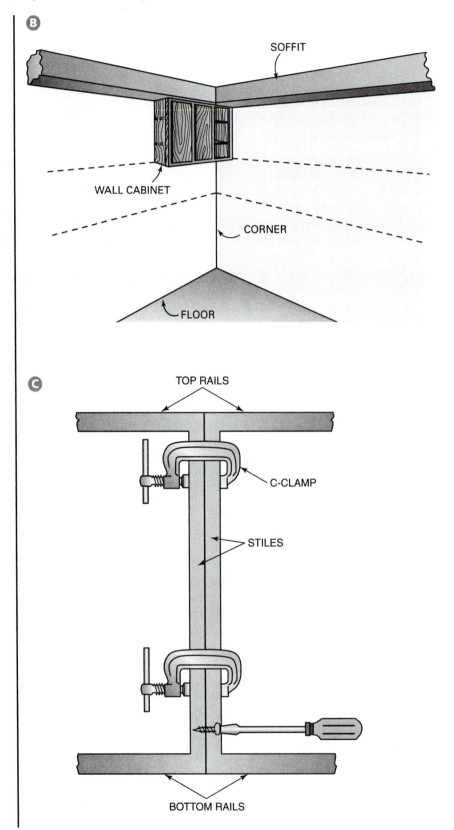

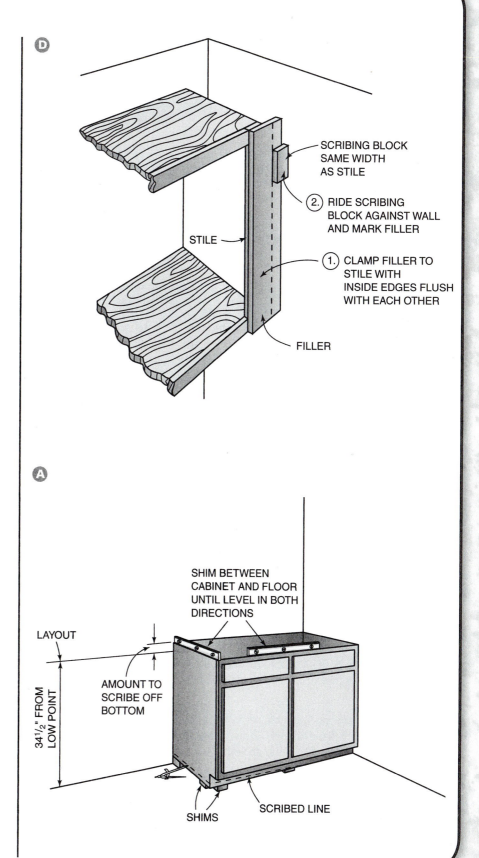

D Procedure for scribing a filler strip at the end of a run of cabinets.

• The space between the top of the wall unit and the ceiling may be finished by installing a soffit.

D

SCRIBING BLOCK SAME WIDTH AS STILE

(2.) RIDE SCRIBING BLOCK AGAINST WALL AND MARK FILLER

STILE

(1.) CLAMP FILLER TO STILE WITH INSIDE EDGES FLUSH WITH EACH OTHER

FILLER

Installing Base Cabinets

A Start the installation of base cabinets in a corner. Shim the bottom until the cabinet top is on the layout line. Then level and shim the cabinet from back to front. If cabinets are to be fitted to the floor, shim until their tops are level across width and depth. This will bring the tops above the layout line that was measured from the low point of the floor. Adjust the scriber so that the distance between the points is equal to the amount the top of the unit is above the layout line. Scribe this amount on the bottom end of the cabinets by running the dividers along the floor.

A

SHIM BETWEEN CABINET AND FLOOR UNTIL LEVEL IN BOTH DIRECTIONS

LAYOUT

$34\frac{1}{2}$" FROM LOW POINT

AMOUNT TO SCRIBE OFF BOTTOM

SHIMS

SCRIBED LINE

Procedures — Installing Manufactured Cabinets (Continued)

 Cut both ends and toeboard to the scribed lines. Replace the cabinet in position. The top ends should be on the layout line. Fasten it loosely to the wall. Install the remaining base cabinets in the same manner. Align and clamp the stiles of adjoining cabinets. Fasten them together. Finally, fasten all units securely to the wall.

Installing Countertops

- After the base units are fastened in position, cut the countertop to length. Fasten it on top of the base units and against the wall.

Scribe the backsplash, limited by the thickness of its scribing strip, to an irregular wall surface. Use pencil dividers to scribe a line on the top edge of the backsplash. Then plane or belt sand to the scribed line.

- Fasten the countertop to the base cabinets with screws up through triangular blocks usually installed in the top corners of base units. Take care not to drill through the countertop. Use screws of sufficient length, but not so long that they penetrate the countertop.

- Exposed cut ends of postformed countertops are covered by specially shaped pieces of plastic laminate. Sink cutouts are made by carefully outlining the cutout and cutting with a saber saw or router. The cutout pattern usually comes with the sink. Use a fine tooth blade to prevent chipping out the face of the laminate beyond the sink. Some duct tape applied to the base of the power tool will prevent scratching of the countertop when making the cutout.

Procedures — Removing Manufactured Cabinets

- Before attempting to remove and/or replace cabinets, first ensure that the contents of the cabinet have been removed.

- If the cabinets have electrical outlets or plumbing connections in them, turn off the utilities before starting.

- If both base and wall cabinets are to be removed, start by removing the base cabinets. Remove any flooring that may be attached to the toeboard.

- Remove any utilities connected to the cabinets.

- Remove all molding and/or decorative accents attached to the cabinets.

- Remove all nails and/or screws along the inside edge of the cabinets.

- Once all the nails and/or screws have been removed, carefully remove the base cabinets.

- When removing wall cabinets, always have at least one helper.

- Once the cabinet has been unsecured from the wall, remove it from the area.

Procedures

Installing Cylindrical Locksets

- To install a cylindrical lockset, first check the contents and read the manufacturer's directions carefully. Many kinds of locks are manufactured that the mechanisms vary greatly. Follow the directions included with the lockset carefully.

Ⓐ However, there are certain basic procedures. Open the door to a convenient position. Wedge the bottom to hold it in place. Measure up, from the floor, the recommended distance to the centerline of the lock. This is usually 36–40 inches. At this height, square a light line across the edge and stile of the door.

Marking and Boring Holes

Ⓐ Position the center of the paper template supplied with the lock on the squared lines. Lay out the centers of the holes that need to be bored. Fold the template over the high corner of the beveled door edge. The distance from the door edge to the center of the hole through the side of the door is called the *backset* of the lock. Usual backsets are $2\frac{3}{8}$ inches for residential and $2\frac{3}{4}$ inches for commercial constructions. Make sure that the backset is marked correctly before boring the hole. One hole must be bored through the side and one into the edge of the door. The manufacturer's directions specify the hole sizes where a 1-inch hole for bolts and $2\frac{1}{8}$-inch hole for locksets are common.

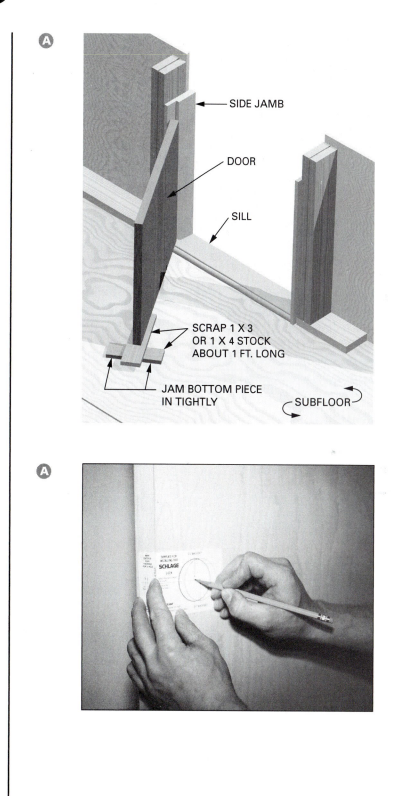

Ⓐ
SIDE JAMB
DOOR
SILL
SCRAP 1 X 3 OR 1 X 4 STOCK ABOUT 1 FT. LONG
JAM BOTTOM PIECE IN TIGHTLY
SUBFLOOR

Ⓐ

Procedures

Installing Cylindrical Locksets (Continued)

- The hole through the side of the door should be bored first. Stock for the center of the boring bit is lost if the hole in the edge of the door is bored first. It can be bored with hand tools, using an expansion bit in a bit brace. However, it is a difficult job. If you are using hand tools, bore from one side until only the point of the bit comes through. Then bore from the other side to avoid splintering the door.

Using a Boring Jig

Ⓐ A **boring jig** is frequently used. It is clamped to the door to guide power-driven **multispur bits**. With a boring jig, holes can be bored completely through the door from one side. The clamping action of the jig prevents splintering.

- After the holes are bored, insert the latchbolt in the hole bored in the door edge. Hold it firmly and score around its faceplate with a sharp knife. Remove the latch unit. Deepen the vertical lines with the knife in the same manner as with hinges. Take great care when using a chisel along these lines. This may split out the edge of the door. Then, chisel out the recess so that the faceplate of the latch lays flush with the door edge.

Ⓑ Use **Faceplate markers**, if available, to lay out the mortise for the latch faceplate. A marker of the appropriate size is held in the bored latch hole and tapped with a hammer. Complete the installation of the lockset by following specific manufacturer's directions.

Ⓐ
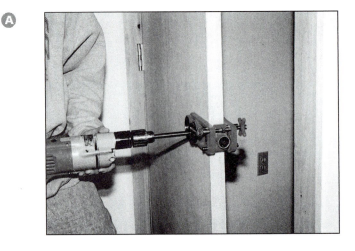

Ⓑ
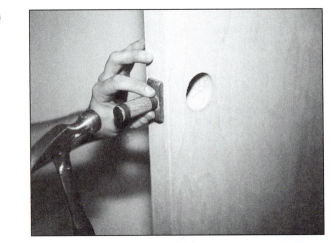

Installing the Striker Plate

- The striker plate is installed on the door jamb, so when the door is closed it latches tightly with no play. If the plate is installed too far out, the door will not close tightly against the stop. It will then rattle. If the plate is installed too far in, the door will not latch.

Ⓐ To locate the striker plate in the correct position, place it over the latch in the door. Close the door snugly against the stops. Push the striker plate in against the latch. Draw a vertical line on the face of the plate flush with the outside face of the door.

- Open the door. Place the striker plate on the jamb. The vertical line, previously drawn on it, should be in line with the edge of the jamb. Center the plate on the latch. Hold it firmly while scoring a line around the plate with a sharp knife. Chisel out the mortise so that the plate lies flush with the jamb. Screw the plate in place. Chisel out the center to receive the latch.

Rekey Locks

- Take the locking part of the lock out by unscrewing it from the inside. Leave the rest of the lock in place.

- Take the knob with the locking mechanism with its proper key to the hardware store. The hardware store person will replace the pins so that they'll fit another key.

- Put the door lock back together.

Ⓐ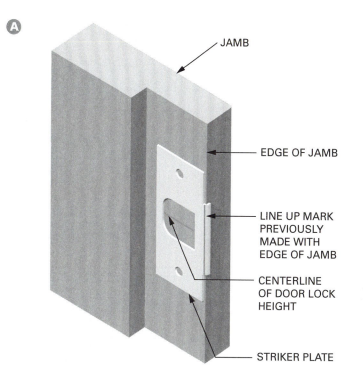

JAMB

EDGE OF JAMB

LINE UP MARK PREVIOUSLY MADE WITH EDGE OF JAMB

CENTERLINE OF DOOR LOCK HEIGHT

STRIKER PLATE

Procedures

Cutting and Fitting Gypsum Board

 A Take measurements accurately to within ¹/₄ inch for the ceiling and ¹/₈ inch for the walls. Using a utility knife, cut the board by first scoring the face side through the paper to the core. Guide it with a *drywall T-square* using your toe to hold the bottom. Only the paper facing needs to be cut.

B Bend the board back against the cut. The board will break along the cut face. Score the backside paper.

- Lifting the panel off the floor, snap the cut piece back quickly to the straight position. This will complete the break.

FROM EXPERIENCE

Cut only the center section of the backside paper, leaving the bottom and top portions. These will act as hinges for the cut piece when it is snapped back into place.

 C To make cuts parallel to the long edges, the board is often gauged with a tape and scored with a utility knife. When making cuts close to long edges, score both sides of the board before the break to obtain a clean cut.

- Ragged edges can be smoothed with a drywall rasp, coarse sanding block, or knife.

A
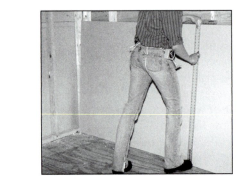

B

C
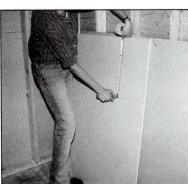

Procedures   Installing Sheet Paneling

Starting the Application

A Mark the location of each stud in the wall on the floor and ceiling. Paneling edges must fall on stud centers, even if applied with adhesive over a backer board, in case supplemental nailing of the edges is necessary.

- If the wall is to be wainscoted, snap a horizontal line across the wall to indicate its height.

- Apply narrow strips of paint on the wall from floor to ceiling over the stud where a seam in the paneling will occur. The color should be close to that of the seams of the paneling. This will hide the joints between sheets if they open slightly because of shrinkage.

- Cut the first sheet to a length about ¼ inch less than the wall height. Place the sheet in the corner. Plumb the edge and tack it temporarily into position.

B Notice the joint at the corner and the distance the sheet edge overlaps the stud. Set the distance between the points of a scriber to the same as the amount the sheet overlaps the center of the stud. Scribe this amount on the edge of the sheet butting the corner.

- Remove the sheet from the wall and cut close to the scribed line. Plane the edge to the line to complete the cut. Replace the sheet with the cut edge fitting snugly in the corner.

- If a tight fit between the panel and the ceiling is desired, set the dividers and scribe a small amount at the ceiling line. Remove the sheet again and cut to the scribed line. The joint at the ceiling need not be fit tight if a molding is to be used.

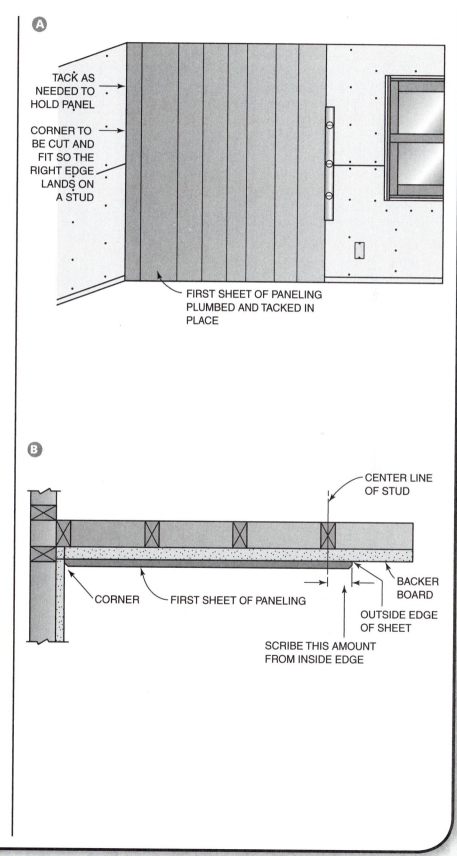

A

TACK AS NEEDED TO HOLD PANEL

CORNER TO BE CUT AND FIT SO THE RIGHT EDGE LANDS ON A STUD

FIRST SHEET OF PANELING PLUMBED AND TACKED IN PLACE

B

CENTER LINE OF STUD

CORNER   FIRST SHEET OF PANELING

BACKER BOARD

OUTSIDE EDGE OF SHEET

SCRIBE THIS AMOUNT FROM INSIDE EDGE

Procedures | Installing Sheet Paneling (Continued)

Wall Outlets

(A) To lay out for wall outlets, plumb and mark both sides of the outlet to the floor or ceiling, whichever is closer. Level the top and bottom of the outlet on the wall out beyond the edge of the sheet to be installed.

- Place the sheet in position and tack. Level and plumb marks from the wall and floor onto the sheet for the location of the opening.

(B) Remove the sheet and cut the opening for the outlet. When using a saber saw, cut from the back of the panel to avoid splintering the face.

Fastening

- Apply adhesive beads 3 inches long and about 6 inches apart on all intermediate studs. Apply a continuous bead along the perimeter of the sheet. Put the panel in place. Tack it at the top when panel is in proper position.

- Press on the panel surface to make contact with the adhesive. Use firm, uniform pressure to spread the adhesive beads evenly between the wall and the panel. Then, grasp the panel and slowly pull the bottom of the sheet a few inches away from the wall.

- Press the sheet back into position after about 2 minutes. Drive nails as needed and recheck the sheet for a complete bond after about 20 minutes. Apply pressure to assure thorough adhesion and to smooth the panel surface.

- Apply successive sheets in the same manner. Panels should touch only lightly at joints.

(A)

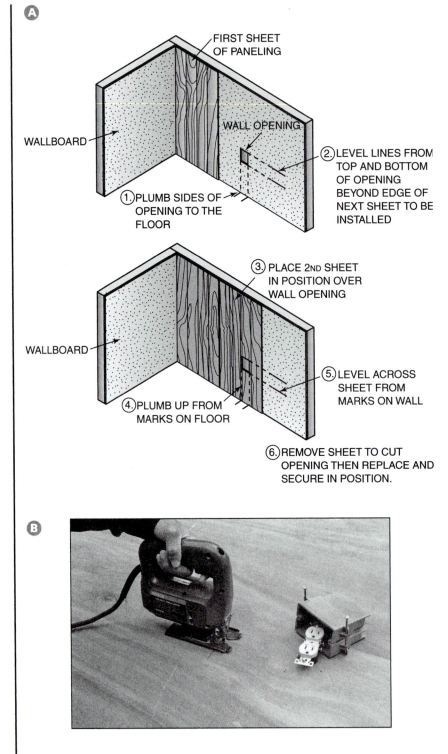

FIRST SHEET OF PANELING

WALLBOARD

WALL OPENING

① PLUMB SIDES OF OPENING TO THE FLOOR

② LEVEL LINES FROM TOP AND BOTTOM OF OPENING BEYOND EDGE OF NEXT SHEET TO BE INSTALLED

③ PLACE 2ND SHEET IN POSITION OVER WALL OPENING

WALLBOARD

④ PLUMB UP FROM MARKS ON FLOOR

⑤ LEVEL ACROSS SHEET FROM MARKS ON WALL

⑥ REMOVE SHEET TO CUT OPENING THEN REPLACE AND SECURE IN POSITION.

(B)

Ending the Application

- Take measurements at the top, center, and bottom. Cut the sheet to width and install. If no corner molding is used, the sheet must be cut to fit snugly in the corner. To mark the sheet accurately, first measure the remaining space at the top, at the bottom, and about the center. Rip the panel about ½ inch wider than the greatest distance.

A Place the sheet plumb with the cut edge in the corner and the other edge overlapping the last sheet installed. Tack the sheet in position so that the amount of overlap is exactly the same from top to bottom. Set the scriber for the amount of overlap and scribe this amount on the edge in the corner.

- Cut close to the scribed line and then plane to the line. If the line is followed carefully, the sheet should fit snugly between the last sheet installed and the corner, regardless of any irregularities.

B Exterior corners may be finished by capping the joint.

- Use a wood block for more accurate scribing of wide distances.

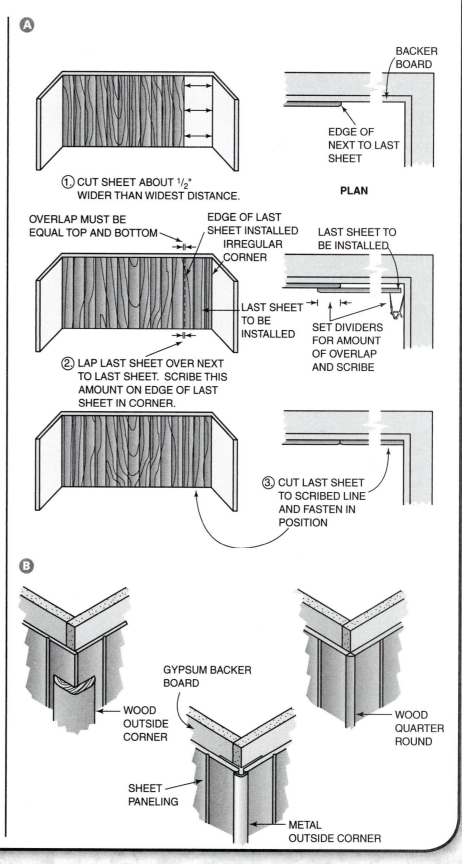

A

① CUT SHEET ABOUT ½" WIDER THAN WIDEST DISTANCE.

BACKER BOARD

EDGE OF NEXT TO LAST SHEET

PLAN

OVERLAP MUST BE EQUAL TOP AND BOTTOM

EDGE OF LAST SHEET INSTALLED IRREGULAR CORNER

LAST SHEET TO BE INSTALLED

LAST SHEET TO BE INSTALLED

② LAP LAST SHEET OVER NEXT TO LAST SHEET. SCRIBE THIS AMOUNT ON EDGE OF LAST SHEET IN CORNER.

SET DIVIDERS FOR AMOUNT OF OVERLAP AND SCRIBE

③ CUT LAST SHEET TO SCRIBED LINE AND FASTEN IN POSITION

B

WOOD OUTSIDE CORNER

GYPSUM BACKER BOARD

WOOD QUARTER ROUND

SHEET PANELING

METAL OUTSIDE CORNER

Procedures

Installing Solid Wood Paneling

Starting the Application

Ⓐ Select a straight board with which to start. Cut it to length, about ¼ inch less than the height of the wall. If tongue-and-groove stock is used, tack it in a plumb position with the grooved edge in the corner.

- Adjust the scribers to scribe an amount a little more than the depth of the groove. Rip and plane to the scribed line.

- Replace the piece and face-nail along the cut edge into the corner with finish nails about 16 inches apart. Blind-nail the other edge through the tongue.

- Apply succeeding boards by blind-nailing only into the tongue. Make sure that the joints between boards come up tightly. Severely warped boards should not be used.

- As installation progresses, check the paneling for plumb. If it is out of plumb, gradually bring back by driving one end of several boards a little tighter than the other end. Cut out openings in the same manner as described for sheet paneling.

Ⓐ

Applying the Last Board

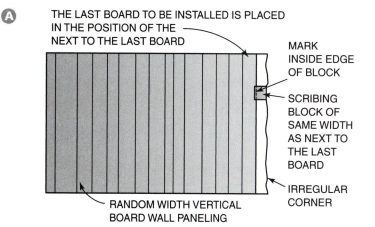

THE LAST BOARD TO BE INSTALLED IS PLACED IN THE POSITION OF THE NEXT TO THE LAST BOARD

MARK INSIDE EDGE OF BLOCK

SCRIBING BLOCK OF SAME WIDTH AS NEXT TO THE LAST BOARD

IRREGULAR CORNER

RANDOM WIDTH VERTICAL BOARD WALL PANELING

A Cut and fit the next to the last board and then remove it. Cut, fit, and tack the last board in the place of the next-to-the-last board just previously removed.

- Cut a scrap block about 6 inches long and equal in width to the finished face of the next-to-the-last board. The tongue should be removed. Use this block to scribe the last board by running one edge along the corner and holding a pencil against the other edge.

- Remove the board from the wall. Cut and plane it to the scribed line. Fasten the next-to-the-last board in position. Fasten the last board in position with the cut edge in the corner.

- Face-nail the edge nearest the corner.

Procedures   Installing Flexible Insulation

A Install positive ventilation chutes between the rafters where they meet the wall plate. This will compress the insulation slightly against the top of the wall plate to permit the free flow of air over the top of the insulation.

• Install the air-insulation dam between the rafters in line or on with the exterior sheathing. This will protect the insulation from air movement into the insulation layer from the soffits.

A

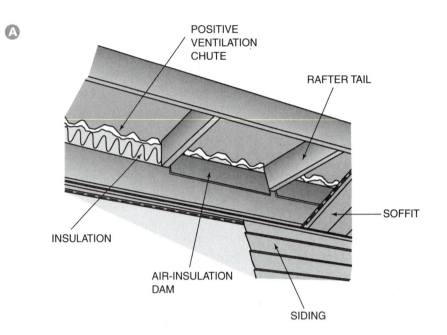

POSITIVE VENTILATION CHUTE

RAFTER TAIL

INSULATION

AIR-INSULATION DAM

SOFFIT

SIDING

B To cut the material, place a scrap piece of plywood on the floor to protect the floor while cutting. Roll out the material over the scrap. Using another scrap piece of wood, compress the insulation and cut it with a sharp knife in one pass.

C Place the batts or blankets between the studs. The flanges of the vapor retarder may be stapled either to cover the studs or to cover the inside edges of the studs as well as the top and bottom plates. A better vapor retarder is achieved with fastening to cover the stud, but the studs are less visible for the installation of the gypsum. Use a hand or hammer-tacker stapler to fasten the insulation in place.

B

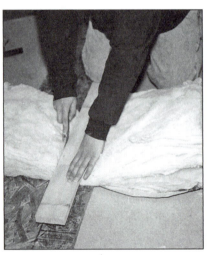

C

D Fill any spaces around windows and doors with spray-can foam. Nonexpanding foam will fill the voids with an airtight seal and protect the house from air leakage. After the foam cures, add flexible insulation to fill the remaining space.

• Install ceiling insulation by stapling it to the ceiling joists or by friction-fitting it between them. Push and extend the insulation across the top plate to fit against the air-insulation dam.

E Flexible insulation installed between floor joists over crawl spaces may be held in place by wire mesh or pieces of heavy-gauge wire wedged between the joists.

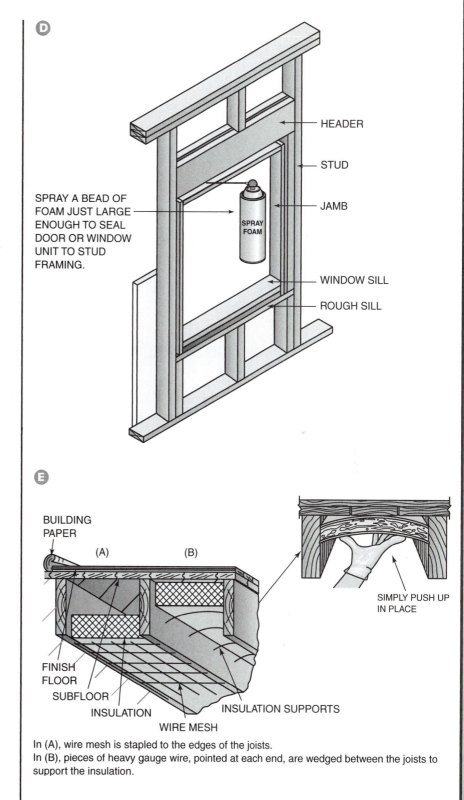

D

SPRAY A BEAD OF FOAM JUST LARGE ENOUGH TO SEAL DOOR OR WINDOW UNIT TO STUD FRAMING.

SPRAY FOAM

HEADER

STUD

JAMB

WINDOW SILL

ROUGH SILL

E

BUILDING PAPER

(A)   (B)

SIMPLY PUSH UP IN PLACE

FINISH FLOOR

SUBFLOOR

INSULATION

WIRE MESH

INSULATION SUPPORTS

In (A), wire mesh is stapled to the edges of the joists.
In (B), pieces of heavy gauge wire, pointed at each end, are wedged between the joists to support the insulation.

Procedures   Installing Windows

A Place window in the opening after removing all shipping protection from the window unit. Do not remove any diagonal braces applied at the factory. Close and lock the sash. Windows can easily be moved through the openings from the inside and set in place.

B Center the unit in the opening on the rough sill with the exterior window casing against the wrapped wall sheathing. Level the window sill with a wood shim tip between the rough sill and the bottom end of the window's side jamb, if necessary. Secure the shim to the rough sill.

• Remove the window unit from the opening and caulk the backside of the casing or nailing flange. This will seal the unit to the building. Replace the unit and nail the lower end of both sides of the window. Next, plumb a side and nail the unit along the sides and top. Check that the sash operates properly. If not, make necessary adjustments.

C Flash the head casing by cutting to length of the flashing with tin snips. Its length should be equal to the overall length of the window head casing. If the flashing must be applied in more than one piece, lap the joint about 3 inches. Slice the housewrap just above the head casing and slip the flashing behind the wrap and on top of the head casing. Secure with fasteners into the wall sheathing. Refasten the house wrap. Tape all seams in housewrap and over window nailing flanges.

CAUTION

Have sufficient help when setting large units. Handle them carefully to avoid damaging the unit or breaking the glass. Broken glass can cut through protective clothing and cause serious injury.

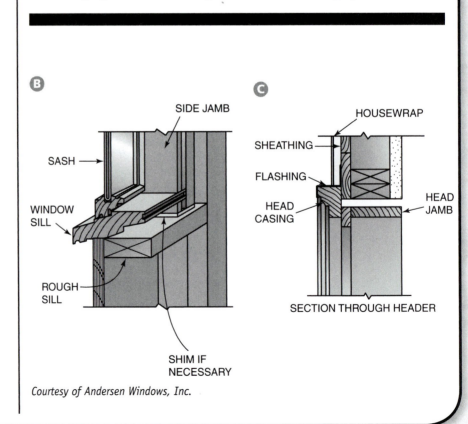

Courtesy of Andersen Windows, Inc.

Procedures   Replacing a Damaged Window Screen

- Remove the damaged window screen from the window frame.

- After placing the screen on a sturdy flat surface (large enough to support the damaged screen), remove the rubber spline, thus allowing the old screen to be separated from the frame

- With the old screen removed, measure the length and width of the screen frame. When cutting the new screen, add 2 inches to the measure of the screen frame.

- Lay the new screen onto the frame, using a screen-rolling tool and starting in one corner press the rubber spline and screen firmly into the spline groove.

- Continue this process all the way around the screen frame.

- Using a utility knife carefully trim the excess screen.

Exterior Carpentry Maintenance

Procedures   Installing Gutters

A On both ends of the fascia, mark the location of the bottom side of the gutter. The top outside edge of the gutter should be in relation to a straight line projected from the top surface of the roof. The height of the gutter depends on the pitch of the roof.

- Stretch a chalk line between the two marks. Move the center of the chalk line up enough to give the gutter the proper pitch from the center to the ends. Snap a line on both sides of the center.

- Fasten the gutter brackets to the chalk line on the fascia with screws. All screws should be made of stainless steel or other corrosion-resistant material. Aluminum brackets may be spaced up to 30 inches on center (OC).

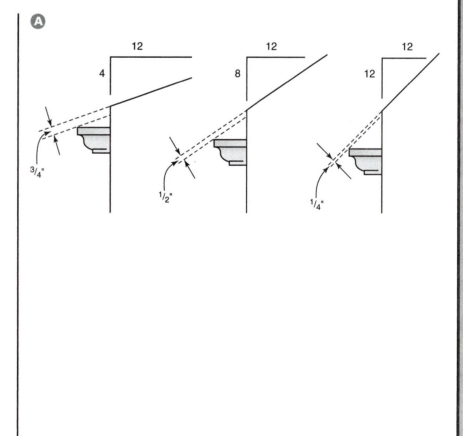

Procedures | Installing Gutters (Continued)

B Locate and install the outlet tubes in the gutter as required, keeping in mind that the downspout should be positioned plumb and square with the building. Add end caps and caulk all seams only on the inside surfaces.

- Hang the gutter sections in the brackets. Use slip-joint connectors to join larger sections. Use either inside or outside corners where gutters make a turn. Caulk all inside seams.

- Fasten downspouts to the wall with appropriate hangers and straps. Downspouts should be fastened at the top and bottom and every 6 feet in between. The connection between the downspout and the gutter is made with elbows and short straight lengths of downspout.

B

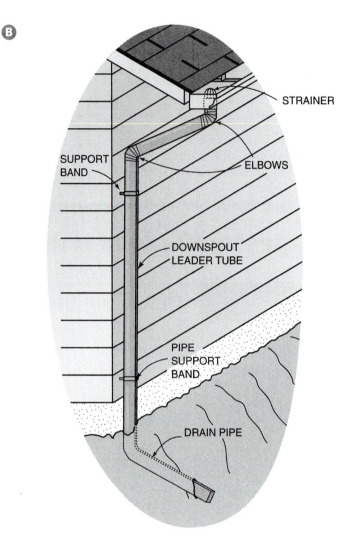

STRAINER

SUPPORT BAND

ELBOWS

DOWNSPOUT LEADER TUBE

PIPE SUPPORT BAND

DRAIN PIPE

C Because water runs downhill, care should be taken when putting the downspout pieces together. The downspout components are assembled where the upper piece is inserted into the lower one. This makes the joint lap such that the water cannot escape until it leaves the bottom-most piece.

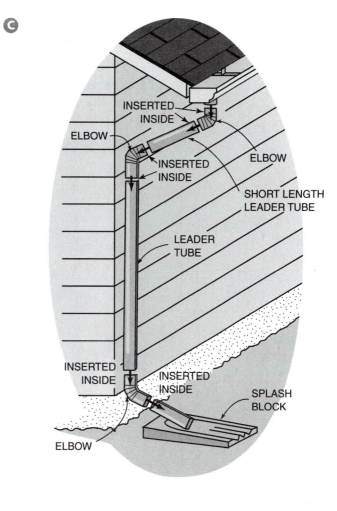

Procedures

Installing Asphalt Shingles

CAUTION

Installation of roofing systems involves working on ladders and scaffolding as well as on top of the building. Workers should always be aware of the potential for falling. Keep the location of roof perimeter in mind at all times.

Ⓐ Prepare the roof deck by clearing sawdust and debris that will cause a slipping hazard.

• Begin underlayment over the deck at a lower corner. Lap the following courses of felt over the lower course at least 2 inches. Make any end laps at least 4 inches. Lap the felt 6 inches from both sides over all hips.

Ⓑ Nail or staple through each lap and through the center of each layer about 16 inches apart. Roofing nails driven through the center of metal disks or specially designed, large head felt fasteners hold the underlayment securely in strong winds until shingles are applied.

• Install metal drip edge along the perimeter on top of the underlayment. This will help prevent blow-offs.

• Prepare the starter course by cutting off the exposure taps lengthwise through the shingle. Save these tabs as they may be used as the last course at the ridge. Install the course so that no end joint will fall in line with an end joint or tab cutout of the regular first course of shingles.

Ⓐ
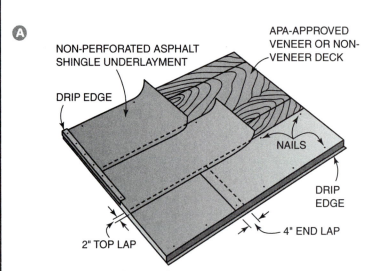

Courtesy of Asphalt Roofing Manufacturers Association

Ⓑ

Courtesy of APA—The Engineered Wood Association

FROM EXPERIENCE

Use a utility knife to cut shingles from the back side. Cut only half way through and then fold and break the shingle to complete the cut. When cutting from the granular top surface, use a hook blade.

C Determine the starting line, either the rake edge or vertical center-snapped lines. To start from the middle of the roof, mark the center of the roof at the eaves and the ridge. Snap a chalk line between the marks. Snap a series of chalk lines from this one, 4 or 6 inches apart, depending on the desired end tab, on each side of the centerline. When applying the shingles, start the course with the end of the shingle to the vertical chalk line. Start succeeding courses in the same manner. Break the joints as necessary, working both ways toward the rakes.

D Starting shingle layout at the rake edge involves placing the first course, with a whole tab at the rake edge. The second course is started with a shingle that is 6 inches shorter; the third course, with a strip that is a full tab shorter; the fourth, with one and one-half tabs removed, and so on. These starting pieces are precut for faster application.

C

SNAPPED LINES PERPENDICULAR TO FASCIA OR PARALLEL TO RAKE FASCIA

METAL DRIP EDGE

FIRST SHINGLE OF EACH COURSE STARTS AGAINST CHALK LINE

STARTER STRIP

D

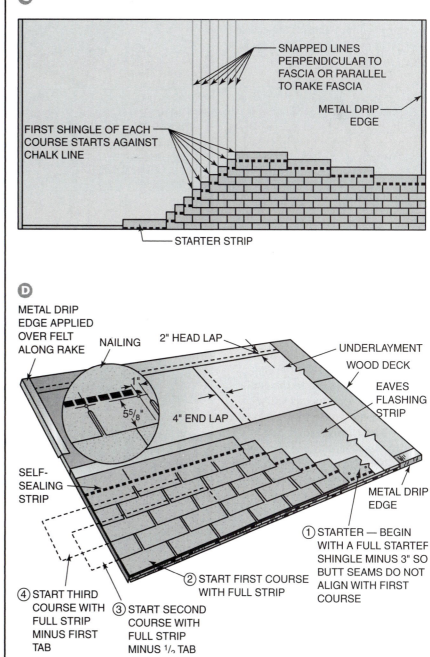

METAL DRIP EDGE APPLIED OVER FELT ALONG RAKE

NAILING

2" HEAD LAP

UNDERLAYMENT

WOOD DECK

EAVES FLASHING STRIP

1"

5⁵/₈"

4" END LAP

METAL DRIP EDGE

SELF-SEALING STRIP

① STARTER — BEGIN WITH A FULL STARTEF SHINGLE MINUS 3" SO BUTT SEAMS DO NOT ALIGN WITH FIRST COURSE

② START FIRST COURSE WITH FULL STRIP

④ START THIRD COURSE WITH FULL STRIP MINUS FIRST TAB

③ START SECOND COURSE WITH FULL STRIP MINUS ¹/₂ TAB

Procedures   Installing Asphalt Shingles (Continued)

E If the cutouts are to break on the thirds, cut the starting strip for the second course by removing 4 inches. Remove 8 inches from the strip for the third course, and so on.

• Fasten each shingle from the end nearest the shingle just laid, which prevents buckling. Drive fasteners straight so that the nail heads will not cut into the shingles. Both ends of the course should overhang the drip edge ¼ to ³⁄₈ inch.

E

NAILING

2" TOP LAP

UNDERLAYMENT

EAVES FLASHING STRIP

1"

⁵⁄₈"

4" END LAP

5"

SEALING STRIP

DRIP EDGE

① STARTER — BEGIN WITH FULL STARTER SHINGLE MINUS 3" SO BUTT SEAMS DO NOT ALIGN WITH FIRST COURSE

④ START THIRD COURSE WITH FULL SHINGLE MINUS 8"

FOURTH COURSE START WITH FULL SHINGLE

② START FIRST COURSE WITH FULL SHINGLE

③ START SECOND COURSE WITH FULL SHINGLE MINUS 8"

SHINGLES OVERHANG EAVES AND RAKE EDGES TO ¼" TO ³⁄₈"

F Install vented ridge cap as per manufacturer's instructions. Cut cap shingles and begin installation from one end. Center the cap shingle over the vented ridge cap. Secure each shingle with one fastener on each side.

• Apply the cap across the ridge until 3 or 4 feet from the end. Then space the cap to the end in the same manner as spacing the shingle course to the ridge. Cut the last ridge shingle to size. Apply it with one fastener on each side of the ridge. Cover the two fasteners with asphalt cement to prevent leakage.

F

VENTED RIDGE CAP

Procedures

Installing Roll Roofing

Roll Roofing with Concealed Fasteners

A Apply 9-inch wide strips of the roofing along the eaves and rakes overhanging the drip edge about $3/8$ inch. Fasten with two rows of nails one inch from each edge spaced about 4 inches apart.

- Apply the first course of roofing with its edge and ends flush with the strips. Secure the upper edge with nails staggered about 4 inches apart. Do not fasten within 18 inches of the rake edge.

- Apply cement only to the edge strips covered by the first course. Press the edge and rake edges firmly to the strips. Complete the nails in the upper edge out to the rakes.

- Apply succeeding courses in a similar manner. Make all end laps 6 inches wide. Apply cement the full width of the lap.

- After all courses are in place, lift the lower edge of each course. Apply the cement in a continuous layer over the full width and length of the lap. Press the lower edges of the upper courses firmly into the cement. A small bead should appear along the entire edge of the sheet. Care must be taken to apply the correct amount of cement.

- To cover the hips and ridge, cut strips of 12 inches × 36 inches roofing. Bend the pieces lengthwise through their centers.

- Snap a chalk line on both sides of the hip or ridge down about 5½ inches from the center. Apply cement between the lines. Fit the first strip over the hip or ridge.

A

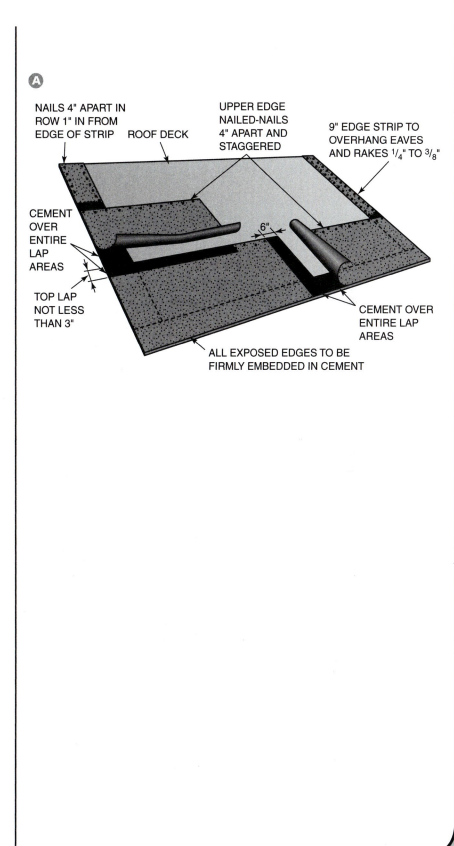

NAILS 4" APART IN ROW 1" IN FROM EDGE OF STRIP

ROOF DECK

UPPER EDGE NAILED-NAILS 4" APART AND STAGGERED

9" EDGE STRIP TO OVERHANG EAVES AND RAKES $1/4$" TO $3/8$"

CEMENT OVER ENTIRE LAP AREAS

6"

TOP LAP NOT LESS THAN 3"

CEMENT OVER ENTIRE LAP AREAS

ALL EXPOSED EDGES TO BE FIRMLY EMBEDDED IN CEMENT

Procedures — Installing Roll Roofing (Continued)

B Press it firmly into place. Start at the lower end of a hip and at either end of a ridge. Lap each strip 6 inches over the preceding one. Nail each strip only on the end that is to be covered by the overlapping piece.

• Spread cement on the end of each strip that is lapped before the next one is applied. Continue in this manner until the end is reached.

Double Coverage Roll Roofing

A Cut the 19-inch strip of *selvage*, nonmineral surface side, from enough double coverage roll roofing to cover the length of the roof. Save the surfaced portion for the last course at the ridge. Apply the selvage portion parallel to the eaves. It should overhang the drip edge by $3/8$ inch. Secure it to the roof deck with three rows of nails.

B Apply the first course using a full width strip of roofing. Secure it with two rows of nails in the selvage portion.

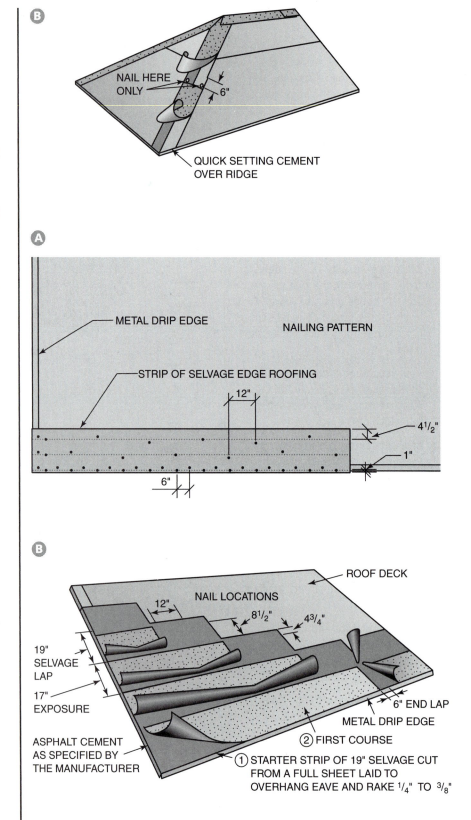

B

NAIL HERE ONLY

6"

QUICK SETTING CEMENT OVER RIDGE

A

METAL DRIP EDGE

NAILING PATTERN

STRIP OF SELVAGE EDGE ROOFING

12"

$4^{1}/_{2}$"

1"

6"

B

ROOF DECK

12"

NAIL LOCATIONS

$8^{1}/_{2}$"

$4^{3}/_{4}$"

19" SELVAGE LAP

17" EXPOSURE

6" END LAP

METAL DRIP EDGE

② FIRST COURSE

ASPHALT CEMENT AS SPECIFIED BY THE MANUFACTURER

① STARTER STRIP OF 19" SELVAGE CUT FROM A FULL SHEET LAID TO OVERHANG EAVE AND RAKE $1/_{4}$" TO $3/_{8}$"

C Apply succeeding courses in the same manner. Lap the full width of the 19-inch selvage each time. Make all end laps 6 inches wide. End laps are made in the manner shown in the accompanying figure. Stagger end laps in succeeding courses.

- Lift and roll back the surface portion of each course. Starting at the bottom, apply cement to the entire selvage portion of each course. Apply it to within ¼ inch of the surfaced portion. Press the overlying sheet firmly into the cement. Apply pressure over the entire area using a light roller to ensure adhesion between the sheets at all points.

D Apply the remaining surfaced portion left from the first course as the last course. Hips and ridges are covered in the same manner shown in the accompanying figure.

- Follow specific application instructions because of differences in the manufacture of roll roofing. Follow specific requirements for quantities and types of adhesive.

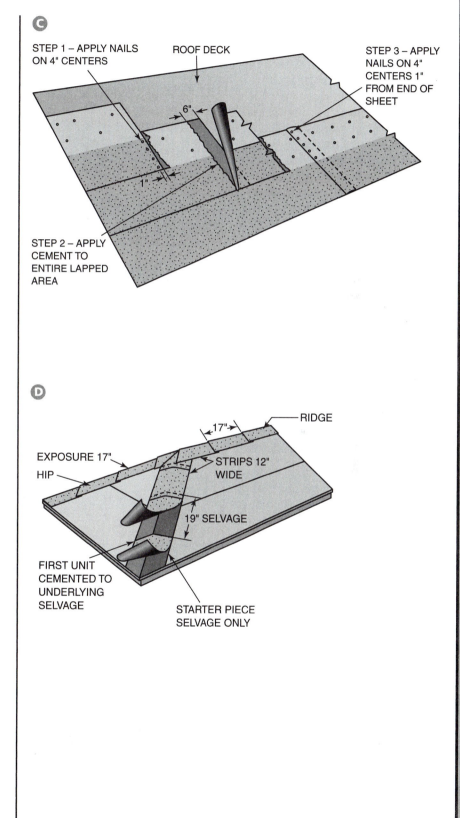

Procedures

Woven Valley Method

Ⓐ Install underlayment and starter strip to both roofs.

• Apply first course of one roof, say the left one, into and past the center of the valley. Press the shingle tightly into the valley and nail, keeping the nails at least 6 inches away from the valley centerline. Cut shingles to adjust the butt ends so that there is no butt seam within 12 inches of the valley centerline.

• Apply the first course of the other (right) roof in a similar manner, into and past the valley.

• Succeeding courses are applied by repeating this alternating pattern, first from one roof and then on the other.

Ⓐ
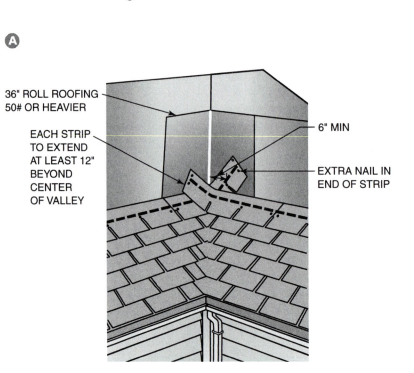

36" ROLL ROOFING 50# OR HEAVIER

EACH STRIP TO EXTEND AT LEAST 12" BEYOND CENTER OF VALLEY

6" MIN

EXTRA NAIL IN END OF STRIP

Procedures

Closed Cut Valley Method

Ⓐ Begin by shingling the first roof completely, letting the end shingle of every course overlap the valley by at least 12 inches. Form the end shingle of each course snugly into the valley. Cut shingles to adjust the butt ends so that there is no butt seam within 12 inches of the valley centerline.

• Snap a chalk line along the center of the valley on top of the shingles of the first roof.

• Apply the shingles of the second roof, cutting the end shingle of each course to the chalk line. Place the cut end of each course that lies in the valley in a 3-inch wide bed of asphalt cement.

Ⓐ
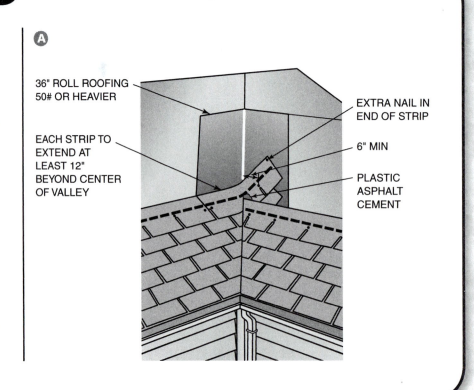

36" ROLL ROOFING 50# OR HEAVIER

EACH STRIP TO EXTEND AT LEAST 12" BEYOND CENTER OF VALLEY

EXTRA NAIL IN END OF STRIP

6" MIN

PLASTIC ASPHALT CEMENT

Procedures | Step Flashing Method

A Snap a chalk line in the center of the valley on the valley underlayment.

- Apply the shingle starter course on both roofs. Trim the ends of each course that meet the chalk line.

- Fit and form the first piece of flashing to the valley on top of the starter strips. Trim the bottom edge flush with the drip edge. Fasten with two nails only in the upper corners of the flashing. Use nails of like material to the flashing to prevent electrolysis.

- Apply the first regular course of shingles to both roofs on each side of the valley, trimming the ends to the chalk line. Bed the ends in plastic asphalt cement. Do not drive nails through the metal flashing. Apply flashing to each succeeding course in this manner.

A

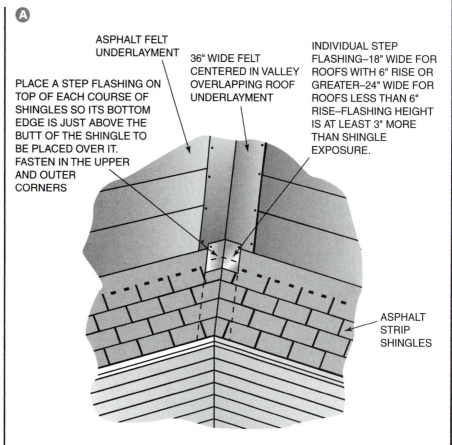

ASPHALT FELT UNDERLAYMENT

36" WIDE FELT CENTERED IN VALLEY OVERLAPPING ROOF UNDERLAYMENT

INDIVIDUAL STEP FLASHING–18" WIDE FOR ROOFS WITH 6" RISE OR GREATER–24" WIDE FOR ROOFS LESS THAN 6" RISE–FLASHING HEIGHT IS AT LEAST 3" MORE THAN SHINGLE EXPOSURE.

PLACE A STEP FLASHING ON TOP OF EACH COURSE OF SHINGLES SO ITS BOTTOM EDGE IS JUST ABOVE THE BUTT OF THE SHINGLE TO BE PLACED OVER IT. FASTEN IN THE UPPER AND OUTER CORNERS

ASPHALT STRIP SHINGLES

Procedures

Installing Horizontal Siding

(A) First determine the siding exposure so that it is about equal both above and below the window sill. Divide the overall height of each wall section by the maximum allowable exposure. Round up this number to get the number of courses in that section. Then divide the height again by the number of courses to find the exposure. These slight adjustments in exposure will not be noticeable to the eye.

(A)

EXAMPLE: Consider the overall dimensions in the accompanying figure. Divide the heights by the maximum allowable exposure, 7 inches in this example. Then round up to the nearest number of courses that will cover that section. Divide the section height by the number of courses to find the exposure.

$40\frac{1}{2} \div 7 = 5.8 \Rightarrow 6$ courses $\qquad$ $40\frac{1}{2} \div 6 = 6.75$ or $6\frac{3}{4}$" exposure

$45\frac{1}{5} \div 7 = 6.5 \Rightarrow 7$ courses $\qquad$ $45\frac{1}{2} \div 7 = 6.5$ or $6\frac{1}{2}$" exposure

$12\frac{1}{2} \div 7 = 1.8 \Rightarrow 2$ courses $\qquad$ $12\frac{1}{2} \div 2 = 6.25$ or $6\frac{1}{4}$" exposure

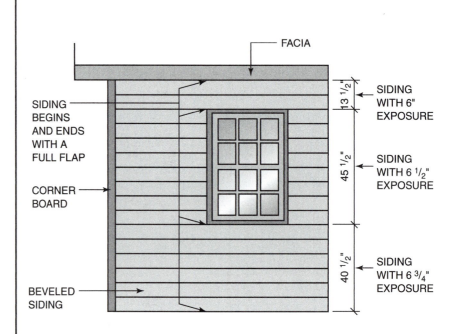

FACIA

SIDING BEGINS AND ENDS WITH A FULL FLAP

CORNER BOARD

BEVELED SIDING

13 1/2" SIDING WITH 6" EXPOSURE

45 1/2" SIDING WITH 6 1/2" EXPOSURE

40 1/2" SIDING WITH 6 3/4" EXPOSURE

B Install a starter strip of the same thickness and width of the siding at the headlap fastened along the bottom edge of the sheathing. For the first course, a line is snapped on the wall at a height of the top edge of the first course of siding.

- From this first chalk line, layout the desired exposures on each corner board and each side of all openings. Snap lines at these layout marks. These lines represent the top edges of all siding pieces.

- Install the siding as per manufacturer's recommendations, staggering the butt joints in adjacent courses as far apart as possible. A small piece of felt paper is used behind the butt seams to ensure the weathertightness of the siding.

C When applying a course of siding, start from one end and work toward the other end. With this procedure, only the last piece will need to be fitted. Tight-fitting butt joints must be made between pieces. Measure carefully and cut it slightly longer. Place one end in position. Bow the piece outward, position the other end, and snap into place. Take care not to move the corner board with this technique. Do not use this technique on cementitious siding.

D Siding is fastened to each bearing stud or about every 16 inches. On bevel siding, fasten through the butt edge just above the top edge of the course below. Do not fasten through the lap. This allows the siding to swell and shrink with seasonal changes without splitting of the siding. Blind-nail cementitious siding by fastening only along the top edge. Blind-nailing is not recommended in high-wind areas.

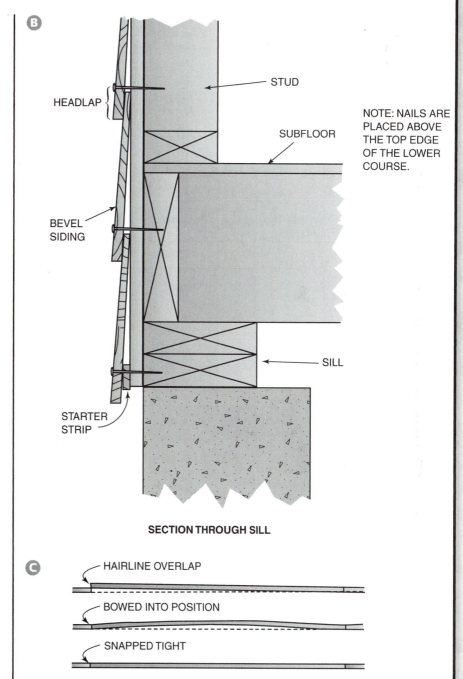

SECTION THROUGH SILL

NOTE: NAILS ARE PLACED ABOVE THE TOP EDGE OF THE LOWER COURSE.

Courtesy of California Redwood Association.

FROM EXPERIENCE

Setting up a comfortable work station for cutting will allow the carpenter to work more efficiently and safely. This will also reduce waste and improve workmanship.

Procedures — Installing Horizontal Siding (Continued)

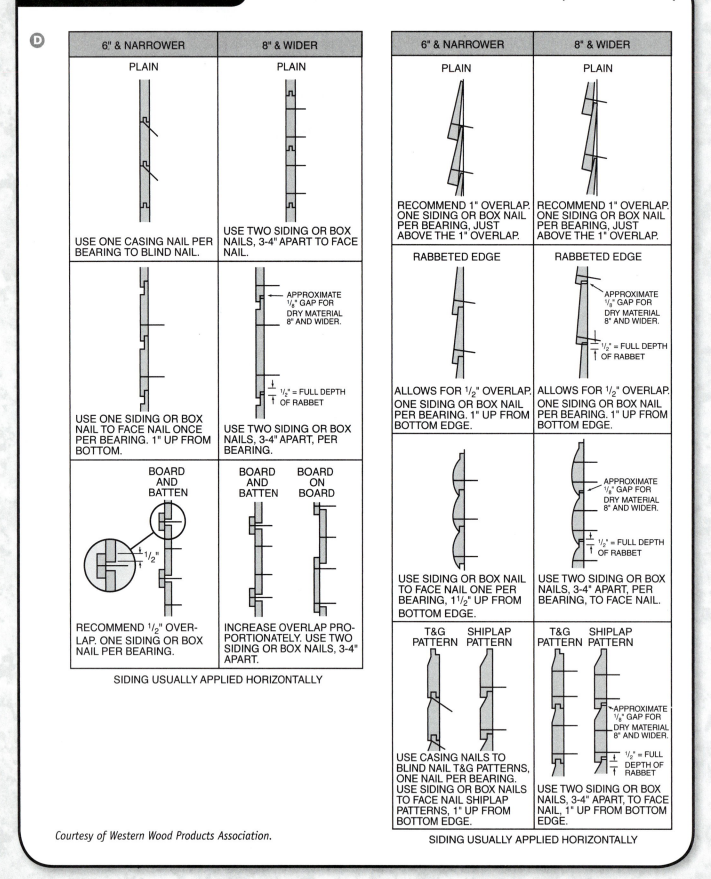

D

6" & NARROWER	8" & WIDER
PLAIN	PLAIN
USE ONE CASING NAIL PER BEARING TO BLIND NAIL.	USE TWO SIDING OR BOX NAILS, 3-4" APART TO FACE NAIL.
USE ONE SIDING OR BOX NAIL TO FACE NAIL ONCE PER BEARING. 1" UP FROM BOTTOM.	APPROXIMATE $1/8$" GAP FOR DRY MATERIAL 8" AND WIDER. $1/2$" = FULL DEPTH OF RABBET. USE TWO SIDING OR BOX NAILS, 3-4" APART, PER BEARING.
BOARD AND BATTEN	BOARD AND BATTEN — BOARD ON BOARD
$1/2$" RECOMMEND $1/2$" OVERLAP. ONE SIDING OR BOX NAIL PER BEARING.	INCREASE OVERLAP PROPORTIONATELY. USE TWO SIDING OR BOX NAILS, 3-4" APART.

SIDING USUALLY APPLIED HORIZONTALLY

6" & NARROWER	8" & WIDER
PLAIN	PLAIN
RECOMMEND 1" OVERLAP. ONE SIDING OR BOX NAIL PER BEARING, JUST ABOVE THE 1" OVERLAP.	RECOMMEND 1" OVERLAP. ONE SIDING OR BOX NAIL PER BEARING, JUST ABOVE THE 1" OVERLAP.
RABBETED EDGE	RABBETED EDGE
ALLOWS FOR $1/2$" OVERLAP. ONE SIDING OR BOX NAIL PER BEARING. 1" UP FROM BOTTOM EDGE.	APPROXIMATE $1/8$" GAP FOR DRY MATERIAL 8" AND WIDER. $1/2$" = FULL DEPTH OF RABBET. ALLOWS FOR $1/2$" OVERLAP. ONE SIDING OR BOX NAIL PER BEARING. 1" UP FROM BOTTOM EDGE.
USE SIDING OR BOX NAIL TO FACE NAIL ONE PER BEARING, $1 1/2$" UP FROM BOTTOM EDGE.	APPROXIMATE $1/8$" GAP FOR DRY MATERIAL 8" AND WIDER. $1/2$" = FULL DEPTH OF RABBET. USE TWO SIDING OR BOX NAILS, 3-4" APART, PER BEARING, TO FACE NAIL.
T&G PATTERN SHIPLAP PATTERN	T&G PATTERN SHIPLAP PATTERN
USE CASING NAILS TO BLIND NAIL T&G PATTERNS, ONE NAIL PER BEARING. USE SIDING OR BOX NAILS TO FACE NAIL SHIPLAP PATTERNS, 1" UP FROM BOTTOM EDGE.	APPROXIMATE $1/8$" GAP FOR DRY MATERIAL 8" AND WIDER. $1/2$" = FULL DEPTH OF RABBET. USE TWO SIDING OR BOX NAILS, 3-4" APART, TO FACE NAIL, 1" UP FROM BOTTOM EDGE.

SIDING USUALLY APPLIED HORIZONTALLY

Courtesy of Western Wood Products Association.

Procedures

(A) Slightly back-bevel the ripped edge. Place it vertically on the wall with the beveled edge flush with the corner similar to making a corner board. Face-nail the edge nearest the corner.

- Fasten a temporary piece on the other end of the wall projecting below the sheathing by the same amount. Stretch a line to keep the bottom ends of other pieces in a straight line.

(B) Apply succeeding pieces by toenailing into the tongue edge of each piece. Make sure that the edges between boards come up tight. Drive the nail home until it forces the board up tight. Make sure to keep the bottom ends in a straight line. If butt joints are necessary, use a mitered or rabbeted joint for weathertightness.

(C) To cut the piece to fit around an opening, first fit and tack a siding strip in place where the last full strip will be located. Level from the top and bottom of the window casing to this piece of siding and mark the piece.

Installing Vertical Tongue-and-Groove Siding

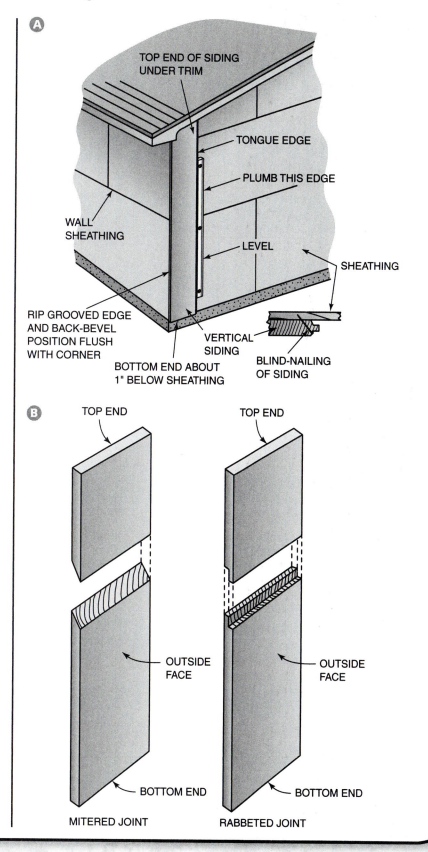

(A)

TOP END OF SIDING UNDER TRIM

TONGUE EDGE

PLUMB THIS EDGE

LEVEL

SHEATHING

WALL SHEATHING

RIP GROOVED EDGE AND BACK-BEVEL POSITION FLUSH WITH CORNER

VERTICAL SIDING

BOTTOM END ABOUT 1" BELOW SHEATHING

BLIND-NAILING OF SIDING

(B)

TOP END

TOP END

OUTSIDE FACE

OUTSIDE FACE

BOTTOM END

BOTTOM END

MITERED JOINT

RABBETED JOINT

Procedures

Installing Vertical Tongue-and-Groove Siding (Continued)

- Next, use a scrap block of the siding material, about 6 inches long, with the tongue removed. Be careful to remove all of the tongue, but no more. Hold the block so that its grooved edge is against the side casing and the other edge is on top of the tacked piece of siding. Mark the vertical line on the siding by holding a pencil against the outer edge of the block while moving the block along the length of the side casing. Remove and cut the piece, following the layout lines carefully. Set this piece aside for the time being. Cut and fit another full strip of siding in the same place as the previous piece. Fasten both pieces in position.

- Continue the siding by applying the short lengths across the top and bottom of the opening as needed.

- **D** Fit the next full-length siding piece to complete the siding around the opening. First tack a short length of siding scrap above and below the window and against the last pieces of siding installed. Tack the length of siding to be fitted against these blocks in the grooves. Level and mark from the top and bottom of the window to the full piece. Lay out the vertical cut by using the same block with the tongue removed, as used previously. Hold the grooved edge against the side casing. With a pencil against the other edge, ride the block along the side casing while marking the piece to be fitted.

- Remove the piece and the scrap blocks from the wall. Carefully cut the piece to the layout lines. Then fasten in position. Continue applying the rest of the siding.

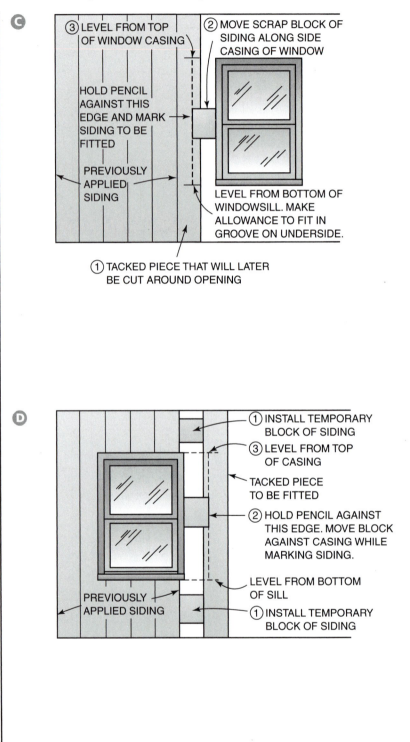

C

③ LEVEL FROM TOP OF WINDOW CASING

② MOVE SCRAP BLOCK OF SIDING ALONG SIDE CASING OF WINDOW

HOLD PENCIL AGAINST THIS EDGE AND MARK SIDING TO BE FITTED

PREVIOUSLY APPLIED SIDING

LEVEL FROM BOTTOM OF WINDOWSILL. MAKE ALLOWANCE TO FIT IN GROOVE ON UNDERSIDE.

① TACKED PIECE THAT WILL LATER BE CUT AROUND OPENING

D

① INSTALL TEMPORARY BLOCK OF SIDING

③ LEVEL FROM TOP OF CASING

TACKED PIECE TO BE FITTED

② HOLD PENCIL AGAINST THIS EDGE. MOVE BLOCK AGAINST CASING WHILE MARKING SIDING.

LEVEL FROM BOTTOM OF SILL

① INSTALL TEMPORARY BLOCK OF SIDING

PREVIOUSLY APPLIED SIDING

Procedures

Installing Panel Siding

(A) Install the first piece with the vertical edge plumb. Rip the sheet to size, putting the cut edge at the corner. The factory edge should fall on the center of a stud. Panels must also be installed with their bottom ends in a straight line. It is important that horizontal butt joints be offset and lapped, rabbeted, or flashed. Vertical joints are either shiplapped or covered with **battens**.

• Apply the remaining sheets in the first course in like manner. Cut around openings in a similar manner as with vertical tongue-and-groove siding. Carefully fit and caulk around doors and windows. Trim the end of the last sheet flush with the corner.

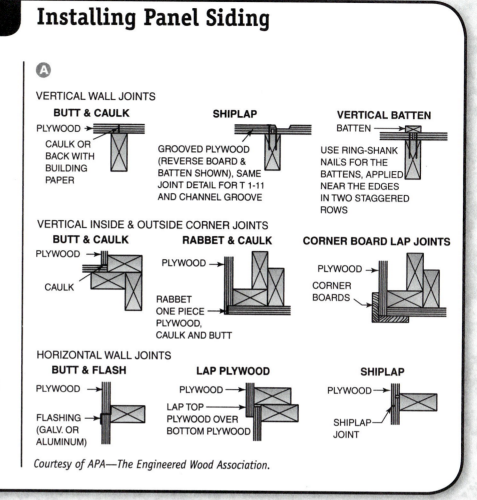

Courtesy of APA—The Engineered Wood Association.

Procedures

Installing Wood Shingles and Shakes

(A) Fasten a shingle on both ends of the wall with its butt about 1 inch below the top of the foundation. Stretch a line between them from the bottom ends. Fasten an intermediate shingle to the line to take any sag out of the line. Even a tightly stretched line will sag in the center over a long distance.

• Fill in the remaining shingles to complete the undercourse. Take care to install the butts as close to the line as possible without touching it and remove the line.

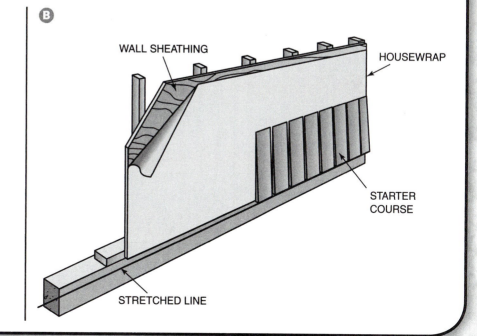

Procedures

Installing Wood Shingles and Shakes (Continued)

B Apply another course on top of the first course. Offset the joints in the outer layer at least 1½ inches from those in the bottom layer. Shingles should be spaced ⅛ to ¼ inch apart to allow for swelling and prevent buckling. Shingles can be applied close together if factory-primed or if treated soon after application.

B

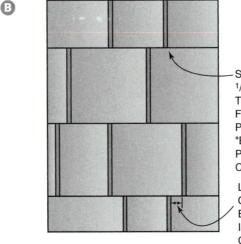

SHINGLES SPACED ⅛" TO ¼" APART. THESE JOINTS ALLOW FOR EXPANSION AND PREVENT POSSIBLE "BUCKING." FACTORY PRODUCTS MAY BE CLOSER.

LEAVE A SIDE LAP OF AT LEAST 1½" BETWEEN JOINTS IN SUCCESSIVE COURSES.

Copyright 2010 Cedar Shake & Shingle Bureau.

C To apply the second course, snap a chalk line across the wall at the shingle butt line. Using only as many finish nails as necessary, tack 1 × 3 straightedges to the wall with their top edges to the line. Lay individual shingles with their butts on the straightedge.

C

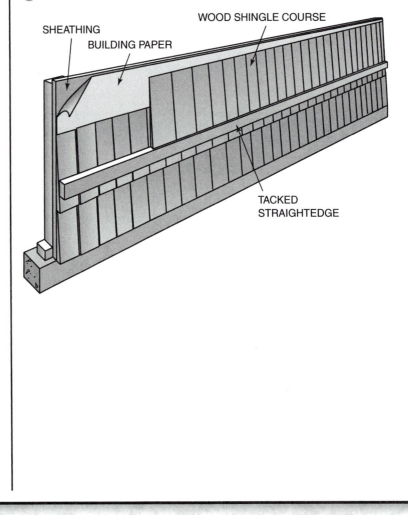

SHEATHING

BUILDING PAPER

WOOD SHINGLE COURSE

TACKED STRAIGHTEDGE

Procedures

Applying Horizontal Vinyl Siding

A Snap a level line to the height of the starter strip all around the bottom of the building. Fasten the strips to the wall with their edges to the chalk line. Leave a ¼-inch space between them and other accessories to allow for expansion. Make sure that the starter strip is applied as straight as possible. It controls the straightness of entire installation.

A

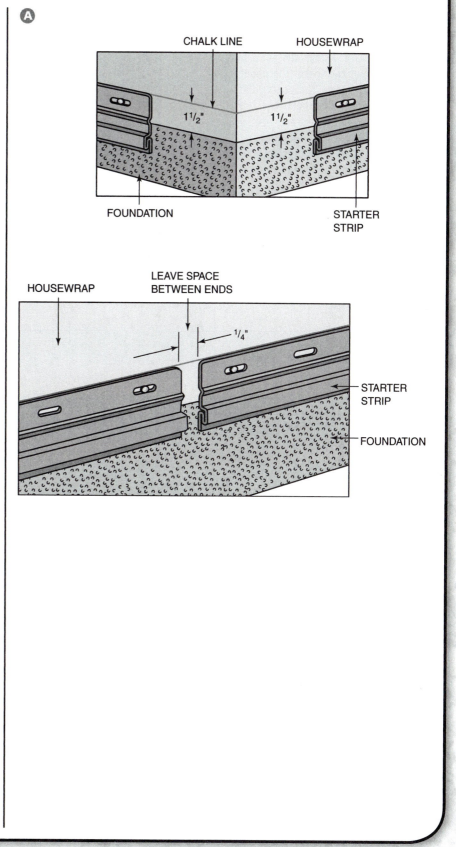

Procedures

Applying Horizontal Vinyl Siding (Continued)

B Cut the corner posts so that they extend ¼ inch below the starting strip. Attach the posts by fastening at the top of the upper slot on each side. The posts will hang on these fasteners. The remaining fasteners should be centered in the nailing slots. Make sure that the posts are straight, plumb, and true from top to bottom.

B

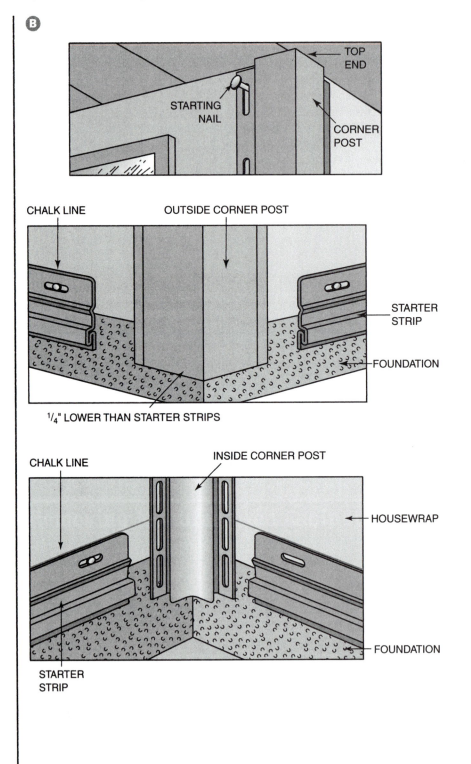

C Cut each J-channel piece to extend, on both ends, beyond the casings and sills a distance equal to the width of the channel face. Install the side pieces first by cutting a ¾-inch notch, at each end, out of the side of the J-channel that touches the casing. Fasten in place.

• On both ends of the top and bottom channels, make ¾-inch cuts at the bends leaving the tab attached. Bend down the tabs and miter the faces. Install them so the mitered faces are in front of the faces of the side channels.

C

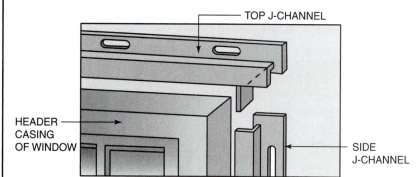

TOP J-CHANNEL

HEADER CASING OF WINDOW

SIDE J-CHANNEL

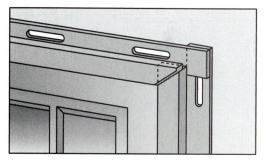

COMPLETE INSTALLATION WITH TOP J-CHANNEL ON OUTSIDE OF SIDE J-CHANNEL

Procedures

Applying Horizontal Vinyl Siding (Continued)

D Snap the bottom of the first panel into the starter strip. Fasten it to the wall. Start from a back corner, leaving a ¼-inch space in the corner post channel. Work toward the front with other panels. Overlap each panel about 1 inch. The exposed ends should face the direction from which they are least viewed.

• Install successive courses by interlocking them with the course below and staggering the joints between courses.

E To fit around a window, mark the width of the cutout, allowing ¼-inch clearance on each side. Mark the height of the cutout, allowing ¼-inch clearance below the sill. Using a special *snaplock punch,* punch the panel along the cut edge at 6-inch intervals to produce raised lugs facing outward. Install the panel under the window and up in the undersill trim. The raised lugs cause the panel to fit snugly in the trim.

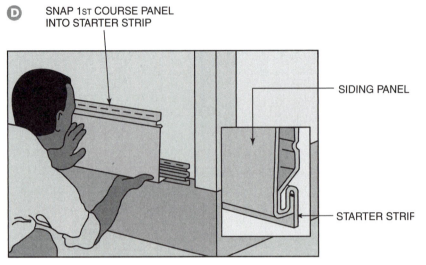

D SNAP 1ST COURSE PANEL INTO STARTER STRIP

SIDING PANEL

STARTER STRIP

DO NOT FORCE PANEL UP OR DOWN WHEN FASTENING

1"

LAP PANELS AT LEAST 1"

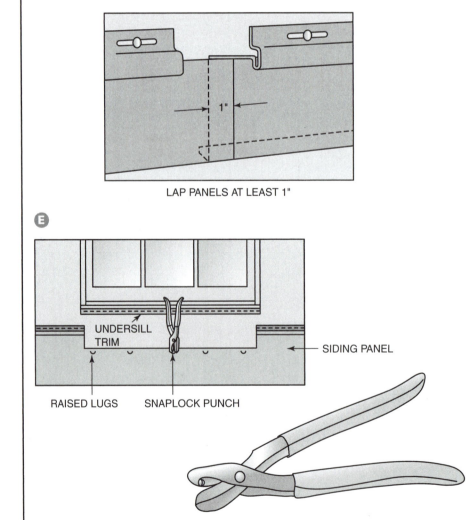

E

UNDERSILL TRIM

SIDING PANEL

RAISED LUGS   SNAPLOCK PUNCH

F Panels are cut and fit over the windows in the same manner as under them. However, the lower portion is cut instead of the top. Install the panel by placing it into the J-channel that runs across the top of the window.

G Install the last course of siding panels under the soffit in a manner similar to fitting under a window. An *undersill trim* is applied on the wall and up against the soffit. Panels in the last course are cut to width. Lugs are punched along the cut edges. The panels are then snapped firmly in place into the undersill trim.

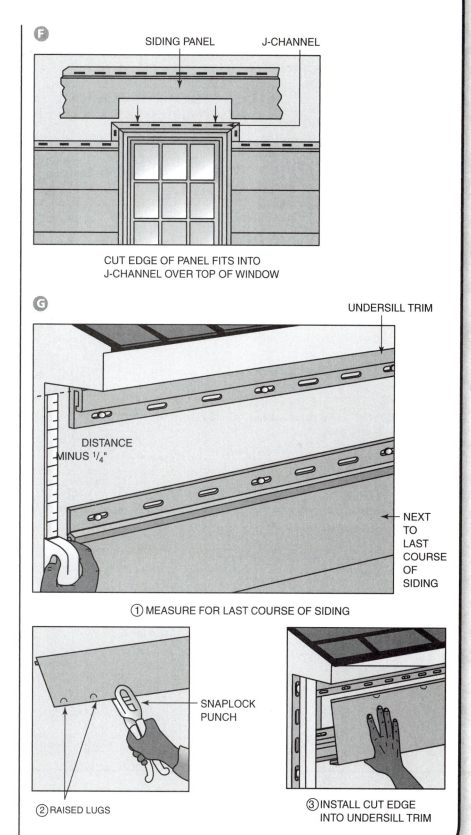

F

SIDING PANEL   J-CHANNEL

CUT EDGE OF PANEL FITS INTO
J-CHANNEL OVER TOP OF WINDOW

G

UNDERSILL TRIM

DISTANCE MINUS 1/4"

NEXT TO LAST COURSE OF SIDING

① MEASURE FOR LAST COURSE OF SIDING

SNAPLOCK PUNCH

② RAISED LUGS

③ INSTALL CUT EDGE INTO UNDERSILL TRIM

Procedures   Applying Vertical Vinyl Siding

Ⓐ Measure and lay out the width of the wall section for the siding pieces. Determine the width of the first and last pieces.

- Cut the edge of the first panel nearest to the corner. Install an undersill trim in the corner board or J-channel with a strip of furring or backing. This will keep the edge in line with the wall surface. Punch lugs along the cut edge of the panel at 6-inch intervals. Snap the panel into the undersill trim. Place the top nail at the top of the nail slot. Fasten the remaining nails in the center of the nail slots.

- Install the remaining full strips making sure that there is ¼-inch gap at the top and bottom. Fit around openings in the same manner as with fitting vertical siding. Install the last piece into undersill trim in the same manner as for the first piece.

Ⓐ

EXAMPLE: What is the starting and finishing widths for a wall section that measures 18 feet to 9 inches for siding that is 12 inches wide?

Convert this measurement to a decimal by first dividing the inches portion by 12 and then adding it to the feet to get 18.75 feet.

Divide this by the siding exposure, in feet: 18.75 ÷ 1 foot = 18.75 pieces.

Subtract the decimal portion along with one full piece giving 1.75 pieces. Next 1.75 ÷ 2 = 0.875, multiplied by 12 gives 10½ inches.

This is the size of the starting and finishing pieces. Thus there are 17 full-width pieces and two 10 ½-inch wide pieces.

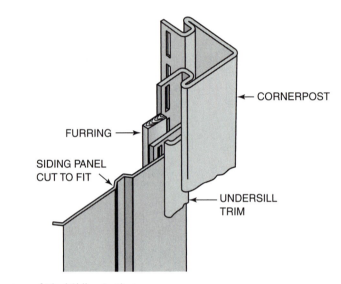

Courtesy of Vinyl Siding Institute.

Review Questions

Select the most appropriate answer.

① True or false? The most common method of classifying wood is by its source.

② List three common hardwoods.

③ List three common softwoods.

④ Define the term "engineer panels."

⑤ List the steps for estimating the amount of drywall needed for a particular application.

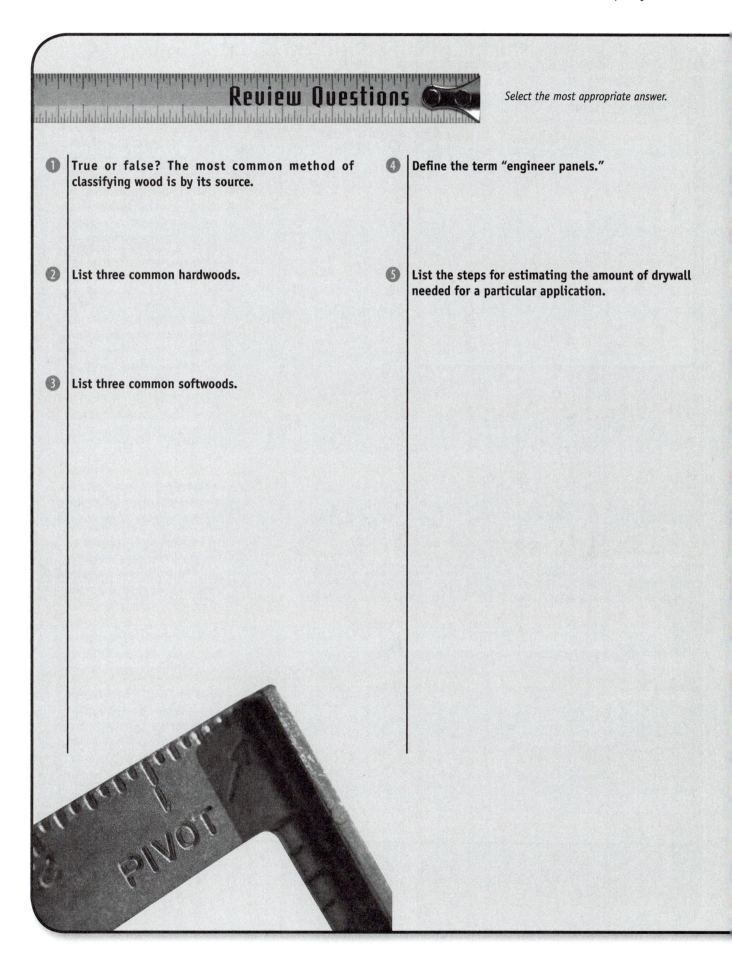

Name: _____

Date: _____

Carpentry

Identify Characteristics of Wood

Upon completion of this job sheet, you should be able to understand how wood is classified.

1. Go to your local lumber yard or building supply center where there are various types of wood. Make a list of the different types of wood available and classify it as either hardwood or softwood.

Name of the Wood	Classification

2. Are the woods available native to your area? (Y/N) If not, where does the wood come from?

3. At your local lumber yard or building supply center, list the available types of engineered panels they make.

Type of Panel	How They Are Made

Instructor's Response:

_____

_____

_____

Chapter 8 Surface Treatments

OBJECTIVES

By the end of this chapter, you will be able to:

Knowledge-Based

- ✪ Identify and select proper surface finishes.
- ✪ Identify and select proper finishing tools for different types of finishes.

Skill-Based

- ✪ Prepare surface and site properly for finishing, including sanding, caulking, and covering exposed surfaces.
- ✪ Apply paint using roller and brush according to manufacturer and job specifications.
- ✪ Apply paint using a paint sprayer according to manufacturer and job specifications.
- ✪ Clean and store paint materials including brushes, rollers, thinners, and spray guns according to manufacturer's specifications and OSHA regulations.

Glossary of Terms

Mask a faux finish used to apply a new color over a dry base coat to create an image or shape

Mural a faux finish used to give the illusion of scenery or architectural elements

Kill spot an inconspicuous area used to start and end wallpaper in a room

Booking the process used to activate the paste

Introduction

Interior surfaces are primarily used to define spaces and hide structural, electrical, and plumbing elements. If these surfaces are constructed using drywall/gypsum board, then they will need a surface treatment to protect the construction. There are an unlimited number of ways to treat or finish drywall/gypsum board surfaces.

These finishes can include paint, wallpaper, texturing, faux finish, or a combination of these applications. The function of the space, along with cost, durability, acoustics, and building codes, is a key factor in the type of surface finish chosen. Whether treating the surface of a new drywall or replacing an existing treatment, achieving a professional result requires proper surface preparation, tool and material selection, and application.

Painting Safety

Before attempting any painting and finishing project, always follow some general safety rules:

- Before performing any painting-related work in areas that are heavily populated, be sure that all the occupants have been notified.
- Never use painting products near an open flame and/or heat source.
- Never smoke while handling painting supplies and/or painting.
- Always use an approved respirator mask when painting or using thinners.
- For indoor projects always provide adequate ventilation before starting the painting project.
- To prevent skin irritation always use vinyl gloves before handling paints and/or thinners.
- Always use painter's goggles to prevent splatters from entering the eye.
- If you should get paint in your eye(s):
  1. Flush your eye(s) for 10 minutes using cool water.
  2. Seek medical attention.
- Always wear a hat to prevent paint from getting into your hair.
- Follow the manufacturer's instructions for handling all epoxy materials, thinners, catalysts, paint removers, and so on.
- When using ladders for a painting project, always make sure that the ladder's safety devices are properly locked into place before using.
- Check the ladder placement and ensure that it is properly leveled before attempting to use.
- Use ladders only for their intended use and always follow the manufacturer's recommendations.

Figure 8-1: Wall with nail holes

Fixing Drywall Problems

Address all drywall problems before attempting to add a treatment to a wall. Drywall can have nail pops, which are visible dimples in the drywall caused by underlying nails. Drywall can also have minor defects caused by small nails used for hanging pictures, thumbtacks, or small dents caused by furniture. In some cases, you may see large holes in the drywall that must be patched properly to ensure a smooth finish after the new wall treatment is complete (Figure 8-1). The steps for fixing a number of drywall problems are given next.

Steps for Fixing Nail Pops

1. If you can remove the nail without damage, do so. Refasten the drywall with drywall screws. Drive the screws until they are recessed but do not break through the paper covering on the drywall.
2. If the nail cannot be removed, drive it back in place. Fill the dent and screw holes with joint compound (Figure 8-2). Allow the compound to dry, then sand smooth.

Steps for Repairing Holes or Scrapes

1. Remove loose drywall plaster and cut away torn paper with a utility knife (Figure 8-3).
2. Roughen the edges of the hole with coarse sandpaper. Wipe dust away from the hole (Figure 8-4).
3. Using a putty knife, fill the area with joint compound. Allow the joint compound to dry, then sand smooth.

Figure 8-2: **Repairing flaws in the wall (Image copyright Mark Stout Photography, 2009. Used under license from Shutterstock.com)**

Figure 8-3: **Utility knife**

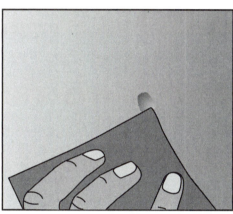

Figure 8-4: **Rough up hole with sandpaper**

Steps for Repairing Cracks

1. Use a utility knife to cut a "V" along the length of the crack. Starting on one side of the crack, cut at an angle from just outside the crack to its center. Repeat this process on the opposite side of the crack to complete the "V" (Figure 8-5).
2. Using a putty knife, fill the area with joint compound. Allow the joint compound to dry, then sand smooth.

Steps for Patching a Large Drywall Hole

1. Mark out a rectangle around the hole with a straightedge or carpenter's rule (Figure 8-6).
2. Use a keyhole saw to cut out the rectangle or score along the lines using a utility knife. Use several more passes to finalize the rectangular cutout with the utility knife (Figure 8-7).

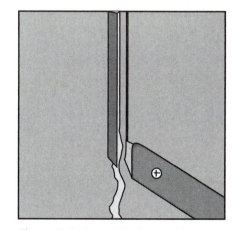

Figure 8-5: **Cut a "V" along the length of the crack**

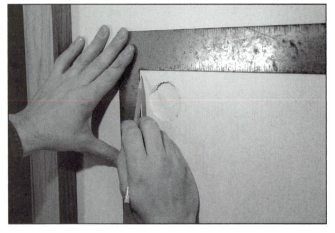

Figure 8-6: **Draw rectangle around hole**

Figure 8-7: **Cut rectangular hole in the wall**

3. Cut a drywall patch 2 inches each direction larger than the hole. Remove the 2-inch perimeter, but leave the facing paper (Figure 8-8).
4. Spread joint compound around the outside edges of the hole and along its inside edges (Figure 8-9).

Figure 8-8: **Cut drywall patch**

Figure 8-9: **Joint compound on patch**

5. Place the patch in position and hold it in place for several minutes while it begins to adhere. Spread more joint compound as needed with a drywall knife. Allow the compound to dry completely (Figure 8-10).
6. Sand, prime, and paint (Figure 8-11).

Painting as a Surface Treatment

A common way of finishing and protecting drywall/gypsum is by painting the surface. The proper sequence for painting a room is from top to bottom. Begin with the ceiling, then paint the walls, the windows, the doors, and the baseboards.

Figure 8-10: **Position patch on hole being repaired**

Surface Preparation

Before starting the actual painting, prepare the surfaces properly to get a professional-looking paint job. To start the paint process, select the proper paint for the type and location of the surface and use of the area.

Types of Paints

There are several types of paint, each designed for a specific purpose, surface, or place in a building.

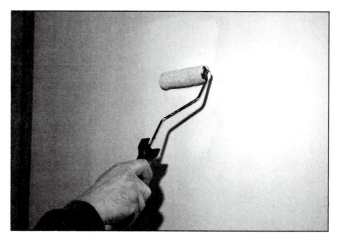

Figure 8-11: **Paint patch**

- Latex paints—water-based, durable, and lower in odor than oil-based products. In addition, latex paints can be cleaned up easily with soap and water. For these reasons, latex is preferred for most interior surfaces.
- Oil paints—durable and resist scraping and wear and tear. Most contain alkyd, a soya-based resin that dries harder than latex. Because these paints are solvent-based, brushes and spills must be cleaned up using paint thinner.
- No or low VOC paints—includes acrylic and oil-based paints. VOCs refer to the volatile organic compounds that are outgassed during application and for years after the paint has dried. Paints and primers with no VOC and low VOC have fewer toxic chemicals that are released during outgassing.

Types of Finishes

The more resin a paint contains, the greater the sheen (or shine) it has after drying. The resin adds durability and makes maintenance/cleaning much easier. However, the shinier the surface, the more pronounced the imperfections of the wall surface and/or the application process. Also, prices increase along with the level of sheen.

- Gloss paints—oil-based and include resin to give them a hard-wearing quality. The higher the gloss level, the higher the shine and easier it is to maintain.
- Flat paints—have no sheen and are inexpensive. Ideal for low-traffic areas such as formal dining rooms and master bedrooms. They provide a beautiful matte coating that hides minor surface imperfections. However, this finish stains easily and is difficult to clean.
- Eggshell paints—provide a smooth finish with a subtle sheen that is slightly glossier than flat. They are washable; durable; and ideal for bedrooms, hallways, home offices, and family rooms.
- Satin paints—a step above eggshell in their ability to be cleaned, providing a nice balance between being washable and having a subtle gloss. They perform and look great in just about any room.
- Semigloss paints—ensure maximum durability. They are commonly used in children's rooms and high-moisture areas such as bathrooms, as well as for trim.
- High-gloss paints—highly reflective and work well for highlighting details, such as trim and decorative molding. They are also the best choice for doors, cabinets, or any area that sees a high volume of abuse.
- Ceiling flats—designed specifically for ceilings. These are usually extra spatter-resistant.

Identify and Select Proper Tools

Selecting the proper tools will save you a lot of time and additional effort. Always choose top-quality tools.

Figure 8-12: **Putty knife (Image copyright Simon Krzic, 2009. Used under license from Shutterstock.com)**

Figure 8-13: **Polyester, nylon, and poly/nylon paint brushes**

Figure 8-14: **White China and Black China bristle brushes**

Types of Finishing Tools

- Pressure washers—blast away many years worth of dirt and grime in a short time and with minimal amount of effort.
- Paint scrapers—efficiently remove the loose and peeling paint from any surface. Many types of paint scrapers are available. Examples include the classic putty knife (Figure 8-12), the 11-in-1 multipurpose tool, and the double-edge wood scraper.
- Power sanders—help with the removal of old peeling paint.

Types of Brushes

Choosing the right brush for your task will make the job easier and provide best results. Determining which paint brush is best for the project is based on the type of paint or finish, the surface to be painted, and personal preference. Two basic categories of brushes are water-based and oil-based paint brushes. Choose the appropriate type of brush for the type of paint, finish, and surface.

Water-Based Paint Brushes

There are three main types of water-based paint brushes (see Figure 8-13).

1. Polyester—these bristle brushes hold and release more paint, which provides a smoother finish. Cleanup is faster and more thorough than with other synthetic brushes.
2. Nylon—these bristle brushes wear longer and are stiffer than any other filament. A nylon brush is well suited for rough surfaces.
3. Poly/nylon—these blends provide longer wear, maximum resiliency, and easy cleanup.

Oil-Based Paint Brushes

There are two basic types of oil-based paint brushes (see Figure 8-14).

1. White China bristles—the best choice for varnishes, polyurethane, and stains. They are finer than Black China bristles and provide a finer finish.
2. Black China bristles—best used with oil-based paint, primer, and enamels.

Brush Sizes and Shapes

Paint brushes are available in various shapes, such as angular, flat, and oval. Sizes range from 1 inch to 4 inches in width (see Figure 8-15).

- Angular brushes—great for angular or narrow surfaces. They are also good to use when the painting surface is hard to reach. They are an excellent choice for an all-purpose brush.
- Flat and oval brushes—used on all surfaces but are best suited for flat surfaces such as wide trim, doors, cutting-in walls, or ceilings.

Figure 8-15: Various brush sizes (Image copyright Baevskiy Dmitry, 2009. Used under license from Shutterstock.com)

Types of Rollers

The most important factor in selecting a paint roller is the surface that is going to be painted. The rule for using almost all roller covers is as follows: The smoother the surface to be painted, the shorter the nap; the rougher the surface, the longer the nap. A high-quality roller cover should have a phenolic core which will not soften in water but will withstand every type of paint solvent.

Roller Covers

Rollers are available in both natural (mohair or lamb's wool) and synthetic materials (nylon, polyester, or a combination of the two). Natural materials are best with oil-based paints, whereas synthetic materials are best with water-based paints. For latex paint, use only synthetic rollers. Natural materials are too absorbent (see Figure 8-16).

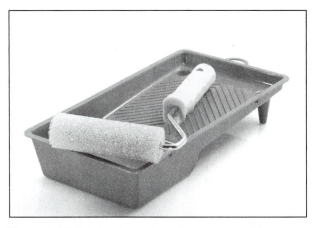

Figure 8-16: Paint roller and pan (Image copyright Vangelis, 2009. Used under license from Shutterstock.com)

- Smooth surface—select a short nap (1/8-inch to 1/4-inch) cover. A longer nap can leave a pronounced "orange peel" effect. Use this cover on smooth plaster, sheet rock, wallboard, smooth wood, Masonite, and Celotex.
- Slightly rough surface—select a medium nap (3/8-inch to 1/2-inch) cover. Longer fibers push the paint into rough surfaces without causing an "orange peel" effect. Use this cover on sand finish plaster, texture plaster, acoustical tile, poured concrete, rough wood, and shakes.
- Rough surface—select a long nap (3/4-inch to 11/4-inch) cover. Longer fibers push the paint into the deep valleys of rough surfaces. Use this cover on concrete block, stucco, brick, Spanish plaster, cinder block, corrugated metal, and asphalt or wood shingles.

Roller Frames

Roller cage frames are available in various styles (see Figure 8-17). U-shaped frames are generally sturdier. When choosing frames, be sure to select those that are sealed on the ends to help keep the paint on the roller, where it belongs.

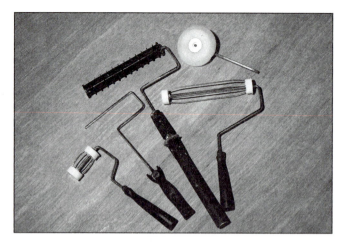

Figure 8-17: **Various roller cage frame types**

Prepare the Surface for Painting

Proper surface preparation is critical to professional, long-lasting results.

Figure 8-18: **Furniture covered with drop cloth**

Steps for Preparing the Surface for Painting

1. Remove as much furniture as possible from the room. Group the heavier items in the center of the room and cover them with drop cloths (Figure 8-18).
2. Mask off the baseboards with 2-inch masking tape and old newspaper or with 12-inch baseboard masking. Use a drop cloth to protect the floor (a canvas drop cloth is less slippery and creates a safer work area).
3. Repair drywall. Drywall problems are relatively easy to spot. Nails sometimes pop out from the drywall. Corners get dented, scraped, or otherwise damaged. Tape can split. Dents, gouges, and holes appear. All drywall problems are relatively easy to fix as was discussed at the beginning of this chapter.
4. Turn off power to the affected switches/outlets at the service panel. Remove switch plates and receptacle plates. Never try to tape or paint over switch plates or receptacle plates. Always remove them. Be sure to tape the screws to the plate so that they are not lost (Figure 8-19).
5. Check around the windowpanes for loose putty. If putty is loose or missing, replace with new putty before starting the painting job.

Figure 8-19: **Switch plate has been removed (Image copyright Lisa F. Young, 2009. Used under license from Shutterstock.com)**

Painting the Ceiling

If the walls are going to be painted, there is no need to worry about protecting them. However, if you are painting only the ceiling, you should protect the walls by taping plastic drop cloths to the top of the walls.

Steps for Painting the Ceiling

1. Paint a 2-inch wide strip with a brush around the edges of the ceiling.
2. Switch to a roller with a 4–5-foot extension pole. Starting at a corner, paint a section about 3 feet square. Use a zigzag pattern to form a "W" pattern on the ceiling. This will disperse the paint evenly on the roller.
3. Fill in this 3-foot section. Paint the section without reloading the roller until you have completely covered the section.
4. Continue to cover the ceiling. Work across the ceiling's shortest dimension in 3-foot square sections. Overlap your strokes while the paint is wet to minimize lap marks (Figure 8-20).

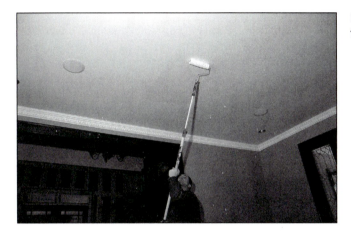

Figure 8-20: **Paint ceiling in small sections**

When moving furniture, be extremely careful to protect yourself and the furniture. Consider the following:

- *If you know you must move large furniture, get a helper.*
- *Use a back support to protect your back.*
- *Walk large, heavy pieces by moving legs to the left and right in small increments.*
- *Turn a piece of carpet upside down and place under the object; this will make it easier to slide across the floor.*
- *Use a dolly to move heavy pieces.*

Painting the Walls

Once the ceiling has been completed, move to the walls. Be sure that the walls are fully prepared to be painted.

Steps for Painting with a Brush and Roller

1. Prime the walls. Proper priming is key to a long-lasting job (see Table 8-1).
2. Paint a 2-inch strip at the ceiling using a brush.
3. Paint 2-inch strips in corners, around windows, doors, cabinets, and baseboards using a brush.
4. Switch to a roller. Paint in a vertical direction using a zigzag pattern. Push the roller upward on the first stroke and then form an "M" pattern to evenly distribute the paint on the roller. Work in 3-foot sections. Fill in the "M" pattern without reloading the roller until you have completely covered the area. Continue in 3-foot sections until the wall is finished. Touch up spots you missed when the paint is wet to help minimize sheen differences (Figure 8-21).

Types of Paint Primers

- Primers/sealers—also known as drywall repair clears (DRCs). These oil- or acrylic-based coatings are designed to penetrate, seal, and

Figure 8-21: **Painting a wall with a roller**

Surface and Paint to be Used	Recommended Preparation
Painting oil or acrylic on new drywall	Use acrylic primer.
Painting oil on new plaster	After proper curing time (approximately 4 weeks), use an oil-based primer.
Painting acrylic on new plaster	Use a diluted first coat of the final paint.
Drywall previously painted with latex or oil	Scrape any loose or flaking paint and sand smooth. Lightly sand entire wall to remove gloss. Use acrylic primer or prep coat.
Drywall previously painted with an intense pigment (very bright or very dark color)	Scrape any loose or flaking paint and sand smooth. Lightly sand entire wall to remove gloss. Use stain-killing primer.
Drywall previously painted with builder's flat	Scrape any loose or flaking paint and sand smooth. Use acrylic primer.
Plaster wall previously painted with latex or oil	Scrape any loose or flaking paint. Lightly sand entire wall to remove gloss. Use acrylic primer or primer/sealer.
Any wall with stains	Use stain-killing primer.
Any wall with mold	Spray affected area thoroughly with distilled white vinegar and allow to dry, then use stain-killing primer. *Note:* This method will kill 82% of the mold. Bleach will not kill mold.

Table 8-1: **Wall Preparation for Painting**

protect the wall while priming the surface to provide a firm bond between the substrate and the previous paint or new paint layer.

- Stain-killing primers—these oil- or acrylic-based primers do not actually remove stains from a surface. They use resin to provide an impenetrable seal between the surface and the final coat of paint. Without a stain-killing primer, many stains and intense colors will continue to bleed through all subsequent coats of paint.

Steps for Painting with a Paint Sprayer

1. Pour the paint through a strainer into the bin or bucket. Be sure to avoid lumps or odd bits of nonpaint material (Figure 8-22).
2. Thin the paint. Do not thin paint more than what is recommended by the manufacturer or it will not cover properly.
3. Be sure that you are covered from head to toe. Wear a long-sleeved shirt, gloves, and dust mask or respirator (Figure 8-23).
4. Start from the top and paint downward in smooth, steady strokes. Start at a corner, work from the top down, and keep your strokes steady and smooth. A lot of paint is going on the surface in a short period of time. It is better to paint several light coats than one heavy one (Figure 8-24).

Faux Finishes

Once the base coat has been applied, the painting process can be taken to the next level by creating a faux finish. Faux (pronounced "foe") is a French word meaning "false." This technique is also called Trompe l'Oeil (pronounced "trome play").

Acrylic- and oil-based primers are available in no and low VOC versions. Paints and primers with no or low VOC have fewer toxic chemicals that are released during outgassing. This creates a safer environment during the painting process and for those who will live/work within the building after the installation is complete.

Figure 8-22: **Paint strainer**

Figure 8-23: **Protected painter**

Again, this is a French phrase meaning "fools the eye." Both terms refer to the technique of using paint to create the illusion of other, more expensive, material.

Tools for Creating Decorative Faux Finishes

An endless variety of tools can be used to create decorative finishes, from specialty tools purchased at home improvement stores to improvised tools that can be found around the home. Some of the most common tools are shown in Figure 8-25(A) through (H). Their uses will be discussed in the following section.

Figure 8-24: **Paint sprayer**

General Categories of Faux Finishes

Five general categories of faux finishes are antiquing or distressing, combing, ragging or sponging, masking, and murals. A faux finish can consist of a single category or a combination of two or more categories.

Antiquing or Distressing

Antiquing or distressing gives the appearance of age or wear. This finish is created by adding color washes or glazes to a dry base coat to create shade variations that mimic aged surfaces. Crackling can also be used to give the appearance of age. It intentionally creates cracks in the top coat to allow an undercoat to show through. It is done by applying a special clear coat that dries very quickly causing the paint layer just below it to crack (see Figure 8-26).

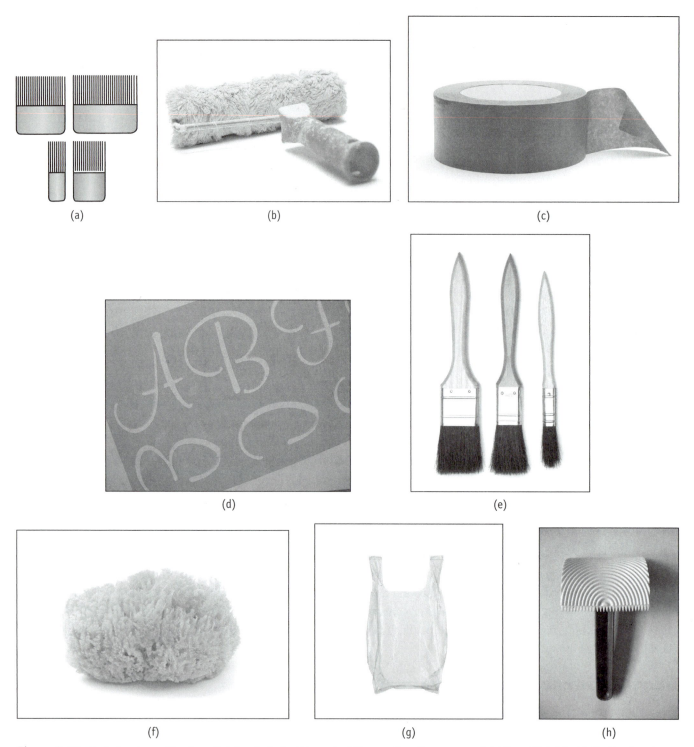

Figure 8-25: **Tools for decorative faux finishes include (A) comb, (B) rag roller (Image copyright Marc Dietrich, 2009. Used under license from Shutterstock.com), (C) painter's tape (Image copyright Todd Taulman, 2009. Used under license from Shutterstock.com), (D) stencil, (E) stipple brush (Image copyright Andrjuss, 2009. Used under license from Shutterstock.com), (F) sponge (Image copyright marylooo, 2009. Used under license from Shutterstock.com), (G) bag (Image copyright Dzarek, 2009. Used under license from Shutterstock.com), and (H) wood grainer**

Combing

Combing can mimic the look of fabric. It is created by adding a top coat of paint to a dry base coat, then dragging a brush or comb through the wet top coat. The comb can be dragged in a hatch or cross-hatch pattern to give the illusion of woven threads in fabric. The two coats of paint usually have a similar hue but different shades (for example, a medium blue base coat with a light blue combed coat on top). This technique can also be used to mimic wood grain. Instead of a comb or brush, a special wood-graining tool is simultaneously rocked and dragged through the wet paint of the top coat (see Figure 8-27).

Ragging or Sponging

Ragging or sponging creates the illusion of marble or stone. The texture is created by adding or subtracting color using a sponge, crumpled rag, plastic bag, or stipple brush. The additive method adds one or more colors to a dry base coat using one of these tools. In the subtractive method, new layers of paint are rolled onto the surface of the dry base coat, then subtracted using one of the tools listed (see Figure 8-28).

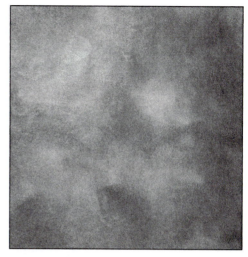

Figure 8-26: **Distressed faux finish (Creatas Images/Jupiterimages)**

Figure 8-27: **Combed faux finish (Creatas Images/Jupiterimages)**

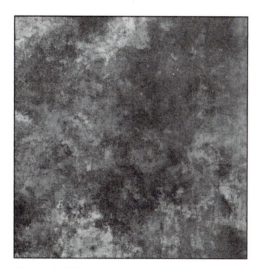

Figure 8-28: **Ragged faux finish (Creatas Images/Jupiterimages)**

Masking

Masking mimics wallpaper or border. A **mask** is placed on a dry base coat and a new color is applied over the mask to create an image or shape. Masking requires either painter's tape or a stencil. Painter's tape can be used to create stripes by masking off a dry base coat at regular intervals, adding a second color, then removing the painter's tape. Squares or diamonds can also be created by varying the placement of the painter's tape.

Murals

Murals give the illusion of scenery or architectural elements. This type of finish is done by an artist, and depicts an image or scene painted on a wall or ceiling. A mural

To keep the top coat from bleeding under the painter's tape, a clear coat can be applied over the tape before adding the top coat. A stencil is a mylar sheet with an image cutout at the center. When the mylar is placed against a surface, it creates a mask around the cutout image. Paint is stippled, or gently "pounced," over the mylar with a sponge or stipple brush to reproduce the image on the surface. Stencils are available in a wide variety of shapes, patterns, and images. They can be used to create a continuous border or an overall pattern (see Figure 8-29).

Figure 8-29: Stenciled faux finish (Image Source Black/Jupiterimages)

can be very simple, like white clouds painted on a blue base coat to create a skyscape, or it can be very complex, like a detailed depiction of a landscape or fantasy scene (like cherubs). Finally, a mural can give the illusion of architectural elements, such as columns, intricate molding, and windows. Murals can be created using regular paint or fresco. Fresco uses wet pigments as paint on uncured plaster (see Figure 8-30).

Figure 8-30: Mural (Image copyright William Casey, 2009. Used under license from Shutterstock.com)

Cleaning and Storing Equipment and Supplies

It is important to clean all paint equipment thoroughly, store equipment and supplies properly, and discard empty containers appropriately.

Steps for Cleaning Paint Rollers

1. Disassemble paint roller.
2. If you are using oil-based paint, submerge cover in solvent.
3. Wash and rinse cover. Wash the roller cover in mild detergent and rinse it in clear water (see Figure 8-31).
4. If you are using oil-based paint, remove paint from frame and hardware using mineral spirits or a natural alternative such as Citrus Solvent; otherwise, use water.
5. Hang roller to dry.

Figure 8-31: Rinse paint roller under tap

Steps for Cleaning Paint Brushes

1. Remove excess paint. Either work it out on a piece of newspaper or run the edge of a scraper along the bristles (Figure 8-32).
2. Use water to clean water-based paints. Water-based paints can simply be rinsed out under a running tap. Work them in a bucket of water first to remove much of the paint buildup. This makes rinsing them under the tap faster and easier. Make sure that the water running through the brush is clear before finishing (Figure 8-33).
3. Use mineral spirits or a natural alternative such as Citrus Solvent to clean oil-based paints. Brushes used in oil-based paints should first be worked in a container of white spirit. Next, shake out the excess paint and work the brush in brush cleaner. Brush cleaner is water-soluble so the brush can then be rinsed under a running tap. Wash the brushes in soapy water and rinse under running water.
4. Shake out the excess water by spinning the brush handle between your palms.
5. Wrap the brush in paper to retain its shape. *Tip*: Paint that has dried into the brush can be removed by soaking it in brush restorer. Really dried paint can be removed using a wire brush after soaking. Do not use the wire brush too harshly. It is easy to damage the bristles on the paint brush. Always brush in the direction of the bristles, working away from the handle.

Figure 8-32: Person removing excess paint from brush

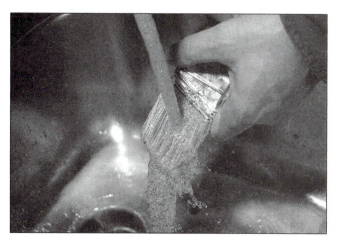

Figure 8-33: Clean brush with water

Steps for Cleaning a Spray Gun

1. Remove any unused paint from the container.
2. Use water to clean water-based paints. Fill container with water and spray until it emerges clear.
3. Use mineral spirits or a natural alternative such as Citrus Solvent to clean oil-based paints. Fill container with white spirit instead of water. When clear of paint, clean the container with a mix of hot water and detergent.
4. Dismantle the machine and clean all parts with a damp cloth.

Proper Paint and Trash Disposal

Water-thinned (latex) and oil-based paints should be used completely. Be sure to reserve a small amount of the paint for touch-ups. Dry, empty paint containers may be recycled in a recycling program. In most states, latex paint can be dried out in the can and disposed of in your household trash. Leave the lid off to show that the paint has hardened.

Finishing Up

Let the paint dry thoroughly. Then remove the masking tape carefully. Use a snap-off razor if necessary to separate the tape from the paint. Pull tape slowly back over itself at a steep angle. No job is complete until the entire area is cleaned up completely and checked for quality. Review all painted areas. Screw all plates back on the wall and check for straightness. Check floors, walls, nonpainted areas, and room entrances for paint spatters or drips and clean as needed. If a client is involved, ask the client to conduct a final assessment of the area with you. Walk through the newly painted areas and point out your work. Note any dissatisfaction and remedy the problem immediately and eagerly. Consider providing the client a small bottle or can of the original paints, clearly labeled, so that they can do minor touch-ups if needed.

Wall Coverings as a Surface Treatment

Another technique for protecting and decorating drywall/gypsum board is by covering the surface with wallpaper. After fixing any drywall problems as discussed at the beginning of the chapter, additional surface preparation is required for a wallpaper application.

Surface Preparation for Wall Coverings

Proper preparation will help the wallpaper adhere to the wall correctly. If the wallpaper does not bond solidly to the wall, seams will be exposed during the drying process and, over time, the paper may begin to peel away from the surface. If, however, the wallpaper bonds too tightly to the surface, it will be extremely difficult to remove and may even damage the surface of the wall.

Primers are used to create a more uniform substrate for its improved bonding with the wallpaper. Primers and sealers also increase the slide of the wallpaper. This allows the wallpaper to slide over the surface of the wall for easier placement during the installation process. However, not all primers allow the wallpaper to slide easily during installation. Always consult manufacturer guidelines for the chosen wallpaper and adhesive. If no guidelines are recommended, use the following information to prepare the wall(s) based on the type of surface being prepared.

Steps for Wall Preparation

1. Remove as much furniture as possible from the room. Group the heavier items in the center of the room and cover them with drop cloths.
2. Protect the floor with a drop cloth (a canvas drop cloth is less slippery and creates a safer work area).
3. Refer to the "Fixing Drywall Problems" section earlier in this chapter. Make all necessary repairs.
4. For existing installations, use a sponge to wash all installation surfaces with a trisodium phosphate (TSP) solution and rinse with water. Allow walls to dry for 24 hours. This will help remove excess oil, grease, and dirt from the surface. Do not wash new, unpainted drywall or plaster surfaces.
5. Turn off power to the affected switches/outlets at the service panel. Remove switch and receptacle plates. Be sure to tape the screws to the plate so that they are not lost.
6. Paint any surfaces such as ceilings or trim before priming or wallpapering the walls (see Table 8-2).
7. Use the manufacturer's guidelines or the following chart to determine and apply the proper primer, sealer, or sizing.

Surface	Recommended Preparation
New drywall	Use primer/sealer.
New plaster	After proper curing time (approximately 4 weeks), use size or a diluted solution of wallpaper adhesive.
Drywall painted with latex or oil	Scrape any loose or flaking paint. Lightly sand entire wall to remove gloss. Use acrylic primer or prep coat.
Drywall painted with an intense pigment (very bright or very dark color)	Scrape any loose or flaking paint. Lightly sand entire wall to remove gloss. Use stain-killing primer.
Drywall painted with builder's flat	Use primer/sealer.
Plaster wall painted with latex or oil	Scrape any loose or flaking paint. Lightly sand entire wall to remove gloss. Use acrylic primer or primer/sealer.
Any wall with stains	Use stain-killer primer.
Any wall with mold	Spray affected area thoroughly with distilled white vinegar and allow to dry; then use stain-killer primer. *Note*: This method will kill 82% of the mold. Bleach will not kill mold.
Wallpaper	Every effort should be made to remove old wallpaper. Puncture or score the surface with a puncture roller or heavy grit sand paper to allow moisture to penetrate. Use a steamer or a damp sponge to loosen the adhesive; then scrape old wallpaper away using a broad knife. If removal is not an option, check all corners and edges to confirm good contact with the wall. Reglue or remove loose wallpaper, spackle and sand where necessary. Use an oil-based primer.
True paper wall covering	Check for proper bond on all old corner and edges; then use size.

Table 8-2: **Wall Preparation for Wallpapering**

Types of Wallpaper Primers

Primer/sealers—also known as drywall repair clears or DRCs. These oil- or acrylic-based coatings are designed to penetrate and seal the wall while priming the surface to provide a firm bond between the substrate and the wallpaper. This type of primer protects the drywall and allows the wallpaper to slide more easily during the installation process.

Primers/sizes—these acrylic coats provide a tacky surface for increased bonding with wallpaper. Stain-killing primers—these oil- or acrylic-based primers do not actually remove stains from a surface. They use resin to provide an impenetrable seal between the surface and the wallpaper. Without a stain-killing primer, many stains and intense colors will bleed or show through wallpaper.

Size or sizing—some wallpaper manufacturers plainly specify the use of sizing for the installation of their products. Size is commonly sold as a powder to be mixed with water according to the manufacturer's instructions. Sizing is used on plaster walls to prevent excessive amounts of wallpaper paste from being absorbed into the plaster during installation. Sizing is occasionally used on drywall for increased bonding between the wall and the covering.

Basic Categories of Wall Coverings

The basic categories of wallpaper are listed here. Generally, wallpaper contains either a paper or a fabric substrate laminated to a paper, vinyl, or fabric decorative face.

True paper has a paper substrate, and the decorative layer is printed directly on this substrate. Most papers are uncoated, but some seal in the decorative inks with a slight top coating. This type of wallpaper is very delicate and difficult to install, and is more suited to areas that will see minimal abuse.

Vinyl-coated paper has a paper substrate. The decorative surface is coated with vinyl or polyvinyl chloride (PVC). This type of wallpaper is durable and is easy to clean and remove. It is resistant to moisture and grease, making it a good option for bathrooms, kitchens, and basements.

Coated fabric has a fabric substrate that has been coated with liquid vinyl or acrylic. The decorative face is printed directly on this substrate. Because this type of wallpaper allows more moisture to "breathe" through the surface, it is best for use in low-moisture rooms that will not compromise the integrity of the adhesive, such as living areas.

Paper-backed vinyl/solid sheet vinyl has a paper substrate laminated to the decorative surface that consists of a solid sheet of vinyl. This type of wallpaper is extremely durable and is easy to clean and remove. Solid sheet vinyl can be used in most areas of the home since it resists moisture and is stain and grease resistant. It is resistant to moisture and grease, making it a good option for bathrooms, kitchens, and basements. However, it is not a good candidate for areas that receive a lot of physical abuse such as mudrooms or storage rooms.

Fabric-backed vinyl has a woven or nonwoven fabric ground (mesh-like textile backing) substrate laminated to a solid vinyl decorative face. This type of wallpaper is extremely durable and is used almost exclusively for commercial and institutional applications.

Specialty may use either a paper or a fabric substrate. The decorative surface may include foils, synthetic fiber flocking, natural elements such as cork or sisal, or other exotic elements. This type of wallpaper is usually expensive and difficult to install. It is usually long-lived, but not easily cleaned and does not withstand abuse. It is used mainly in highly decorative, low-traffic installations.

Anaglypta and *supaglypta* have a paper substrate. Anaglypta has a pulp surface and supaglypta has a cotton fiber surface. The surface is deeply embossed and left uncoated so that it can be painted after installation. Because this is an extremely thick wallpaper, it is a good choice for covering up wall imperfections. The look can be quickly changed and adapted with a fresh coat of paint. The embossing is easily damaged with sharp objects or furniture. Depending on the surface paint used, this wallpaper can be used in any room that does not see a high volume of abuse. *Environmental-friendly paper* has a paper substrate often from recycled material. The surface, or face of the wallpaper, is generally an organic or recycled material (rice paper, grass, wood, or newspaper). It improves interior air quality because it has no or low VOC and no vinyl. It often comes prepasted with toxin-free adhesives. Because this type of covering is breathable, air bubbles and mildew are minimized. Check with the manufacturer to determine if the paper you choose will work well in high humidity locations such as bathrooms and kitchens. Continual improvements have made ecofriendly wallpaper competitive with traditional wallpaper choices, but this option remains somewhat more expensive and may not be suitable for high abuse areas such as mudrooms and storage rooms.

> *Remove masking tape before beginning the wallpapering process.*

Determining the Type and Amount of Wall Coverings

It can be tricky to estimate the amount of wallpaper needed for a project. Extra time spent measuring and planning will greatly reduce the risk of coming up short at the end.

1. Calculate the square footage of the space. Multiply the height by the width of each wall to find the overall square footage. Multiply the height by the width of each door, window, fireplace, etc., then subtract this total from the overall square footage to find the final coverage area. Add 10 percent to the total for waste. Round your total up to the nearest foot (see Figure 8-34).
2. Determine the starting place where you will begin (and end) hanging the wallpaper. Since it is virtually impossible to wallpaper an entire room and get the final strip to fit perfectly, plan to start (and end) the wallpaper in an inconspicuous area, for example, in the corner behind the entrance door, above a door, or window that is not directly across from the entry. This area is known as the **kill spot**.
3. Determine the number of strips of wallpaper to cover the space. This can be accomplished by using the width of the wallpaper to mark off the number of strips it will take to cover the entire area. For example, if the wallpaper is 27-inch wide, begin at the kill spot and measure out horizontally 27 inches. Place a small mark on the wall, and measure another 27 inches from this mark. Number each strip marked on the wall consecutively, for example, 1, 2, 3, ... Continue around the room until you come back to the kill spot (see Figure 8-35).
4. Determine the repeat or drop in the pattern. This information is usually determined by the manufacturer and included on the label of the wallpaper itself. It refers to the linear space on the roll between identical pattern elements. The amount of vertical space between the pattern repeat will be important when aligning the strips of wallpaper horizontally. Multiply the amount of pattern drop by the number of strips used to cover the space (see Figure 8-36). For

> *If the trial layout created in Step 2 leaves awkward areas such as long narrow strips (less than 3 inches wide) next to a door or corner, adjust your starting point or the width of your starting strip to eliminate the awkward areas. Be sure to mark the new strip areas on the wall. When adjusting the width of the starting strip, always trim from the kill spot side of the strip.*

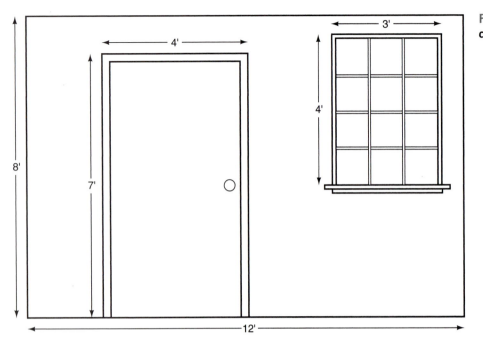

Figure 8-34: **Wall diagram with calculations**

REPEAT THE FOLLOWING PROCESS FOR EACH WALL THAT WILL BE COVERED
OVERALL WALL DIMENSIONS: 8 Ft × 12 Ft = 96 Ft
DIMENSIONS OF DOOR: 4 Ft × 7 Ft = 28 Ft
DIMENSIONS OF WINDOW: 4 Ft × 3 Ft = 12 Ft
96 Ft − 28 Ft − 12 Ft = 56 Ft
56 Ft + 10% (OR 5.6 Ft) = 61.6 Ft
ROUND 61.6 Ft UP TO 62 Ft

example, if you determine in Step 3 that it will take 22 strips to complete the room, and the wallpaper has a 6-inch drop; multiply 22 × 6. Divide this total by 12 to convert the results to feet (22 inches × 6 inches = 132 inches) (132/12 = 11 feet). This will give you the total amount of wallpaper that will be lost when aligning each strip to match the pattern. Round your total up to the nearest foot (see Figure 8-36).

5. Consult the dealer or refer to the wallpaper label to find the total square footage contained on each roll. Most American double rolls of wallpaper cover about 70 square feet. Combine the totals from the final coverage area (Step 1) with the pattern drop loss (Step 4) and divide by the number of square feet contained in one roll of wallpaper. For example, if the final coverage area including 10 percent waste is 220 square feet, and the pattern drop loss is 11 feet (220 feet + 11 feet = 231 feet) (231 feet ÷ 70-feet roll coverage = 3.3 rolls). Round up to the next whole roll. So the room in the example would require four rolls.

6. Check the label or with the distributor to find out if the wallpaper chosen is prepasted. If it is not, use the wallpaper manufacturer's guidelines to determine the type and amount of paste you will need.

Figure 8-35: **Measuring width of strips on wall (Image copyright mearicon, 2009. Used under license from Shutterstock.com)**

Table 8.3 provides a checklist of tools that are necessary when wallpapering (see also Figure 8-37).

Vinyl smoother (recommended) or wallpaper smoothing brush
☐ Metal square or yard stick
☐ Vinyl-to-vinyl wallpaper paste
☐ Ladder
☐ Thirty-inch water tray
☐ Bucket
☐ Seam roller
☐ Sponge
☐ Eight-inch broad knife
☐ Carpenter pencil
☐ Scissors
☐ Painter's masking tape
☐ Snap-off razor and extra pack of blades
☐ Level or plumb boÇb
☐ Paint roller with 3/8-inch nap cover (if the paper is not prepasted)
☐ Metal square or yard stick
☐ Basswood table or a sheet of plywood and sawhorses to create a cutting surface

Table 8-3: **Tools Used for Wallpaper Installation**

Figure 8-36: **Aligning the pattern repeat in two strips of wallpaper**

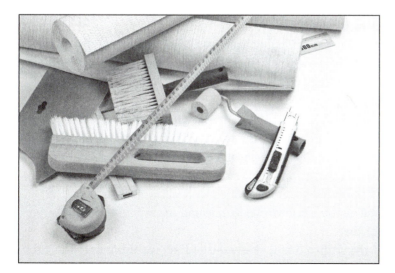

Figure 8-37: **Wallpaper tools (Image copyright Dmitry Melnikov, 2009. Used under license from Shutterstock.com)**

Applying Wallpaper

The installation of wallpaper is not a difficult task to complete as long as the facility maintenance technician follows a few basic steps, which are listed next. However, it should be noted that all manufactures' instructions and recommendations should be followed.

1. Read and follow all instructions included with the wallpaper. Failure to follow these guidelines could void and warranty and cause the wallpaper to fail.
2. Using the markings created during the trial layout, start in the most visible area (usually directly across from the kill spot). With a pencil, extend the vertical mark of first strip from floor to ceiling using a level or plumb bob. A proper vertical line will make hanging the wallpaper much easier and greatly enhance the final appearance of the job (see Figure 8-38).
3. Measure the height of the wall and add 4 inches to the measurement for trim allowance. Unroll the wallpaper face down on the cutting surface. Mark your measurements using a pencil.
4. Use a metal square or straight edge to make a straight cut across the paper. If the wallpaper is a straight match, you can continue cutting out the remainder of the strips. Keep the trial layout in mind for areas that do not require a full strip (below a window or above a fireplace) (see Figure 8-39).
5. If the wallpaper has a drop match pattern:
   A. Turn the strip face up and unroll the next strip face up to the left of the first.
   B. Extend the top of the second strip until the pattern aligns across the two strips.
   C. Use the bottom of the first strip as the measuring line to cut the second strip.
   D. Label the back of this strip "left," and save for later use. This strip will be longer than the first to allow for the drop match.
   E. Unroll the next strip face up to the right of the first.
   F. Extend the top of the second strip until the pattern aligns across the two strips.
   G. Use the bottom of the first strip as the measuring line to cut the second strip.
   H. Label the back of this strip "right." This strip will be longer than the first to allow for the drop match.
   I. Continue using Steps E through G to cut out the remaining strips to the kill spot. Then, use Steps A through C to finish the other half of the area around to the kill spot. Before installing each strip, be sure to use it as a guide for cutting the next strip.
6. If the paper is not pre-pasted, use a paint roller to apply the paste to the back of the wallpaper according the wallpaper and paste instructions. Roll the paste across the narrow expanse of the paper, then along the length of the strip to the top and bottom edges to evenly distribute the paste. Then follow points D and E in Step 7 to allow the paper to relax before installation.
7. If the wallpaper is prepasted, consult the wallpaper's guidelines for booking. **Booking** is the process used to activate the paste. Manufacturer's guidelines vary for soaking and relaxing times. But the general process is as follows:
   A. Roll the measured and cut strip face inward.
   B. Place the rolled strip in the water tray (filled with warm water unless otherwise specified) and gently swish it side to side and allow it to soak for the prescribed amount of time.
   C. Grasp the end of the strip and pull slowly out of the tray, unwinding the strip face down (see Figure 8-40).

Figure 8-38: **Creating an initial plumb line**

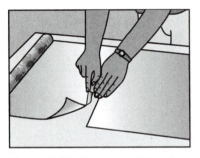

Figure 8-39: **Using a straight edge to cut the wallpaper**

D. Once completely out of the water, fold half of the wallpaper back over itself to the approximate center of the strip so that the pasted sides are together. Fold the other half back over itself to meet and overlap the opposite edge by a few inches. *Do not* crease the paper at the "folds;" it should create a small loop (see Figure 8-41).

E. Starting with the folded ends, roll the booked wallpaper loosely together to form a scroll and allow it to relax for the time indicated in the guidelines.

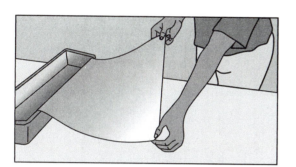

Figure 8-40: **Removing wallpaper from the tray**

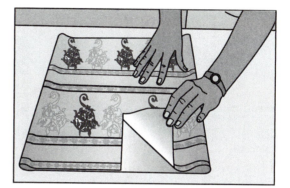

Figure 8-41: **Booking wallpaper**

8. Unroll the wallpaper. Make sure the pattern is facing up. Separate the edges of the pasted side only on the top half. While allowing a few inches to extend past the ceiling/trim line, align the wallpaper with the vertical line marked on the wall and firmly press the exposed portion of the paste to the wall. Although it is tempting, do not try to align the edge of the wallpaper with the ceiling/trim. Allow for overhang to be trimmed later for a better fit and more professional appearance (see Figure 8-42).

9. Once the wallpaper is in position, use a vinyl smoother or wallpaper brush to gently work out wrinkles and air bubbles. Be careful not to squeeze excessive amounts of paste out during this process. It could leave the wallpaper with too little adhesive for a proper bond. If a wrinkle is not coming out, pull the far edge (the edge farthest away from the plumb line or previous strip) of the wallpaper away from the wall and smooth back down with the vinyl smoother (see Figure 8-43).

10. Place the 8-inch broad blade firmly along the trim line at the ceiling/trim (see Figure 8-44). Use a snap-off razor to trim the excess by cutting along the edge of the broad blade. The broad blade should always be placed between the installed wallpaper and the blade. This will hold the paper in place and protect the wallpaper from damage if the knife slips. Snap off old blades after a few cuts. A dull blade will tear the wallpaper.

11. Once the top half is smoothed and trimmed, unbook the bottom half, smooth, and trim.

12. Book the next strip and install next to the first being careful to align the pattern and butting the edges snugly together (minute shrinkage will occur during drying, so any gaps at this point will increase once the project is complete) (see Figure 8-45).

13. Use the seam roller on the abutting seams being careful not to force too much adhesive out of the joint, which could cause poor adhesion. *Note:* Do not use a seam roller on anaglypta or supaglypta wallpaper (see Figure 8-46).

Figure 8-42: **Align the first strip with the plumb line**

Wallpaper rarely matches perfectly throughout the entire length of the seam due to minor paper stretching and wall imperfections. Use eye level to align the pattern since mismatches at the ceiling or floor will be less noticeable.

Figure 8-43: **Using a wallpaper brush to smooth out air bubbles**

Figure 8-44: **Trimming seam allowances**

Figure 8-45: **Aligning the pattern and butting the seams together**

Figure 8-46: **Rolling the edges with a seam roller**

14. Use a bucket of warm water and a sponge to clean paste off of face of the wallpaper before it dries. Replace the warm water in the water tray when it becomes cold and/or slimy.

15. Work in one direction around to the kill spot. Then, starting from the first strip installed, work in that direction to the kill spot. At the kill spot, place the broad knife vertically in the center of the overlapping area. Use the broad knife as a guide to make a double cut through the top and bottom strips simultaneously. Remove the waste from the top cut; then peel back the newly installed paper to reveal the waste on the bottom. Remove the waste and smooth the wallpaper back into place. Because of the double cut, the two edges should line up perfectly (see Figure 8-47).

Wallpapering Corners and Around Obstacles

Corners (inside or outside) are never straight and plumb. They can be crooked from floor to ceiling and bulge or sink in the middle. Also, walls can shift slightly over time. To keep the wallpaper from twisting or showing seam gaps at corners and to ensure that the paper continues to be plumb and straight after turning a corner, use the following steps.

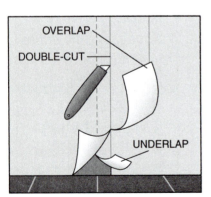

Figure 8-47: **Making a double cut through two strips of wallpaper (redraw illustration)**

Wallpapering an Inside Corner

1. Wrap and smooth the paper folding it around the corner.

2. Using painter's masking tape, tape a carpenter pencil along the length of the broad knife's edge. This will create a guide approximately ¼-inch wide (see Figure 8-48).

3. Place the broad knife vertically in the corner with the pencil against the wall that has just been papered. Use a snap-off razor to cut the wallpaper as you slide the broad knife down the length of the corner. Use scissors to

finish cutting the paper at the top and bottom overlap areas (see Figure 8-49).

4. Peel the remainder of the strip from the wall. If the strip is less than 4 inches, discard the strip and begin with a new strip. If the strip is more than 4-inch wide, rebook it to keep the adhesive wet during the next step.

5. Measure the width of the next piece (either the piece just cut or a full strip of wallpaper). Use this measurement to create a plumb line on the wall after the corner.

6. Hang the cut piece or new strip on the wall using the plumb line as the placement guide. Smooth the wallpaper over the ¼-inch overlap into the corner.

7. Peel the top piece back slightly in the corner. Apply vinyl-to-vinyl wallpaper paste to the overlap area and smooth the paper back into place.

8. Trim the excess from the top and bottom as usual. You may need to peel the corner out slightly and use scissors to make the final cut in the corner; then smooth the corner back down.

9. This will create a ¼-inch overlap at each corner to help mask wall movement and to ensure that the paper will come out of the corner plumb. This will create a slight pattern mismatch in the corner; this is normal and will be less obvious once the job is complete (see Figure 8-50).

Figure 8-48: **Using a pencil and broad knife as a guide for cutting (Courtesy Debbie Standiford)**

Figure 8-49: **Finish cutting the paper at the top and bottom overlap areas.**

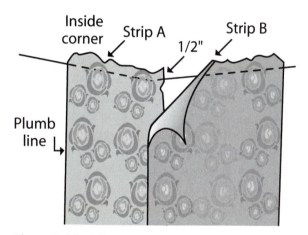

Figure 8-50: **Wallpapering around an inside corner**

Wallpapering an Outside Corner

1. Hang and smooth paper around an outside corner as usual. Cut a vertical slit in the top and bottom trim allowances at the corner to allow the wallpaper to wrap around the corner. Carefully remove all bubbles that tend to form at the peak of the corner.

2. Check the edge of the paper coming out of the corner for plumb. If it is plumb, continue hanging wallpaper from there (see Figure 8-51).

3. If the edge of the wallpaper coming out of the corner is not plumb, use the following steps:

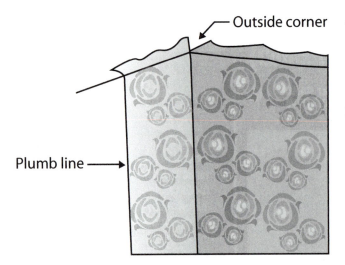

Figure 8-51: Wallpapering around an outside corner that is plumb

A. Measure the distance from the last strip to the corner and add 1 inch. Cut the width of the strip that will go around the corner to this length. Install and smooth this strip. Cut a slit in the top and bottom trim allowances to allow the wallpaper to wrap around the corner smoothly.

B. Measure the width of the cut strip (or a new strip if the remainder is less than 4 inches). Add ¼ inch to this measurement and create a new plumb line on the wall after the corner.

C. Install the cut strip (or new strip) starting with the plumb line and working backward toward the corner. The edge should fall ¼ inch short of the corner. If the edge were installed too close to the corner, it might fray with the abuse that outside corners usually endure (see Figure 8-52).

D. Peel back the top layer at the corner and use vinyl-to-vinyl wallpaper adhesive where the paper overlaps.

E. Trim the allowances from the top and bottom of the wall as usual.

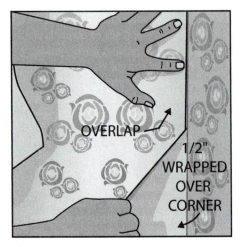

Figure 8-52: Wallpapering around an outside corner that is plumb

Wallpapering Around Obstacles

Obstacles might include anything from windows, doors, built-in shelves, and fireplaces. This section will specifically discuss windows, but the same general principle will apply to most obstacles you will encounter.

1. Hang the wallpaper as normal aligning the edge with the previous strip. Use the vinyl smoother to smooth the paper to the vertical edge of the window.

2. In the area that overlaps the window, use scissors to cut a diagonal slit from the approximate center to the upper corner of the window trim face. Use a snap-off razor to carefully continue this diagonal slit from the face to the trim to the wall (along the depth of the trim) (see Figures 8-53 and 8-54).

Figure 8-53: Using scissors to cut diagonally to the edge of window trim

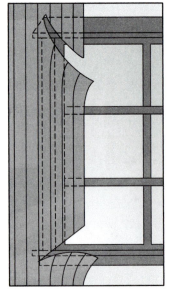

Figure 8-54: Cut lines for wallpapering over window

3. Smooth the top of wallpaper up to the ceiling and down to the top of the window.

4. Again, in the area that overlaps the window, use scissors to cut a diagonal slit from the approximate center to the lower corner of the window trim face. Use a snap-off razor to carefully continue this diagonal slit from the face to the trim to the wall (along the depth of the trim).

5. This area may be more complicated because of the many angles created by the window sill. Take your time and use minidiagonal slits for each corner you encounter.

6. Smooth the lower portion of wallpaper up to the bottom of the window and down baseboard.

7. Trim around the window by placing the broad knife against the trim as a guide. Trim into the wall, not the window facing.

8. Use a bucket of warm water and a sponge to clean paste off windows and trim before it dries.

Wallpapering Around Outlet and Switches

Although only outlets are referred to in this section, both outlets and switches use the same technique.

1. Hang the wallpaper as normal aligning the edge with the previous strip. Use the vinyl smoother to smooth the paper to the vertical edge of the outlet.

2. Use a snap-off razor to make an X-cut over the outlet. The size of the X should be just enough to allow the paper continue smoothly over the outlet. Do not extend the edges of the X to the corners of the outlet box yet.

3. Continue smoothing the entire strip onto the wall. Trim the allowances from the top and bottom of the wall.

4. Use scissors to extend each line of the X into the corners of the outlet box. Trim the excess either by running a snap-off razor along the inside edge of the outlet box or by allowing a ½-inch overhang and wrapping the edges into the outlet box (see Figure 8-55).

5. Use a bucket of warm water and a sponge to clean paste off the face of outlet before it dries.

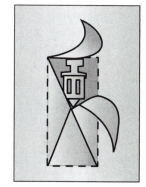

Figure 8-55: **X-cut and trim lines for and outlet**

Finishing Up

Drain water tray and bucket, rinse adhesive from tools with warm water, and dry. Review the area for problems or edges that might need more paste. Screw all plates back on the wall and check for straightness. Check floors, walls, and room entrances for paste spatters or drips and clean as needed. Remove drop cloths and replace furniture. If a client is involved, ask the client to conduct a final assessment of the area with you after the wallpaper has completely dried.

Texture as a Surface Treatment

The final technique we will discuss for protecting and decorating drywall/gypsum board is to create a surface texture. After fixing major drywall problems as discussed at the beginning of the chapter, walls should be primed before being textured. Primer is used to create a more uniform substrate for improved bonding with the texture. Use the following information to prepare the wall(s) based on the type of surface being prepared.

Steps for Wall Preparation

1. Remove as much furniture as possible from the room. Group the heavier items in the center of the room and cover them with drop cloths.
2. Protect the floor with a drop cloth (a canvas drop cloth is less slippery and creates a safer work area).
3. Refer to the "Fixing Drywall Problems" section earlier in this chapter. Repair major defects such as jutting nails; minor defects will be concealed by the texture.
4. For existing installations, use a sponge to wash all installation surfaces with a Trisodium Phosphate (TSP) solution and rinse with water. Allow walls to dry for 24 hours. This will help remove excess oil, grease, and dirt from the surface. Do not wash new, unpainted drywall or plaster surfaces.
5. Turn off power to the affected switches/outlets at the service panel. Remove switch and receptacle plates. Be sure to tape the screws to the plate so that they are not lost.
6. Paint any surfaces such as ceilings or trim before priming the walls (see Table 8-4).

Surface and Location	Recommended Preparation
Bathrooms and basements	Use oil-based primer in high-humidity areas.
Kitchens	Use oil-based primer in areas that may accumulate excessive grease.
New drywall	Use acrylic primer.
New plaster	Use oil-based primer.
Drywall painted with latex or oil	Scrape any loose or flaking paint. Lightly sand entire wall to remove gloss. Use acrylic primer.
Plaster wall painted with latex or oil	Scrape any loose or flaking paint. Lightly sand entire wall to remove gloss. Use acrylic primer.
Any wall with stains or an intense pigment	Scrape any loose or flaking paint. Lightly sand entire wall to remove gloss. Use stain-killer primer.
Any wall with mold	Spray affected area thoroughly with distilled white vinegar and allow to dry, then use stain-killer primer. *Note*: This method will kill 82% of the mold. Bleach will not kill mold.
Wallpaper	Every effort should be made to remove old wallpaper. Puncture or score the surface with a puncture roller or heavy grit sand paper to allow moisture to penetrate. Use a steamer or a damp sponge to loosen the adhesive; then scrape old wallpaper away using a broad knife. The moisture in drywall texture will loosen wallpaper adhesive. It is not recommended that texture be applied over wallpaper.

Table 8-4: **Wall Preparation for Texturing**

Texture Mediums

- All-purpose drywall joint compound—or drywall mud, comes premixed or in powder form. It is generally considered the easiest texture for beginners.
- Gypsum plaster—gypsum, lime, and sand. It is more expensive than joint compound, but it creates a tougher surface that is more resistant to abuse.
- Stucco—cement, lime, and sand.
- Texture paint—about as thick as pancake batter or wet plaster.
- Environmental-friendly plaster—made from natural clays and recycled aggregates. Installation is similar to other textures but may require several applications to build up the surface slowly to prevent cracks.
- Sand—30 or 70 mesh (the larger the mesh the larger the grain) white quartz sand can be added to any texturing medium for a more rustic, old-world appearance.

Types of Textures

Application techniques and the resulting variety of textures created are limited only by the imagination. However, texture is most often applied by using a blower, trowel, or paint roller. The resulting textures generally fall within the following categories:

Figure 8-56: Popcorn texture on drywall (Image copyright Denis Babenko, 2009. Used under license from Shutterstock.com)

- Popcorn texture—can be sprayed or can be rolled on with a roller. The roller method creates peaks or stalactites to give this texture its signature cottage cheese appearance. It was commonly used for ceilings during the 1960s and 1970s (see Figure 8-56).
- Orange peel texture—must be sprayed on the wall using a thin mixture. The resulting spatters resemble the peel of an orange (see Figure 8-57).
- Knockdown texture—begins by applying the orange peel texture. The peaks that are formed during the application are then "knocked down" using a trowel (see Figure 8-58).
- Skip trowel—is rolled on using a paint roller. The surface is then troweled like the knockdown texture. Depending on the thickness of the texture and the amount of pressure used during troweling, this surface can be very subtle or very rough (see Figures 8-59 and 8-60).
- Spanish texture—also called a knife texture. It is applied randomly with a trowel. The trowel marks are left in the texture to give it an old-world appearance (see Figure 8-61).
- Sand texture—can be applied with a sprayer, roller, or brush. Sand is added to texture paint and applied to the wall. This is an easy texture to apply, but distributing the sand evenly across the surface can be tricky (see Figure 8-62).
- Stamp texture—is applied using a paint roller; then it is stamped with an object to give the desired results. The objects used for stamping can be anything that transfers a pattern into the texture medium ranging from a foam image stamp to a stipple brush to a human hand (see Figure 6-63).

Figure 8-57: Orange peel texture on drywall (Image copyright Simon Krzic, 2009. Used under license from Shutterstock.com)

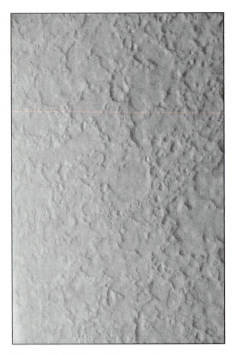

Figure 8-58: Knockdown texture on drywall (Image copyright Kanwarjit Singh Boparai, 2009. Used under license from Shutterstock.com)

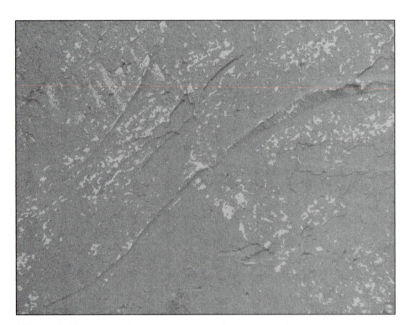

Figure 8-59: Subtle skip trowel texture (Image copyright Dewitt, 2009. Used under license from Shutterstock.com)

Figure 8-60: Rough skip trowel texture (Image copyright Zastolskiy Victor Leonidovich, 2009. Used under license from Shutterstock.com)

Figure 8-61: Spanish texture on drywall (Image copyright Cousin_ Avi, 2009. Used under license from Shutterstock.com)

- Drag texture—is applied using a paint roller; then an object such as a comb or wallpaper brush is dragged across the surface to create the desired effect (see Figure 8-64).
- Stencil texture—the base texture is applied using a paint roller or trowel and allowed to dry. Then a mud stencil is placed on the wall and texture is applied over the stencil. When the stencil is removed, the texture leaves a relief of the stencil design on the wall (see Figure 8-65).

Table 8-5 provides a list of tools that are necessary when texturing drywall.

❏	Texturing medium
❏	Drop cloths
❏	Painter's masking tape
❏	Ladder electric drill with ribbon attachment
❏	Depending on the application
❏	Texture blower or spray gun and compressor
❏	Paint roller, cover, and tray
❏	Trowel

Table 8-5: **Tools for Texturing Drywall**

Figure 8-62: **Sand texture on drywall (Image copyright spe, 2009. Used under license from Shutterstock.com)**

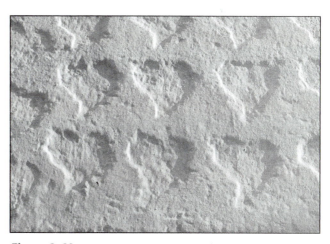

Figure 8-63: **Stamp texture on drywall (Image copyright Carsten Medom Madsen, 2009. Used under license from Shutterstock.com)**

Spraying or Blowing Texture on Drywall

1. Spraying or blowing texture can be extremely messy. Carefully mask all surfaces that will not be textured by using painter's masking tape and plastic drop cloths. Cover the floor with plastic drop cloths. Tape the edges of the drop clothes together to create a secure seal.
2. Cover clothing and shoes; always use goggles and a mask.
3. Turn off electricity at the circuit panel; then remove all outlet and switch covers. Protect the switches and outlets by tucking newspaper around them.
4. If you are using all-purpose drywall joint compound (premixed or dry), add water according to manufacturer's instructions and mix them thoroughly to achieve the proper consistency for spraying.
5. Attach the spray gun to the compressor. Adjust the spray nozzle and the air pressure to achieve the desired results. Generally, a ³⁄₈-inch opening will work

Figure 8-64: Texture relief using stencil and plaster (Courtesy Debbie Standiford)

Figure 8-65: Drag texture on drywall (Image copyright Thoma, 2009. Used under license from Shutterstock.com)

well at 60 psi. As you decrease the air pressure, you will need to increase the nozzle opening.

6. If you are using a knockdown technique:
    A. Wait 10–20 minutes before trying to trowel the surface, or you will smear the texture.
    B. Drag a trowel lightly across the surface at a low angle to flatten the peaks created by the sprayer. Scrape excess texture back into the container each time you raise the trowel from the surface.
7. Allow the texture to dry for at least 24 hours.
8. Paint the surface with a semigloss paint (the higher gloss reduces the amount of paint that is absorbed by the texture). Take special care to coat all surfaces in deep texture applications.
9. Remember, the key to a professional texture is consistency.

Rolling or Troweling Texture on Drywall

1. Mask all surfaces that will not be textured by using masking tape and plastic drop cloths. Cover the floor with plastic drop cloths. If you are texturing only one wall, mask off 12 inches of the adjacent walls using masking tape and newspaper.
2. Turn off electricity at the circuit panel; then remove all outlet and switch covers. Protect the switches and outlets by tucking newspaper around them.

3. When texturing an entire room, it is difficult to texture around inside corners without damaging the texture on the previous wall. Texture facing walls and allow them to dry. Then mask off 12 inches of these textured walls at the corners using masking tape and newspaper before texturing the adjoining walls. This will protect the texture on each wall at the corner.

4. Mix powdered all-purpose drywall compound using manufacturer's guidelines. If you are using ready mixed all-purpose drywall compound:
   A. Remove half of the mud and transfer it to an airtight container.
   B. Thin the remaining mud with one cup of water, and mix thoroughly with the electric drill using the ribbon attachment. Add more water in small amounts until the mud has the consistency of pancake batter or a milkshake (somewhat thicker for a knockdown texture).

5. If you are using a trowel, start at one edge of the wall and begin applying texture.
   A. Use one edge of the trowel to scoop up about a cup of mud.
   B. Set the mud edge of the trowel against the wall. The clean edge of the trowel should be raised away from the wall several inches.
   C. Sweep the trowel across the wall in the desired pattern. Play with the pressure and motion of the trowel in a small area until get the look you want. If you make a mistake while the mud is still wet, simply scrape the mud off the wall with the trowel and try again.
   D. Work in small areas from one side of the wall to the other. Always keep a wet edge to the mud.
   E. Use less texture at corners and trim.

6. If you are using a roller, start at one edge of the wall and begin applying texture.
   A. Fill the paint tray with mud and place the lid on the remainder.
   B. Roll texture onto the roller brush and roll the texture onto the wall. Two factors will determine the look of the texture: the nap or type of roller brush used and the speed of the application. The slower the brush is rolled across the surface, the higher the peaks in the texture.
   C. If you are using a knockdown technique, drag a trowel lightly across the surface at a low angle to flatten the peaks created by the roller. Scrape excess texture back into the container each time you raise the trowel from the surface. You may need to apply texture to a small area then trowel before moving on, or have one person apply texture while a second person trowels.
   D. Work in a small area and play with application techniques to achieve the results you want.
   E. Always keep a wet edge on the mud.
   F. Use less texture at corners and trim.

7. Allow the texture to dry for at least 24 hours.

8. Paint the surface with a semigloss paint (the higher gloss reduces the amount of paint that is absorbed by the texture). Take special care to coat all surfaces in deep texture applications.

9. Remember, the key to a professional texture is consistency.

Finishing Up

Clean the paint sprayer according to manufacturer's instructions, rinse tools with warm water, and dry. Review the area for inconsistencies. Allow the texture to dry completely. Screw all plates back on the wall and check for straightness. Check floors, walls, and room entrances for spatters or drips and clean as needed with warm water. Remove drop cloths and replace furniture. If a client is involved, ask the client to conduct a final assessment of the area.

Other Specialty Finishes

This chapter has discussed three common types of finishing techniques for drywall. However, there are many more ways to treat and protect a surface. These specialty finishes are typically more labor intensive and costly. It is worth noting, however, that almost anything relatively flat, from tile, to leather to precious metals, can be applied to a surface for protection and decoration.

Review Questions

1. List three types of paints commonly used in residential and light commercial.

2. List the three main types of water-based paint brushes.

3. List the two basic types of oil-based paint brushes.

4. List the steps for preparing a surface for painting.

5. List the steps for fixing a nail pop.

6. List the steps for repairing a small hole in drywall.

PIVOT

Name: _____

Date: _____

Painting

Identify and Select Proper Surface Finish

Upon completion of this job sheet, you should be able to demonstrate your ability to identify and select the proper surface finish and explain the selection process.

Describe the surface being painted:

What needs to be painted:	Bedroom	Bathroom	Living room	Dining room
	Kitchen	Hall	Doors	Trim

1 Are surfaces exposed to continual washing or scrubbing? Y/N

If yes, what is the best type of paint to use?

_____

_____

2 Does the surface need to stand up to washing? Y/N

If yes, what is the best type of paint to use?

_____

_____

3 If the room to be painted is a bathroom or kitchen, what is the best type of paint to use?

_____

_____

4 If you are painting a high-traffic area, such as a door or trim, what is the best type of paint to use?

_____

_____

5 If you need to hide minor imperfections, what is the best type of paint to use?

_____

_____

6 If you are painting a ceiling, why is it important to use ceiling paint?

_____

_____

Instructor's Response:

_____

_____

_____

Name: _____

Date: _____

Painting

Select Proper Finishing Tool

Upon completion of this job sheet, you should be able to demonstrate your ability to select the proper finishing tools and explain the selection process.

1 Why is it important to use high-quality tools?

_____

_____

2 If it is necessary to remove loose paint from the surface to be painted, what tool(s) should be used to perform this task?

_____

_____

3 Paint brushes are selected based on the type of finish being used and the surface being painted. List the two types of brushes and the characteristics of each type.

_____

_____

4 Rollers are an alternative to using brushes. Explain the different types of rollers and the types of surfaces they would be used on.

_____

_____

_____

Instructor's Response:

_____

_____

_____

Name: _____

Date: _____

Painting

Properly Preparing Surface for Painting

Upon completion of this job sheet, you should be able to inspect the painting surface and determine what preparations need to be done to the surface before painting.

Inspect the following, indicating whether inspected surfaces need to be repaired prior to painting.

Repair	Replace
Walls	
Baseboard	
Crown molding	
Window trim	
Doors	
Door jambs	

If the surface can be repaired, explain what process will be done to repair it. If the surface needs to be replaced, fill out a work order for the work to be done. Preparation checklist:

Tasks Completed

1 Surface repaired or replaced

2 Furniture moved from room or covered

3 Wall plates and receptacles removed

4 Exposed surfaces covered with caulking

5 Drop cloth or other protective material in place

Instructor's Response:

_____

_____

_____

Name: _____

Date: _____

Painting

Applying Paint by Using a Brush, Roller, or Sprayer

Upon completion of this job sheet, you should be able to paint the surface using a brush, roller, or sprayer.

1 What does the term "cut an area" during the painting process mean?

_____

_____

2 Is the ceiling being painted? If yes, describe the process that you will use to paint the ceiling.

_____

_____

3 Explain the process for painting the walls.

_____

_____

4 Why is it suggested that you paint in a zigzag or "M" pattern?

_____

_____

5 If you are using a sprayer, why is it important to pour the paint into a bucket through a strainer?

_____

_____

6 If you are using a sprayer, why do paint manufacturers recommend thinning the paint first?

_____

_____

Instructor's Response:

_____

_____

_____

Name: _____

Date: _____

Painting

Cleanup and Storing Equipment

Upon completion of this job sheet, you should be able to demonstrate the ability to clean up and store all equipment.

Tasks
Completed

1. Disassemble paint roller.

2. Clean roller and frame in a cleaning solvent.

3. Hang roller to dry.

4. Clean paint brushes in a cleaning solvent.

5. Wrap brushes in paper.

6. If you are using sprayer, remove any unused paint from sprayer.

7. Fill sprayer with appropriate solvent and spray until what emerges is clear.

Instructor's Response:

_____

_____

_____

Chapter 9 Plumbing

Introduction

Today society relies extensively on the use of plumbing systems to maintain its current standard of living. In a residential setting they are commonly used to transport drinking water to a home while conveying the waste products produced away from the home. Therefore it is essential that the plumbing systems be properly maintained to ensure the health and welfare of the people the system services.

Plumbing Safety

It is the responsibility of everyone to ensure that a worksite is safe. Although there are general safety procedures to be followed when working on a jobsite, each trade will have its own set of safety rules to be followed to ensure the safety of the workers and help prevent property damage. If anyone is creating a safety issue, that person should be stopped from doing that and immediately corrected. The following are a few of the safety rules that should be followed as well as some of the dangers of working with or around plumbing.

- Installing an incorrect piping product in a potable water system could contaminate the drinking water supply.
- Installing thin-wall pipe not designed for pressure in a pressure system can cause injury and property damage.
- SV gasket lubricant can irritate the eyes; a material safety data sheet (MSDS) should be reviewed and kept in the work location.
- Solvent welding requires the use of glue and, in most instances, a solvent primer and cleaner.
- These chemicals should be included in a company's Hazardous Communication Program. Refer to an MSDS for information about proper use, dangers, and medical treatment.
- PVC can shatter if overpressurized or damaged.
- Not all fittings are safe for potable drinking water systems. Incorrect fitting selection for drinking water systems can cause serious illness to a consumer.
- When working with copper, wear gloves to avoid cuts from the sharp tube ends. Clean and bandage cuts immediately to avoid infection.
- Heating polyethylene (PE) piping with a torch creates dangerous fumes and, although this practice may be common, it should be avoided.
- All plastic piping is flammable and, when ignited, causes toxic fumes. Never use an open flame near plastic piping or install plastic piping in areas designed for metal pipe applications.
- Cast iron pipe is heavy, so proper lifting procedures must be used to eliminate back injury.
- Always have two people to carry long pipes, and be sure that both people are on the same side of the pipe when carrying.
- Always read the MSDS before using joint compound and cutting oil. Know all safety precautions and medical treatments in case of accidental consumption of oil or compound.

• When handling and storing flexible gas tubing, always protect the outer sheathing from being nicked by sharp objects. Keep ends of tubing covered so debris cannot enter the tubing.

Plumbing Tools

Before any repairs can be made to a plumbing system, some basic tools will be required. These tools can be purchased in most home improvement stores, and they range in price from a few to a couple hundred dollars. However, like most trades and jobs, quality tools do have a tendency to perform better and last longer.

• Locking tape measure—used to make measurements for plumbing fixtures (e.g., sinks) (Figure 9-1).
• Angled jaw pliers—used to make adjustments for valves and fixtures (Figure 9-2).
• Fourteen-inch pipe wrench—used to loosen and tighten metal pipe and fittings (Figure 9-3).

Note: Typically when tightening fittings on a steel pipe, a second pipe wrench is required: one to hold the pipe and one to turn the fitting.

• Hacksaw—used for cutting metal and plastic pipes (Figure 9-4).

Figure 9-1: **Tape measure**

Figure 9-2: **Angled jaw pliers**

Figure 9-3: **Pipe wrench**

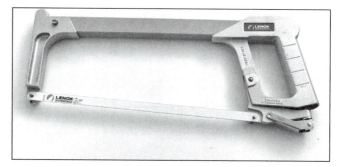

Figure 9-4: **Hacksaw**

- Plunger—used to unclog (unstop) a fixture (sink, toilet, etc.) (Figure 9-5).
- Six- and/or eight-inch adjustable wrench—used to loosen and tighten valves and fittings (Figure 9-6).
- Closet auger—used to unstop toilet fixture by removing the foreign object that is causing the problem (Figure 9-7).

Figure 9-5: Plunger

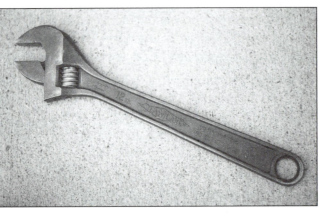

Figure 9-6: **Adjustable wrench**

Figure 9-7: **Closet auger**

Figure 9-8: **Electric spin drain cleaner**

- Electric spin drain cleaner—used to unstop kitchen sinks and lavatory. Typically a ¼-inch cable is used for kitchen sinks (Figure 9-8).
- Torpedo level—used to ensure that fixtures are installed correctly (level) and waste piping systems have the necessary slope (Figure 9-9).
- Midget copper cutter—used for cutting copper tubing (Figure 9-10a).
- Plastic pipe cutter—used for cutting plastic tubing (Figure 9-10b).
- Toolbox—used to store all the plumbing tools (Figure 9-11).
- Miter box—used to cut various precise angles (Figure 9-12).
- Reamer—used to remove burs from piping (Figure 9-13).

Figure 9-9: **Torpedo level**

Figure 9-10a: **Copper tubing cutter**

Figure 9-10b: **Plastic pipe cutter**

Figure 9-11: **Person carrying toolbox** (Image copyright mates, 2009. Used under license from Shutterstock.com)

Figure 9-12: **Miter box**

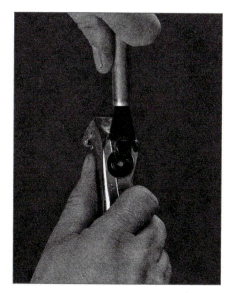

Figure 9-13: **Reamer**

Piping

Pipe is used to bring potable water and gas into a building and to allow sewage and wastewater to drain from a building.

Types of plastic pipes include:

- Polyvinyl chloride (PVC) pipe—available in various sizes and thicknesses (schedules). It is most commonly used in wastewater applications (see Figure 9-14).

Figure 9-14: **PVC pipe**

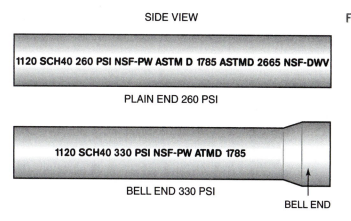

SIDE VIEW

1120 SCH40 260 PSI NSF-PW ASTM D 1785 ASTMD 2665 NSF-DWV

PLAIN END 260 PSI

1120 SCH40 330 PSI NSF-PW ATMD 1785

BELL END 330 PSI

BELL END

- Chlorinated polyvinyl chloride (CPVC) pipe—a yellowish-white flexible pipe used in water distribution systems. In residential facilities, this pipe is identified as SDR 11 (Figure 9-15).
- Acrylonitrile butadiene styrene (ABS) pipe—a black plastic pipe typically used in wastewater venting systems (Figure 9-16).
- PE pipe—used only for exterior water services. This type of pipe is not well suited for use in places where it receives direct sunlight (Figure 9-17).

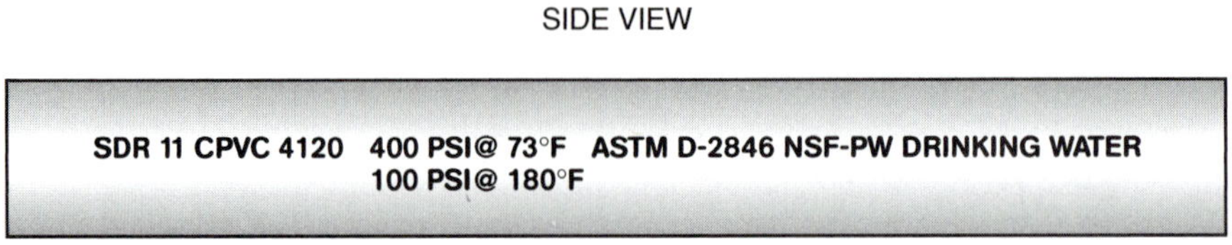

SIDE VIEW

SDR 11 CPVC 4120   400 PSI@ 73°F   ASTM D-2846 NSF-PW DRINKING WATER
100 PSI@ 180°F

Figure 9-15: CPVC pipe

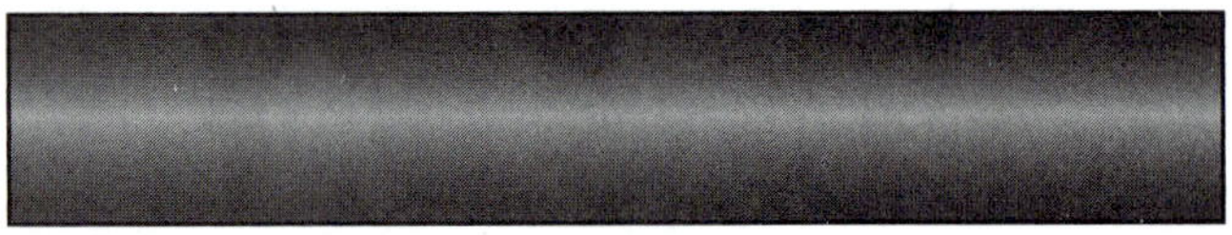

SIDE VIEW

Figure 9-16: ABS pipe

ASTM F-628 ABS DWV NOT FOR PRESSURE

Figure 9-17: PE pipe

SIDE VIEW

← UNROLL TUBING

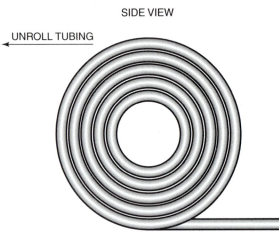

Figure 9-18: PEX pipe

- Cross-linked polyethylene (PEX) pipe—typically a whitish-colored pipe used for water distribution systems (Figure 9-18).

  Types of metal pipes include:

- Cast iron pipe—often used in residential wastewater systems because it is a much quieter pipe for draining applications than plastic (see Figure 9-19).
- Galvanized and black steel pipe—used for residential gas supply piping (Figure 9-20).

  Currently, three types of copper tubings—types K, L, and M—are used for domestic water (see Figure 9-21).

- Type K copper tubing—a thick wall tubing used primarily for underground water service. It can be purchased either in soft rolls or in 20-foot stock lengths.
- Type L copper tubing—a thin wall tubing used in above-ground installations. It can be purchased either in soft rolls or in 20-foot stock lengths.
- Type M copper tubing—a hard thin wall tubing that is sold only in 20-foot stock lengths. It should be used only for indoor applications in which the tubing can be easily accessed (Figures 9-22 and 9-23).

NO HUB ASTM A888

SERVICE WEIGHT ASTM A74

Figure 9-19: Cast iron pipe

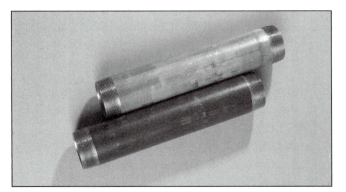

Figure 9-20: Black pipe

Copper Type	Potable Water	DWV	Underground	Available in Roll
DWV		✓		
Type M	✓	✓		
Type L	✓	✓	✓	✓
Type K	✓	✓	✓	✓

Figure 9-21: **Types and basic uses of copper pipe**

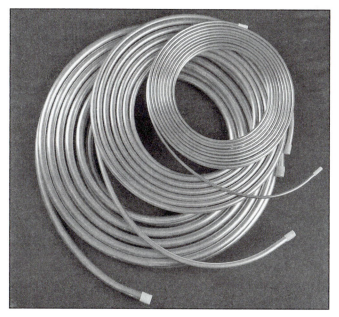

Figure 9-22: **Copper tubing**

Figure 9-23: **Unrolling copper tubing**

Pipe Fittings

Regardless of the type of piping materials used in a plumbing system, the basic components (fittings) used to connect the piping together into a useable system are the same (just made from different materials). Some of the more common fittings used are as follows (Figure 9-24):

- Ninety-degree elbow—used to change the direction of a pipe by 90°.
- Forty-five-degree elbow—used to change the direction of a pipe by 45°. It is typically used in pairs to offset a section of pipe around an obstacle.
- Tee—used to create a branch line.
- Cap—used to terminate a section of pipe. A cap is a fitting that fits around the outside of a pipe.
- Coupling—used to connect two pieces of pipe having the same diameter.
- Plug—used to terminate a section of pipe. A plug is a fitting that screws into the inside of a section of pipe.
- Union—used to connect two pieces of pipe together, while allowing for them to be disconnected without being cut.
- Reducer—used to reduce a section of pipe to a smaller diameter.
- Adapter—used to connect pipes of different sizes or materials together.

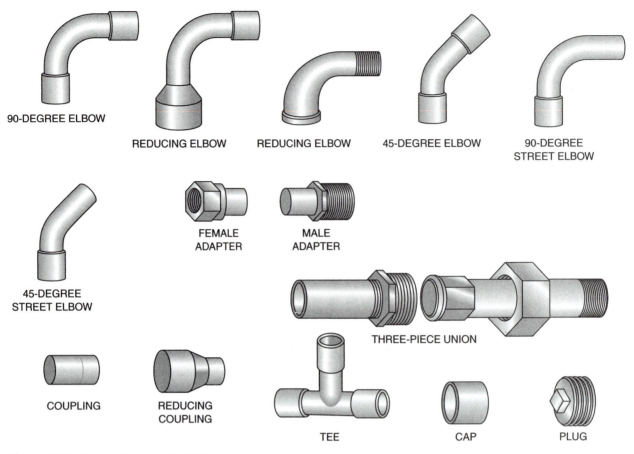

Figure 9-24: **Commonly used pipe fittings**

Piping Support and Hangers

In order for a plumbing system to work properly it must be correctly sized and then properly supported. Typically, the type of support used for a particular section of piping depends on the kind of pipe used, the position of the pipe, and the purpose of that pipe. In general, five common types of plumbing supports are typically used for residential settings (see Figures 9-25 and 9-26).

1. Clamp—used to anchor a section of pipe to an architectural support.
2. Hanger strap—made from a flexible material containing predrilled holes that are equally spaced throughout the length of the hanger.
3. U-bolts—typically used to secure a section of pipe to a structural steel member.
4. Wire hanger—similar to a U-bolt. This type of pipe hanger is typically used on smaller lighter pipe.
5. Pipe staples—typically installed using a staple gun. This type of hanger is used to secure piping to an architectural support.

Measuring and Cutting Pipes

To correctly measure copper tubing, always measure it from the centers of two fittings or from the end of the pipe run to the center of a fitting (Figures 9-27 and 9-28).

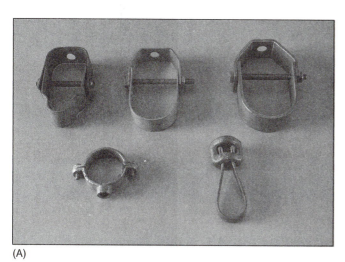

(A)

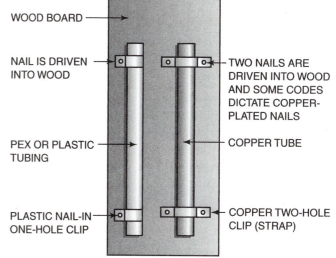

WOOD BOARD

NAIL IS DRIVEN INTO WOOD

TWO NAILS ARE DRIVEN INTO WOOD AND SOME CODES DICTATE COPPER-PLATED NAILS

PEX OR PLASTIC TUBING

COPPER TUBE

PLASTIC NAIL-IN ONE-HOLE CLIP

COPPER TWO-HOLE CLIP (STRAP)

(A)

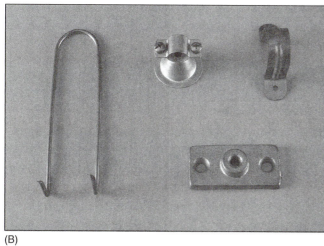

(B)

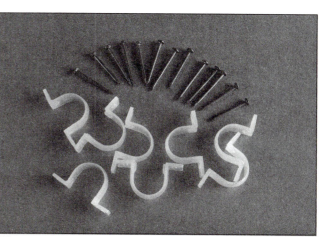

(B)

(C)

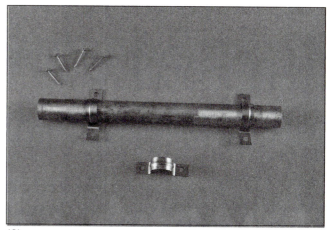

(C)

Figure 9-25: **Various types of support and hangers**

Figure 9-26: **Clips and straps**

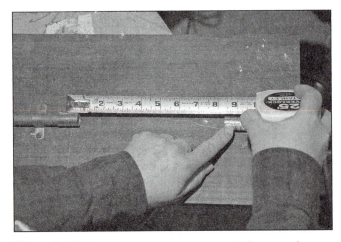

Figure 9-27: **Center-to-center measurement of copper pipe**

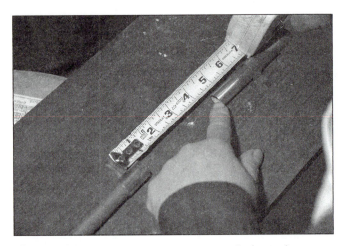

Figure 9-28: **Center-to-end measurement of copper pipe**

Figure 9-29: **Miter box**

Although copper tubing can be cut using either a hacksaw or a tubing cutter, it is recommended that when possible a tubing cutter should be used. A tubing cutter will ensure that the end of the copper tubing is square. However, if copper tubing is cut with a hacksaw, it is recommended that a miter box or fixture be used (Figure 9-29).

Using a Tubing Cutter

To cut copper tubing using a tubing cutter, do the following:

1. Use a locking tape measure to measure and mark the copper tubing to the desired length (Figure 9-30).
2. Place copper tubing into fixture (Figure 9-31).
3. Unwind the pipe cutter so that the pipe can be inserted.
4. Once the cutter has been positioned so that the wheel sits on the pipe where it needs to be cut, tighten the pipe cutter.
5. Tighten the cutter ¼ turn and then rotate it around the pipe before tightening it again.
6. After tubing has been completely cut, remove any burs inside the tubing with a round file (Figure 9-34).

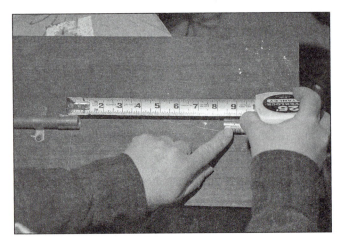

Figure 9-30: **Measuring copper tubing**

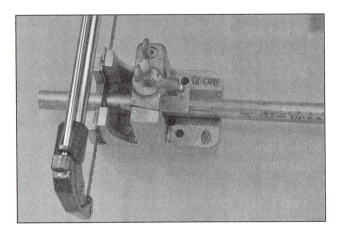

Figure 9-31: **Copper tubing in fixture**

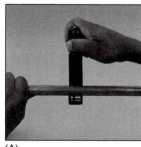
(A)

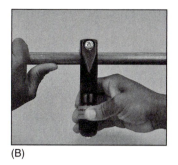
(B)

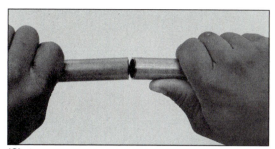
(C)

Figure 9-32: **(A)–(C) The proper procedure for using a tubing cutter**

Figure 9-33: **Cutting copper tubing using a hacksaw**

Using a Hacksaw

To cut copper tubing with a hacksaw do the following:

1. Use a locking tape measure to measure and mark the copper tubing to the desired length (Figure 9-30).
2. Place copper tubing into fixture (Figure 9-31).
3. Start the cutting process by pulling and pushing the hacksaw back and forth (Figure 9-33).
4. After tubing has been completely cut, remove any burs inside the tubing with a round file (Figure 9-34).

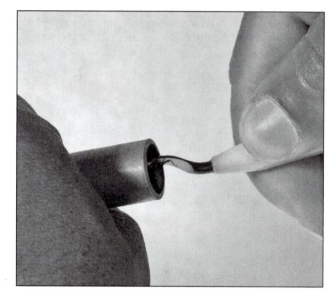

Figure 9-34: **Remove burs from copper tubing**

Unclogging Pipes

One of the most frequent types of calls that a facilities maintenance technician will receive is regarding either a clogged lavatory or toilet. The following situations can usually be resolved without the assistance of a professional by following a few basic steps.

Using a Plunger to Unclog a Kitchen Sink or Lavatory

A kitchen sink or lavatory can be unclogged using a plunger. Care should be taken when unclogging a sink that has a disposal attached using a plunger. The pressure generated by the plunger can damage the flange gasket or tailpiece of the disposal.

Figure 9-35: Sink strainer (Image copyright Joe Gough, 2009. Used under license from Shutterstock.com)

The following steps should be used when unclogging a kitchen sink or lavatory using a plunger:

1. Remove the sink's strainer or plug (Figure 9-35).
2. If the sink does not already have water in it, fill the sink to the halfway mark.
3. Place the plunger's tuber globe over the drain carefully making sure that the entire drain opening is covered by the plunger (Figure 9-36).
4. Using forceful strokes plunge the sink drain at least 15 times before removing the plunger to see if the sink will drain.
5. If the sink is still clogged, repeat Steps 3 and 4.
6. Once the sink has been unclogged, run hot water down the drain for several minutes. This will help clean out anything remaining in the system.

Figure 9-36: Positioning plunger over drain opening

Using a Drain Cleaner to Unclog a Kitchen Sink or Lavatory

When using liquid drain cleaner to unclog a kitchen sink or lavatory, do the following steps:

1. Before using a liquid drain cleaner, carefully read and follow all instructions located on the package.
2. When handling liquid drain cleaner, care should be taken *not* to spill it on any of the surrounds or on human skin. If you do come in contact with the drain cleaner, refer to the packaging or consult a doctor immediately.

A Green Approach to Unclogging a Kitchen Sink or Lavatory

In some cases, a clog can be removed using a more natural, environmental-friendly approach. This approach uses baking soda and vinegar. When baking soda is mixed with vinegar, a chemical reaction is produced in which carbon dioxide is released.

1. Remove any contents from the sink.
2. Pour approximately ¼ cup of baking soda into the drain opening.
3. Pour approximately one cup of vinegar into the drain opening.
4. Cover the drain opening for 15 minutes with a lid.
5. Uncover the drain opening and test the drain.
6. If the drain is still clogged, repeat Steps 2–5.
7. Rinse the drain with hot water.
   Note: A plunger may be required to help loosen the clog.

Using a Plunger to Unclog a Toilet

Do the following when using a plunger to unclog a toilet:

Figure 9-37: Plunger covering drain opening

1. Insert the plunger into the toilet bowl, fully covering the drain opening (Figure 9-37).

2. Pushing down and pulling up the handle of the plunger, vigorously plunge the toilet 15–20 times.
3. Lift the plunger from the drain opening to see if the toilet drains.
4. If the toilet is still clogged, repeat Steps 1–3.

Using a Drain Cleaner to Unclog a Toilet

A liquid drain cleaner should be used on a toilet only as a last resort because of its possible impact on the environment. However, if you use a drain cleaner to unclog a toilet, follow these steps:

1. Before using a liquid drain cleaner, carefully read and follow all instructions located on the package. Also, before using a liquid drain cleaner on a toilet, make sure that the cleaner is safe to use with porcelain.
2. When handling liquid drain cleaner, care should be taken *not* to spill it on any of the surrounds or on human skin. If you do come in contact with the drain cleaner, refer to the packaging or consult a doctor immediately.
3. After the toilet unclogs, flush the toilet several times to check the flow and remove the drain cleaner.

Cleaning a Slow-Draining Lavatory

If a lavatory is slow draining, try running hot water into the sink for about 10 minutes. Running hot water can sometimes loosen contaminates that might be causing the sink to drain slowly. If using hot water does not completely open the drain, try using an environmentally safe cleaner (baking soda and vinegar—see the previous section on using a more natural approach).

While unclogging a lavatory with a cleanout, do the following:

1. Place a bucket under the sink to catch wastewater.
2. Using a wrench, open the cleanout (Figure 9-38).
3. Use a screwdriver to probe around and pull out the clog (Figure 9-39).
4. Replace the cleanout cover and gasket.
5. Run hot water into the sink for 10 minutes.

> *If the clog is not removed using a screwdriver, then a sink snake can be used to unclog the drain.*

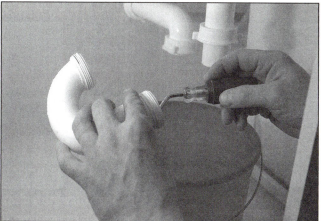

Figure 9-38: **Removing a sink trap using angled jaw pliers**   Figure 9-39: **Probing with a screwdriver**

Figure 9-40: **Place a bucket under a sink**

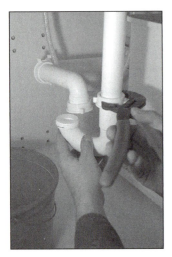

Figure 9-43: **Replacing a sink trap using angled jaw pliers**

Figure 9-44: **Using a plunger in bathtub to unclog drain**

Using a Sink Snake to Unclog a Lavatory

Another way to try unclogging a lavatory is with a cleanout and a sink snake. This can be accomplished as follows:

1. Place a bucket under the sink to catch wastewater (Figure 9-40).
2. Using a wrench, open the cleanout (Figure 9-41).
3. Carefully push the snake into the cleanout opening, moving the tape back and forth to help it navigate the drain pipe (Figure 9-42).
4. When the snake reaches the clog, twist and push it until the clog is removed.
5. Replace the cleanout cover and gasket (Figure 9-43).
6. Run hot water into the sink for 10 minutes.

Figure 9-41: **Removing a sink trap using angled jaw pliers to unclog a drain using a sink snake**

Figure 9-42: **Using a sink snake**

Using a Plunger to Unclog a Bathtub

A plunger can also be used to unclog a bathtub drain:

1. Place the plunger's tuber globe over the drain, carefully making sure that the entire drain opening is covered by the plunger (Figure 9-44).
2. Using forceful strokes, plunge the sink drain at least 15 times before removing the plunger to see if the sink will drain.
3. If the sink is still clogged, repeat Steps 3 and 4.
4. Once the sink has been unclogged, run hot water down the drain for several minutes. This will help clean out anything remaining in the system.

Using a Drain Cleaner to Unclog a Bathtub

When using drain cleaners to unclog a bathtub drain, the following steps should be taken:

1. Before using a liquid drain cleaner, carefully read and follow all instructions located on the package.
2. When handling liquid drain cleaner, care should be taken *not* to spill it on any of the surrounds or on human skin. If you do come in contact with the drain cleaner, refer to the packaging or consult a doctor immediately.

A Green Approach to Unclogging a Bathtub

Using environmental-friendly cleaners to unclog a bathtub drain is another option:

1. Pour approximately ¼ cup of baking soda into the drain opening.
2. Pour approximately one cup of vinegar into the drain opening.
3. Cover the drain opening for 15 minutes with a lid.
4. Uncover the drain opening and test the drain.
5. If the drain is still clogged, repeat Steps 2–5.
6. Rinse the drain with hot water.

Using a Toilet Auger to Unclog a Toilet

1. Loosen the setscrews on the auger and push the cable into the drain, moving it back and forth, until the clog is reached (Figures 9-45 and 9-46).
2. Tighten the set screws on the toilet auger.
3. While pushing on the toilet auger, crank on the auger clockwise until the obstruction is cleaner.
4. Remove the toilet auger from toilet.
5. Test the flow of the toilet by flushing it several times.

Figure 9-45: **Set screws on toilet auger**

Figure 9-46: **Pushing auger cable into drain opening**

Caulking

Caulk is primarily used to seal cracks caused by mating parts (recall the discussion in Chapter 4, page 53). However, in plumbing, caulk is used mostly for waterproofing, for example, the crack produced by a sink and the mating wall.

Applying Caulk

1. Remove all dust, dirt, and any other grime from around area to apply caulk.
2. If water or a solvent is used to clean the area to be caulked, make sure the area is dry.
3. Insert caulk tube into caulk gun (Figure 9-47).
4. Using a knife, remove the tip of the caulk, cutting away as little as possible. The amount of tip removed will affect the size of the bead of caulk that the caulk gun

applies. Make sure the tube of caulk does not contain a second seal. To check, stick a nail into the hole made and cut off the tip (Figure 9-48).

5. Hold the caulk at about a 5–10° angle, in the direction of travel. The direction of travel is the direction in which you will be pulling the caulk gun (Figure 9-49).

6. After applying the caulk to the crack, use your finger to lightly work the caulk into the crack. (*Note*: Keeping your finger wet will help ensure a smoother caulk bead (see Figure 9-50).)

7. Use a wet (moist) towel to wipe away all excess caulk.

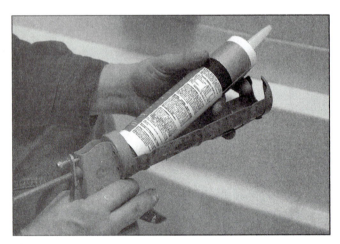

Figure 9-47: **Caulk tube and gun**

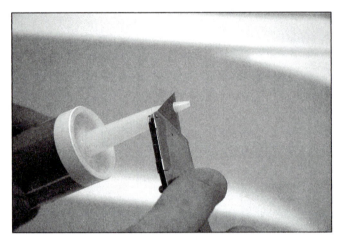

Figure 9-48: **Caulk tube with cut angle on end**

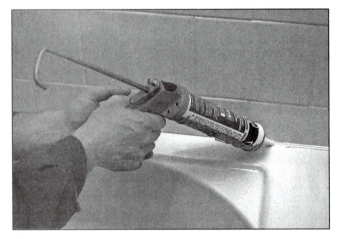

Figure 9-49: **Applying caulking**

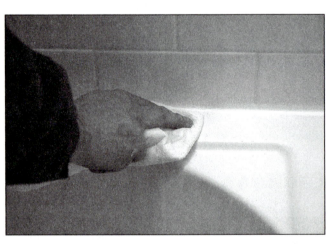

Figure 9-50: **Smoothing caulking with your finger**

Selecting a Caulk

The four most commonly used caulk today are as follows:

1. Acrylic latex caulk—fast-drying all-purpose caulk designed for dry applications. It is well suited for painting and is easily cleaned up using soap and water.

2. Vinyl latex caulk—designed for wet applications and a good choice for showers and tubs.

3. Silicone caulk—long-lasting, mildew-resistant, watertight adhesive caulk that does not discolor. Typically, silicone caulk cannot be painted or cleaned using soap and water.
4. Butyl rubber caulk—designed for use in outdoor applications and can be used to fill large joints or cracks.

Plumber's Putty

Plumber's putty is a pliable sealant used to seal pipe joints and fittings that can be used for quick emergency repairs (Figure 9-51). To apply plumber's putty, follow these steps:

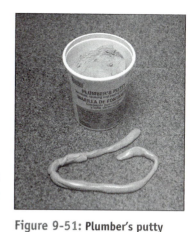

Figure 9-51: **Plumber's putty**

1. Clean the surface of the fitting or fixture where the putty is to be applied.
2. Apply a bead of plumber's putty to the mating part. In the case of a sink drain, apply the putty under the rim of the flange and place the drain into the drain outlet (see Figure 9-52).
3. Tighten the fitting, causing the plumber's putty to spread.

Assembling Pipes Using PVC Cement

See page 302 for a procedure for assembling plastic pipe.

Figure 9-52: **Applying caulk on drain outlet**

Soldering a Copper Fitting onto a Piece of Copper Pipe

See pages 303–306 for procedures for soldering.

National and Local Plumbing Codes

Because local plumbing codes vary from location to location in the United States, it is recommended that you review all local plumbing codes before attempting any plumbing installation and/or repair. The national plumbing code can be accessed at http://emarketing.delmarlearning.com/downloads/StayCurrent_02_05.pdf. Local plumbing codes can be obtained by contacting the state code and administration department.

Adjusting the Temperature of a Water Heater

The U.S. Department of Energy states that a temperature at the tap of 90°F is adequate for most household applications. However, to obtain this temperature the water leaving the tank should be no less than 130°F. This temperature prevents dangerous bacterial growth.

When using a propane torch, always keep a fire extinguisher handy. Always read and carefully follow all directions on the propane torch. Always carefully read the instructions on the fire extinguisher before using the propane torch.

Testing the Water Temperature

1. At one of the sinks in the structure serviced by the hot water tank, turn on the hot water. Be sure to leave the water on for a few moments, making sure the tap water has reached its full temperature.
2. Fill a glass with the hot water.
3. Place a thermometer in the glass of hot water.
4. Remove the thermometer and read the recorded temperature.

Adjusting the Temperature of an Electric Water Heater

The water temperature of an electric water heater is usually controlled by a thermostat located on the side of the water heater. To reset or set the temperature of an electric water heater, simply set the thermostat to the desired temperature.

In some cases, the thermostat may be concealed; if this is the case, do the following:

1. Turn off power to the water heater.
2. Using a screwdriver, remove the covering, revealing the thermostat.
3. Adjust the thermostat to the desired temperature.

Some electric water heaters have multiple thermostats. If your water heater has multiple thermostats, both thermostats must be set on the same temperature (Figure 9-53).

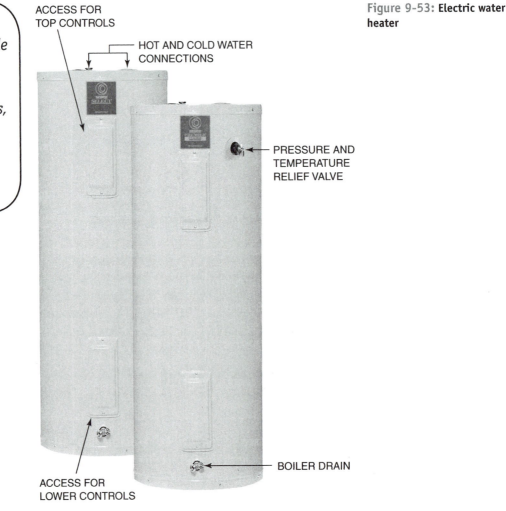

Figure 9-53: Electric water heater

ACCESS FOR TOP CONTROLS

HOT AND COLD WATER CONNECTIONS

PRESSURE AND TEMPERATURE RELIEF VALVE

BOILER DRAIN

ACCESS FOR LOWER CONTROLS

4. Replace the cover.
5. Turn on electricity.

Adjusting the Temperature of a Gas Water Heater

Most gas water heaters have the thermostat located on the side of the water heater; it can be adjusted by simply setting the thermostat to the desired temperature (Figure 9-54). In a few cases, the thermostat may be concealed. If this is the case, then the following steps should be followed:

1. Using a screwdriver, remove the thermostat cover.
2. Adjust the thermostat to the desired temperature.
3. Replace the thermostat cover.

Basic Water Heater Replacement

When you are replacing a water heater, stick with what is already there. If you have an electric water heater, replace it with an electric water heater unless you are willing to run gas lines and exhaust vents. If you are replacing a gas water heater, replace it with a gas water heater unless you are willing and able to install new electrical service.

1. Turn off the gas or electricity to the heater.
2. Drain the heater. Opening a hot water faucet will let air into the system (Figure 9-55).
3. Disconnect the water lines (Figure 9-56).
4. Move the new heater to its location by "walking" it or using an appliance cart, dolly, or hand truck. Position the new heater so the piping—particularly a gas vent pipe—will reach most easily.
5. If you removed the shutoff valve, replace it (Figure 9-57).
6. Install the water lines and pressure-relief line (Figure 9-58).

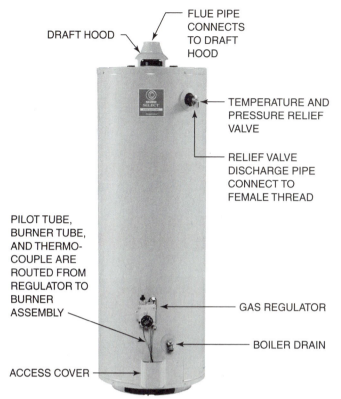

Figure 9-54: **Gas water heater**

Follow manufacturer's instructions for a specific step-by-step list on how to install specific water heaters such as gas or electric heater.

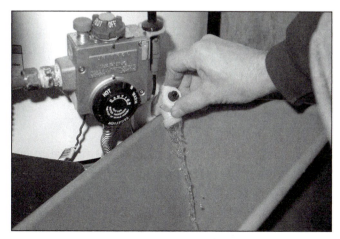

Figure 9-55: **Drain the water heater**

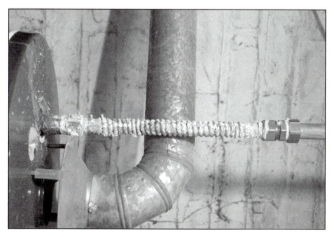

Figure 9-56: **Disconnect water lines**

Figure 9-57: **Water heater shutoff valve**

Figure 9-58: **Install water lines**

Figure 9-59: **Flapper assembly package**

Plumbing Leaks

Plumbing leaks can cause thousands of dollars worth of damage if left unchecked. To check a plumbing system for leaks, the facilities maintenance technician should periodically check under sinks, around toilets, and in crawl spaces. Once a leak has been detected, correct the problem or consult a plumber.

One common source of plumbing leaks is the flapper assembly in a toilet. If the flapper is leaking, then the defective flapper should be replaced. A new flapper can be obtained at your local plumbing supply. To replace the flapper in a toilet, consult the instructions on the flapper package (Figure 9-59).

Shower Seals

If the bathroom floor is getting wet after a shower, then possibly the shower door seal should be replaced. Shower seals can be obtained at your local plumbing supply. To replace the shower seal, consult the instructions located on the packaging of the shower seal.

Repairing a Faucet

Several types, makes, and models of sink faucets are available on the market today. The most common types of sink faucets available are single and double handled. Faucets manufactured today are either a compression or a washerless type.

Compression faucets regulate the flow of water by applying pressure onto a rubber washer located within the faucet, while washerless faucets use a cartridge, ball, or disk to control the flow of water (Figure 9-60).

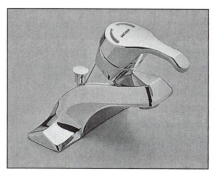

(A)

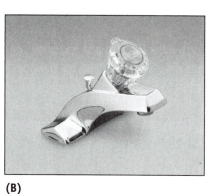

(B)

(C)

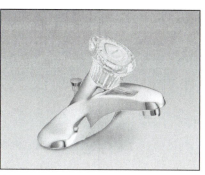

(D)

Figure 9-60: **Various faucet types**

Before repairs are made to a faucet, always close the sink drain to prevent parts of the faucet from going down the drain. In addition, never use a pipe wrench on a polished fixture without first applying tape to the pipe wrench jaws. Applying tape to the pipe wrench jaws will prevent the wrench jaws from damaging the fixture's finish. Finally, shut off all water supplies to the faucet.

Because a number of types and styles of faucets are available on the market today, facilities maintenance technicians should consult the manufacturer's documentation at the local plumbing supply when repairing a sink faucet. Only manufacturer-approved parts should be used. In addition, always read and follow the faucet manufacturer's instructions.

Procedures

Joining Plastic Pipe

- Mark the pipe at the appropriate point with a pencil.

Ⓐ Cut the pipe using either a hacksaw or a tubing shear.

- Remove the burrs from both the inside and the outside of the pipe.

- Apply primer, if required, to both the male and female portions of the joint.

Ⓑ Apply cement to both the male and female portions of the joint.

Ⓒ Insert the male end of the fitting into the female end and rotate the pipe ¼ turn.

- Hold the pipe and fitting together for approximately 1 minute to prevent the pipe from pulling out of the fitting.

FROM EXPERIENCE

When working with plastic pipes, always try to dry-fit the piping arrangement before cementing. Once a joint is cemented, you don't get a second chance!

CAUTION

Follow all safety guidelines provided on the primer and cement containers. Plastic primers and cements should only be used in well-ventilated areas as the fumes from these chemicals are hazardous to your health.

Ⓐ

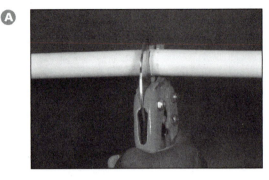

Photo by Bill Johnson.

Ⓑ

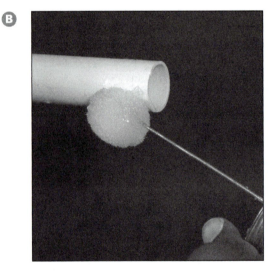

Photo by Bill Johnson.

Ⓒ

Photo by Bill Johnson.

Procedures

Soldering

- Properly cut and ream the sections to be joined. Refer to the cutting and reaming procedures.

- Using sand cloth or steel wool and the correct size pipe brush, clean the male and female portions of the joint being soldered.

(A) Using a flux brush, apply flux to the male portion of the joint.

- Insert the male portion of the joint into the female end.

- Before connecting the acetylene regulator to the tank, quickly open and close the stem on the tank using the refrigeration-service wrench. This will blow any particulate matter from the opening of the tank.

(B) Mount the acetylene regulator and torch kit to the tank, making sure that the connections are tight.

(C) Making certain that the valve on the torch handle is closed, open the stem valve on the acetylene tank ½ to 1 turn using the service wrench. Flip the ratchet on the service wrench so the tank can be closed quickly in the event of an emergency.

(A)

Photo by Bill Johnson.

(B)

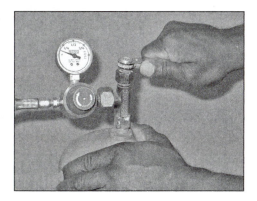

Photo by Bill Johnson.

(C)

Photo by Bill Johnson.

Procedures   Soldering (Continued)

D Using a soap bubble solution, leak-check the regulator, hose, and torch assembly, making certain that no acetylene is leaking from the kit. Tighten any leaking connections.

E Set the regulator on the tank to the middle range to start. It can always be adjusted later on.

• Open the valve on the torch handle most of the way and ignite the fuel with the striker.

• Adjust the flame using the regulator on the tank until the desired flame intensity is obtained. A proper flame will have a bright blue inner cone and a lighter blue outer cone. The hottest portion of the torch flame is the tip of the inner cone.

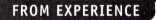

FROM EXPERIENCE

The size of the flame and the amount of heat generated by the torch are directly related to the size of the torch tip used. Larger size pipes will require more heat and will therefore require a larger torch tip. Residential applications typically require the use of an A-3 or A-5 tip for soft soldering.

D

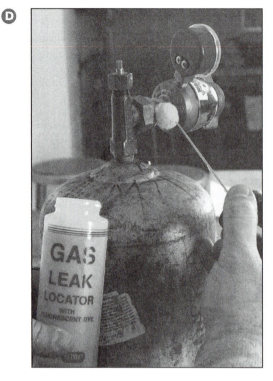

Photo by Bill Johnson.

E

Photo by Bill Johnson.

F Apply heat to the joint, placing the tip of the inner cone on the surface of the joint. The flux will begin to melt and flow into the joint. Be sure to keep the flame moving to heat the entire joint. After heating the joint for a short period of time, apply the solder to the joint. The solder should be melted by the heat of the copper, not by the heat of the torch flame. If the solder does not immediately begin to flow, remove the solder and heat the joint a little more.

F

Photo by Bill Johnson.

FROM EXPERIENCE

Overheating the joint prior to introducing solder will cause the solder to run off the joint instead of sticking to it. If this should occur, the pipes should be recleaned and fluxed to ensure a good solder joint.

Procedures

Soldering (Continued)

G When the solder begins to flow, feed enough solder into the joint to completely fill it. You should use no more solder than is needed to fill the joint. Using too much solder will result in buildup on the inside of the pipe. Using too little solder will result in leaks.

- To extinguish the torch, simply close the valve on the torch handle. When the torch is no longer needed, close the stem on the acetylene tank and bleed off any acetylene from the hoses by opening the valve on the torch handle.

- While the joint is still hot, it is a good practice to wipe the joint with a rag. This removes excess solder and improves the appearance of the solder joint. Applying a small amount of flux to the joint while the pipe is hot also helps clean the joint.

G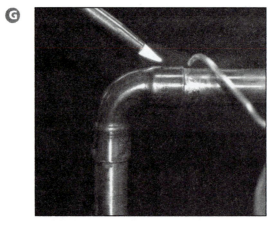

Photo by Bill Johnson.

FROM EXPERIENCE

The amount of soft solder used on any given joint should not exceed a length equal to twice the diameter of the pipe.

FROM EXPERIENCE

Molten solder will flow toward the source of heat. To help ensure that the solder flows into the joint, apply heat to the base of the female portion of the joint.

Review Questions

1. List three types of plastic pipes commonly used in plumbing.

2. List the two main types of metal pipes used in residential and commercial plumbing.

3. List the three types of copper tubings used in residential and commercial plumbing.

4. List the steps for cutting copper tubing with a hacksaw.

5. List the steps for using a plunger to unclog a kitchen sink.

6. List the steps for unclogging a kitchen sink using a natural approach.

7. List the steps for unclogging a lavatory with a cleanout.

8. List the steps for using a drain cleaner to unclog a bathtub.

Name: _____

Date: _____

Plumbing

Tool Identification

Upon completion of this job sheet, you should be able to identify and know the use of any tools used for plumbing tasks at your facilities.

1 Inspect your facility. Identify and list the use of any tools used for plumbing tasks.

Tool	Use

2 Where are the tools stored?

3 Are instructions available for using the tools correctly? If not, do you think they should be?

Instructor's Response:

_____

_____

_____

Name: _____

Date: _____

Plumbing

Pipe Identification

Upon completion of this job sheet, you should be able to identify and know the use of pipes used at your facility.

Pipe Type	Use

Inspect your facility. Identify and list the uses of various types of pipes.

Instructor's Response:

_____
_____
_____

Name: _____

Date: _____

Plumbing

Pipe Identification

Upon completion of this job sheet, you should be able to identify the correct piping material for a particular task.

Match the pipe material to the application.

A. Copper   ____ **1** Gas pipe connected to a central heating unit.

B. Cast iron   ____ **2** Piping connecting a submergible potable pump to a storage tank.

C. Black steel   ____ **3** Primarily used for drain, waste vent application (black in color).

D. Galvanized steel   ____ **4** Yellowish pipe used for potable hot water.

E. PVC   ____ **5** Sold in hard and soft versions and used in most plumbing applications and used by the HVAC industry.

F. ABS   ____ **6** Used primarily for commercial DWV and storm drainage systems.

G. Polyethylene   ____ **7** White flexible plastic-type tubing used for potable water and heating systems.

H. PEX   ____ **8** Black steel pipe lined with a protective coating that allows it to be used for potable water applications.

I. CPVC   ____ **9** Used for cold water applications and DWV.

Instructor's Response:

_____

_____

_____

Name: _____

Date: _____

Plumbing

Fitting Identification

Upon completion of this job sheet, you should be able to identify fittings used in plumbing correctly.

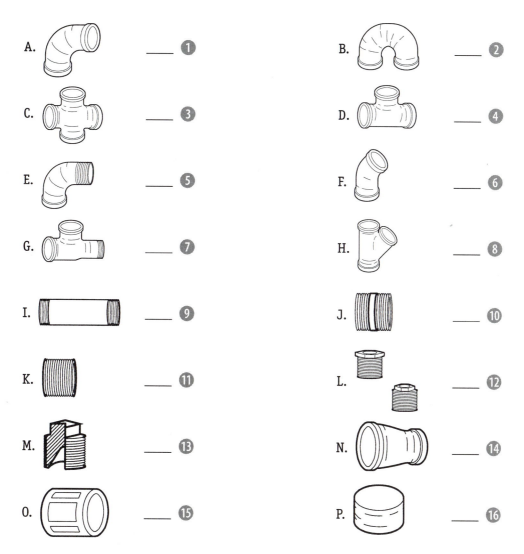

For each fitting, write the name under the picture.

Instructor's Response:

_____

_____

_____

Chapter 10

Heating, Ventilation, and Air-Conditioning Systems

OBJECTIVES

By the end of this chapter, you will be able to:

Knowledge-Based

- Explain the importance of properly installing air filters
- List the three common types of furnaces used

Skill-Based

- Perform general maintenance procedures including:
- General maintenance on a furnace
- Tightening and/or replacing belts
- Adjusting and/or replacing pulleys
- Replacing filters on HVAC units
- Maintain the heat source on gas-fired furnaces.
- Perform general maintenance of hot water or steam boilers.
- Perform general maintenance of an oil burner and boiler.
- Repair and replace electrical devices, zone valves, and circulator pumps.
- Light a standing pilot.
- Perform general maintenance of a chilled water system.
- Clean coils.
- Lubricate motors.
- Follow systematic diagnostic and troubleshooting practices.
- Maintain and service condensate systems.
- Replace through-the-wall air conditioners.

Glossary of Terms

Carbon monoxide a poisonous, colorless, odorless, tasteless gas generated by incomplete combustion

Carbon dioxide by-product of natural gas combustion that is harmful and can even cause death

Introduction

The primary function of heating, ventilation, and air-conditioning (HVAC) systems is to provide healthy and comfortable interior conditions for occupants. This chapter on HVAC will provide a practical description and overview of the various HVAC equipment and systems used in both residential and commercial buildings.

HVAC Safety

As stated earlier, it is the responsibility of everyone to ensure that a worksite is safe. When working on a job site, one should behave in a safe and professional manner. Accidents on the job often result from carelessness. It is very important for workers to be aware of their surroundings at all times and to evaluate the immediate area for possible safety hazards. The following are a few of the safety rules that should be followed when working with or around HVAC:

- Never work on electric circuits while standing on a wet floor or when not wearing rubber-soled boots. Shocks received when standing in a wet location are quite often deadly as the current passes through the heart, causing it to stop pumping. Water and electricity do not mix! Stay dry and stay safe.
- Should a wire come loose from inside an air-conditioning system and come in contact with the casing of the equipment, electric shock can result by simply touching the surface of the unit.
- Never leave tools or other materials on the top platform of the ladder. If the ladder is moved by another individual, the object can fall, causing injury.
- Never horseplay while at work.
- Always be aware of your surroundings and potential hazards.
- Dress properly for work wearing long pants, long sleeved shirts, and work boots.
- Remove metallic jewelry, as it is a good conductor of heat and electricity.
- Use safety glasses, ear plugs, and gloves for additional protection from dangerous conditions on the job site.
- Power tools and equipment should be grounded to protect against electric shock.
- Electric shock occurs when the body becomes part of an electric circuit.
- Always de-energize electric circuits before working on them.
- Ground wires and prongs should never be cut or disconnected.
- The GFI de-energizes a circuit when a current leak to ground is sensed.
- Fire extinguishers are classified by the types of fires they are designed to be used on.
- Fire extinguisher use: Pull, Aim, Squeeze, Sweep (PASS).
- Always use tools for the tasks they are intended to perform.
- Handle and use chemicals according to the manufacturer's directions.
- Be prepared for injuries on the job and have a first aid kit handy.
- OSHA, NFPA, and ANSI help ensure safety in the work place.

Perform General Furnace Maintenance

There are three common types of furnaces: gas, electric, and oil. Because each furnace manufacturer has a different set of specifications, the following are general guidelines for performing furnace maintenance. For more detailed instructions on maintaining a furnace, read the manufacturer's specifications or call a qualified repair person. Some common problems associated with gas furnaces are outlined in Table 10-1 shown below.

Problem	Possible cause	Solution
Furnace won't run		
	1. No power	1. Check for blown fuses or tripped circuit breakers at main entrance panel, at separate entrance panel, at separate entrance panel, and on or in furnace; restore circuit.
	2. Switch off	2. Turn on separate power switch on or near furnace.
	3. Motor overload	3. Wait 30 minutes; press reset button. Repeat it if necessary.
	4. Pilot light out	4. Relight pilot.
	5. No gas	5. Make sure gas valve to furnace is fully open.
Not enough heat		
	1. Thermostat set too low.	1. Raise thermostat setting 5°.
	2. Filter dirty	2. Clean or replace filter.
	3. Blower clogged	3. Clean blower assembly.
	4. Registers closed or blocked.	4. Make sure all registers are open; make sure they are not blocked by tugs, drapes, or furniture.
	5. System out of balance.	5. Balance system.
	6. Blower belt loose or broken.	6. Adjust or replace belt.
Pilot won't light		
	1. Pilot opening blocked.	1. Clean pilot opening.
	2. No gas	2. Make sure pilot light button is fully depressed; make sure gas valve to furnace is fully open.
Pilot won't stay lit		
	1. Loose or faulty thermocouple.	1. Tighten thermocouple nut slightly, if no results, replace thermocouple.
	2. Pilot flame set too low.	2. Adjust pilot so flame is about 2 inches long into thermocouple.
Furnace turns on and off repeatedly		
	1. Filter dirty	1. Clean or replace filter.

Table 10-1: **Common Problems with Gas Furnaces**

Problem	Possible cause	Solution
Blower won't stop running		
	1. Blower control set wrong.	1. Reset thermostat from ON to AUTO.
	2. Limit switch set wrong.	2. Reset limit switch for stop-start cycling.
Furnace noisy		
	1. Access panels loose.	1. Mount and fasten access panels correctly.
	2. Belts sticking, worn, or damaged.	2. Spray squeaking drive belts with belt dressing; replace worn or damaged belts.
	3. Blower belts too loose or too tight.	3. Adjust belt.
	4. Motor and/or blower needs lubrication.	4. If motor and blower have oil ports, lubricate.

Table 10.1: **(Continued)**

Tightening Belts

Loose belts on an air distribution system can cause the following:

- Insufficient airflow
- Evaporator coil freezing
- Inadequate cooling

If a belt is slipping, the inside surfaces of the pulleys will become polished to a near-mirror finish. If such is the case, the pulleys must be replaced. See the "Replacing Pulleys" section on page 325 for more on this. If the pulleys are not polished, proceed to adjust the belts:

1. Make certain that the power to the blower is off and that the blower itself has come to a complete stop.
2. Open the blower access panel/service door on the furnace. Check manufacturer's specifications to locate the blower access panel/service door.
3. With a pencil, mark the position of the motor mounts on the furnace and the bolt positions on the motor base (Figure 10-1).

Never attempt to stop rotating equipment or machinery with your hands. Severe personal injury can result.

Do not use a screwdriver or other similar object to pry the belt off the pulleys. Doing so can result in the slippage of the tool, causing severe personal injury.

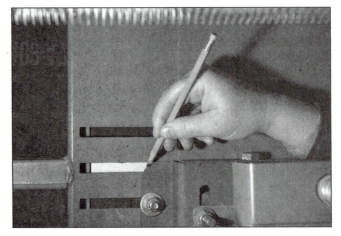

Figure 10-1: **Mark position of motor mounts**

4. Using the proper size wrench, loosen the motor mount bolts so that the motor and the motor mount can move freely. Do not completely remove these bolts. Make certain to leave the motor secured to the motor mount itself (Figure 10-2).
5. Gently move the motor closer to the blower shaft and pulley to further loosen the belt (Figure 10-3).

Figure 10-2: **Loosen the motor mounts**

MOTOR IS ADJUSTED TOWARD COMPRESSOR
FOR BELTS TO BE INSTALLED.

Figure 10-3: **Adjust the motor to remove the belt**

6. Slide the belt off the pulleys.
7. Turn the belt inside out and inspect the underside for any cracks, missing pieces of the belt material, or other signs of excessive belt wear. (See Figures 10-4, 10-5A, and 10-5B.)
8. If the belt is worn, it needs to be replaced. Refer to the "Replacing Belts" section for more information on this. If the belt is in good condition, go to step 9.
9. With the belt removed, inspect the interior surfaces of the pulleys. The interior surfaces of the pulleys should look rough and should not be shiny. Shiny, polished surfaces on the pulleys are an indication that the belts have been slipping. Belt slippage will cause premature wear. Polished pulleys should be replaced. See the section on "Replacing Pulleys" for more on this (see Figure 10-6).

Figure 10-4: **Technician inspecting the belt**

10. If both the belt and pulleys are in good condition, position the motor close to the blower pulley and replace the belt on the pulleys.
11. Gently push the motor away from the blower pulley to increase the belt tension.
12. When the motor mount is slightly past the pencil markings that were made earlier, begin to tighten the motor mounts to the chassis. Make certain that the new position of the motor mount on the chassis is parallel to the original markings to help ensure proper pulley alignment.

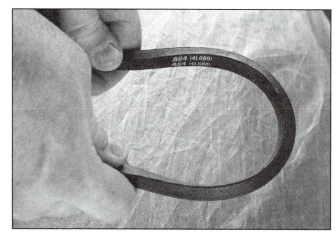

Figure 10-5A: **A good belt**

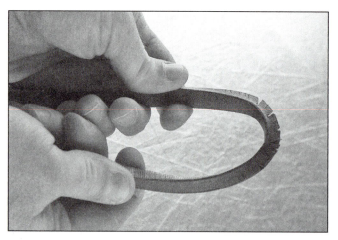

Figure 10-5B: **This belt is cracked and needs to be replaced**

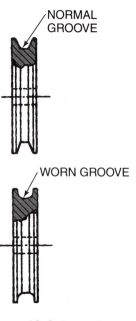

Figure 10-6: **Comparison between normal and worn pulleys**

13. Check the belt tension by placing your thumb and fingers on the opposite sides of the belt and gently squeezing them together. Basically, you are trying to squeeze the two opposite sides of the belts together. There should be some play in the belts, but no more than one inch of deflection (see Figure 10-7).

14. If the sides of the belts can be pushed in more than 1 inch, loosen the motor mounting bolts again and repeat steps 11 through 13.

Figure 10-7: **Check the belt tension**

Do not use a screwdriver or other similar object to replace the belt on the pulleys. Doing so can result in the slippage of the tool, causing severe personal injury.

Replacing Belts

If the belt is broken, damaged, or worn, do not reinstall that belt on the system. Damaged and worn belts should be replaced immediately to help ensure the satisfactory, continued operation of the system.

1. Obtain the belt information from the old belt. If the information on the old belt cannot be read, refer to the "Estimating Belt Sizes" section (Figure 10-9).

2. Obtain a new, exact replacement for the old belt.

Figure 10-8: **Tension gauge**

Figure 10-9: **Belt information**

> *Ideally, technicians should use a belt tension gauge to ensure that the belt tension on the belts is correct (see Figure 10-8).*

3. If the belt has been removed from the pulleys and is damaged, proceed to step 6. If the belt simply broke, continue with step 4.
4. With a pencil, mark the position of the motor mounts on the furnace and the bolt positions on the motor base.
5. Using the proper size wrench, loosen the motor mount bolts so that the motor and the motor mount can move freely. Do not completely remove these bolts and make certain to leave the motor secured to the motor mount itself.
6. Position the motor close to the blower pulley and replace the belt on the pulleys.
7. Gently push the motor away from the blower pulley to increase the belt tension.
8. When the motor mount is slightly past the pencil markings that were made earlier, begin to tighten the motor mounts to the chassis. Make certain that the new position of the motor mount on the chassis is parallel to the original markings to help ensure proper pulley alignment.
9. Check the belt tension by placing your thumb and fingers on the opposite sides of the belt and gently squeezing them together. Basically, you are trying to squeeze the two opposite sides of the belts together. There should be some play in the belts, but no more than 1 inch of deflection.
10. If the sides of the belts can be pushed in more than 1 inch, loosen the motor mounting bolts again and repeat steps 7 through 9.

> *Do not attempt to adjust the pitch on a variable-pitch pulley to tighten the belt tension. Doing so will change the rotation proportions of the drive assembly and change the speed at which the blower turns.*

> *Obtain spare belts and keep them on or near the equipment. In the event the belt breaks in the future, you'll have the system backup quickly.*

Estimating Belt Sizes

There will be times that you will not be able to read the information on an old belt. This can be due to excessive amounts of dirt, age, or simply the destruction of the belt itself. In order to determine important belt information, follow these steps.

1. Measure the center-to-center distance (in inches) between the motor shaft and the blower shaft (Figure 10-10).
2. Multiply the measurement in step 1 by 2.
3. Measure the diameters of the drive pulley and the driven pulley (Figure 10-11).
4. Add the two diameters in step 3 together.
5. Divide the result in step 4 by 2.

Figure 10-10: **Measure center to center**

Figure 10-11: **Measure pulley diameters**

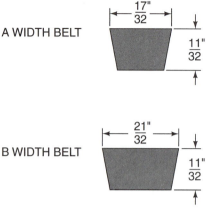

Figure 10-12: **A- and B-width belts**

If the belt sits deep into one or both pulleys, determine the "effective" diameter of the pulley, which is the diameter of an equivalent pulley if the belt was resting at the outer edge of the pulley.

6. Multiply the result from step 5 by 3.14.
7. Add the result from step 6 to the result from step 2. This gives you the approximate length of the required belt.
8. Measure the width of the old belt. An "A" belt has a width of 17/32 inch, while a "B" belt has a width of 21/32 inch (Figure 10-12).
9. The results from steps 7 and 8 provide the type of belt and the approximate length of the required belt.

Sample calculation: Estimate the length of a belt that is used to connect an 8-inch pulley to a 10-inch pulley that is installed on motor and blower shafts that are 16 inches apart.

Here is the step-by-step solution, which corresponds to the steps in the original procedure:

1. The center-to-center distance between the motor shaft and the blower shaft is 16 inches.
2. 16 × 2 = 32 inches
3. Pulley diameters = 8 inches and 10 inches
4. 8 inches + 10 inches = 18 inches
5. 18 inches ÷ 2 = 9 inches
6. 9 inches × 3.14 = 28.26 inches
7. 28.26 inches + 32 inches = 60.26 inches = 60 inches

Adjusting Pulleys

Quite often, the cause for belt slippage, breakage, and premature wear is misaligned pulleys. To check pulley alignment:

1. Make certain that the system is off and all rotating equipment has come to a complete stop. Never attempt to stop rotating equipment by hand. Severe personal injury can result.
2. Remove the access panel on the blower compartment. Check the manufacturer's specifications to locate the blower access panel/service door.
3. Place a straightedge such as a wooden ruler against the faces of both the drive and the driven pulleys (Figure 10-13).

4. The straightedge should touch all four sides of the pulleys:
   a. outside edge of the drive pulley
   b. outside edge of the driven pulley
   c. inside edge of the drive pulley
   d. inside edge of the driven pulley
5. This will determine not only if the pulleys are lined up, but also if they are parallel to each other.
6. If the pulleys are not properly aligned but are parallel to each other, either the drive pulley or the driven pulley will have to be repositioned on the respective shaft. See the section on "Repositioning Pulleys."
7. If the pulleys are not parallel to each other, the motor mount or the blower mount must be adjusted to correct this situation.

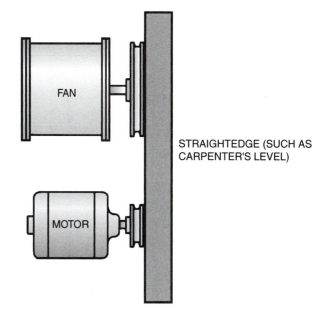

Figure 10-13: **Pulleys must be aligned properly**

Replacing Pulleys

If you have determined to replace one or more pulleys, the original pulley must be removed from the shaft. The shaft may be dirty and/or rusty, making this project potentially very time consuming. Here are some tips to accomplish this.

1. Clean the shaft completely.
2. Spray the shaft with a loosening agent (rust remover). See Figure 10-14.
3. Allow the loosening agent to seep into the joint between the shaft and the hub of the pulley.
4. Completely remove the set screw that holds the pulley to the shaft. This may be a square set screw or an Allen key (Figure 10-15).
5. Be sure to place the set screw in a place where it will not be lost.
6. Spray loosening agent in the set screw hole and allow it to seep into the space between the pulley and the shaft.

Figure 10-14: **Spray the pulley hub and shaft**

Figure 10-15: **Use Allen wrench to loosen set screw**

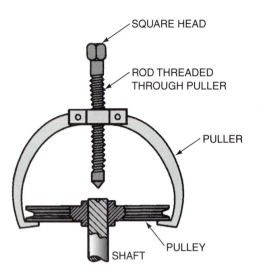

SQUARE HEAD

ROD THREADED THROUGH PULLER

PULLER

PULLEY

SHAFT

Figure 10-16: **A pulley puller**

7. If the above tips do not help in the pulley removal process, a pulley puller should be used (Figure 10-16).
8. Once the pulley has been removed, clean the shaft completely.
9. Obtain the new pulley and position the new pulley on the shaft so that it is perfectly aligned with the other pulley. Refer to the previous section on "Aligning Pulleys" for more on this.
10. Once the pulley has been properly positioned, tighten the pulley securely to the shaft.

Repositioning Pulleys

There are times when the pulley is in good shape, but needs to be repositioned on the shaft. Use the procedures, steps, and tips in the previous two sections to loosen, reposition, align, and retighten the pulley on the shaft.

Replacing Filters on HVAC Units

The most important thing you can do to keep your air conditioner operating efficiently is to routinely replace or clean its filter(s). Clogged, dirty filters restrict normal airflow, which can cause unfiltered air to carry dirt directly into the evaporator coil. Change air-conditioning filters monthly.

1. Make certain that the system is off and all rotating equipment has come to a complete stop. Never attempt to stop rotating equipment by hand. Severe personal injury can result.
2. Remove existing filter(s) from the system (Figure 10-17).
3. Inspect the channel that holds the filters to be sure that the channels are in good shape and that the filter is supported on at least two sides.
4. Measure the filter channel (Figure 10-18).
5. Make certain that the replacement filter is the same size as the filter channel, not necessarily the size of the filter that came out of the unit. (Someone may have put the wrong size filter in the unit.)

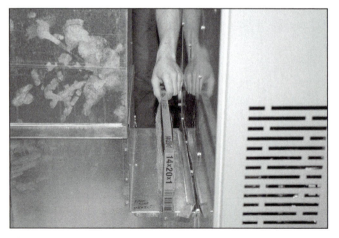

Figure 10-17: **Remove the filter**

Figure 10-18: **Measure the filter rack**

6. Obtain the correct size filter.
7. Locate the arrow on the edge of the filter (Figure 10-19).
8. Install the filter in the channel with the arrow on the filter pointing in the direction of airflow, which is toward the blower.
9. Mark the edge of the filter with the date and your initials (Figure 10-20).
10. Once the filter has been installed, inspect the filter and the channel to be certain that no air is able to bypass the filter.
11. Seal any and all air leaks to prevent/eliminate air bypass.
12. Make certain that any filter channel covers are replaced and secured.

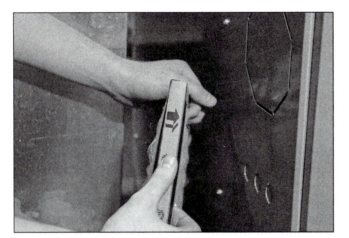

Figure 10-19: **Directional arrow on the filter**

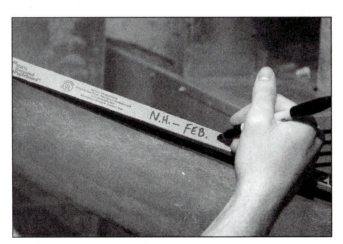

Figure 10-20: **Mark or initial the filter**

Maintaining the Heat Source on Gas-Fired Furnaces

Servicing fossil fuel systems should be done by trained professionals, but there are a number of things that the maintenance technician can do to help ensure that the equipment remains in good working order.

Notes to keep in mind about filter replacement:

- *Air that bypasses the filter will cause dirt and dust to accumulate on the coils, blowers, air distribution system components, supply registers, and ultimately end up back in the occupied space.*
- *Be sure to have an ample supply of filters on hand.*
- *As an absolute minimum, change filters at the beginning of the heating and cooling seasons. It is recommended, however, to change them every month.*
- *If the filters are metal permanent-type filters, they can be cleaned.*
- *If it has been determined that air filters are missing or too small, be sure to visually inspect the return side of the evaporator coil and make certain it is free from dirt and dust.*

1. Typically, the fuel-burning portion of the system does not need to be adjusted each year. So, for best results, do not change any of the current settings on the system.

2. Perform a visual inspection of the burners, pipes, and manifold arrangement. Make certain that these components are free from dirt, dust, and rust. If it is found that there is excessive rust and rust-related damage, call for professional service immediately.

3. If excessive dirt or rust is present, the gas manifold may be carefully removed for cleaning. Follow these steps:

   a. Close the manual gas valve that feeds gas to the appliance.

   b. Disconnect the manifold, making certain to carefully disconnect any components that are attached to the piping (Figure 10-21).

   c. Blow out the manifold with pressurized air, making certain to wear the appropriate personal protection equipment to protect yourself from airborne particles (Figure 10-22).

Figure 10-21: **Disconnect the gas manifold**

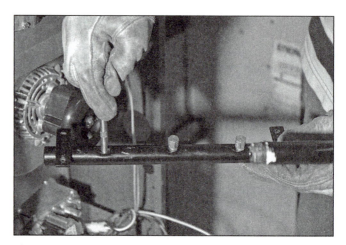

Figure 10-22: **Clean the manifold**

   d. After the cleaning is complete, reassemble the manifold.

   e. Make certain that all piping connections are tight.

   f. Open the manual gas valve.

   g. Restart the system.

4. Observe the burners when they are lit. The flames should be bright blue with slightly orange tips. If the flames are yellow or are blue with yellow tips, **carbon monoxide**, a poisonous, colorless, odorless, tasteless gas generated by incomplete combustion is present. If yellow flames and/or tips are noticed, seek the assistance of a professional immediately (Figure 10-23A).

5. When burning, the flames should rest just above the burners. Flames that are too high above the burner indicate that there is too much air being introduced. Call for help (Figure 10-23B).

6. Flames should be uniform. Erratic flames may be an indication of a system in need of professional adjustment (Figure 10-23C).

7. Schedule a combustion test/analysis on an annual basis, before the beginning of the heating season.

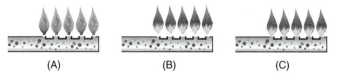

(A)   (B)   (C)

Figure 10-23: **Proper and improper flames**

Perform General Maintenance of a Hot Water or Steam Boiler

Only experienced contractors should work on boilers. If any of the following conditions are present, a professional should be called in to examine, troubleshoot, and remedy the situation:

1. Water accumulates on the floor around the boiler.
2. Water drips from the pressure relief valve.
3. Heat source fails to energize after one attempt to reset/restart the system has failed.
4. Boiler fails to operate after water has been added to the system.
5. Individual zones fail to heat after attempts to bleed air from the system have failed.
6. Individual zones fail to heat after attempts to troubleshoot zone valves have failed.

Perform General Maintenance of an Oil Burner and Boiler

Oil burners typically require regular service to ensure continued satisfactory system operation. Here is a list of items that must be addressed as well as some suggestions for keeping oil-fired heating systems in tip-top shape.

1. Schedule a combustion analysis before the start of the heating season. If the oil-fired equipment is used to supply domestic hot water year-round, this should be done more frequently. As shown in Figure 10-24, the combustion analysis test should include:
   a. Smoke test
   b. Carbon monoxide level
   c. Carbon dioxide (a by-product of natural gas combustion that is harmful and can even cause death) level
   d. Stack temperature
   e. Draft test

2. Keep the oil tank as full as possible. The more oil there is in the tank, the less likely that condensation will form and accumulate in the oil. Water mixed with the oil can result in combustion and operational problems.

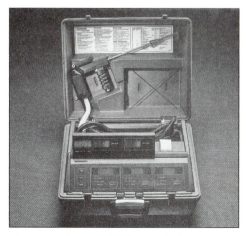

Figure 10-24: **Combustion analysis kit**

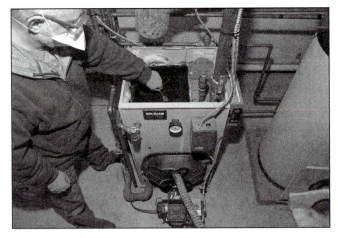

Figure 10-25: **Boiler being cleaned**

Figure 10-26: **A gasket**

(a)

(b)

Figure 10-27: **Changing the oil filter**

3. Clean the heat exchanger on the equipment. To do this, follow the manufacturer's recommendations. The steps involve:
   a. Disconnect the flue pipe connection.
   b. Use a boiler brush and vacuum to clean the spaces between the boiler sections.
   c. Wear a protective dust mask (Figure 10-25).
4. Replace the oil filter.
   a. Make certain that the oil valve line is in the closed position.
   b. Unscrew the existing oil filter.
   c. Replace the filter, making certain that the filter canister gasket is in place (Figure 10-26).
   d. Make certain that the new filter is tight.
   e. Dispose of the oil filter as you would any other hazardous material (Figure 10-27).
5. Check and clean the flue pipe.
   a. Be sure to wear a protective dust mask.
   b. Make certain that a high-quality (filtering) vacuum is used (Figure 10-28).
   c. Be sure to reassemble the flue pipe when finished.

Figure 10-28: **Cleaning the flue**

d. Make certain that the flue pipe is sloped upward toward the chimney.

e. Inspect the chimney if possible and remove any obstructions or debris from the chimney.

6. Check the oil tank for water accumulation.

7. Inspect the area around the unit for traces of oil.

8. Visually inspect the oil lines for damage.

9. Make certain that fill pipe and vent caps are in place.

10. Check for oil leaks in the tank area.

11. Check for unusual oil odors.

12. Inspect the sight glass (steam boilers only).

a. If the water level is low, add water to the system via the feed valve.

b. If the system is losing water at a very fast rate, call for service (Figure 10-29).

Figure 10-29: **Sight glass on a steam boiler**

13. Remove the burner from the unit.

a. Place a drop cloth or other barrier between the boiler and the floor.

b. Make certain that the main oil line is closed.

c. Disconnect the oil line to the burner, making certain that you have rags on hand for any oil droplets (Figure 10-30).

d. Disconnect power to the boiler.

e. Disconnect the wiring connections to the burner, making certain to discharge any capacitors (Figure 10-31).

f. Place the burner on the floor or work bench.

14. Inspect the combustion chamber.

a. Look for cracks in the refractory.

b. Look for and clean up any soot build-up (Figure 10-32).

Figure 10-30: **Disconnect the oil line from the burner**

Figure 10-31: **Disconnect the power supply**

Figure 10-32: **Inside of a combustion chamber**

15. Replace the nozzle (Figure 10-33) and check the firing assembly. You will need to access the manufacturer's guidelines for this, as each manufacturer and each oil burner model has different procedures for accessing/removing the firing assembly.
    a. Make certain that the new nozzle is an exact replacement for the existing part. Be sure to have plenty of spares on hand.
    b. Be careful to not damage the electrode porcelains.
    c. Inspect porcelains for damage.
16. Clean the cad cell (on systems that are equipped with them).
    a. Open the top of the oil burner (Figure 10-34).
    b. Locate the cad cell.
    c. Wipe the cell down to remove accumulated dirt (Figure 10-35).
17. Clean the transformer springs.
    a. Double check to make certain there is no power being supplied to the unit.
    b. Open the top of the oil burner.
    c. Clean the transformer springs (Figure 10-36).

Figure 10-33: **Replacing the nozzle on the oil heating system**

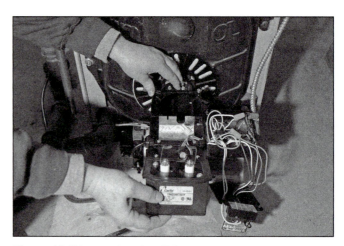

Figure 10-34: **Opening the oil burner**

Figure 10-35: **Cleaning the cad cell**

Figure 10-36: **Cleaning the springs**

Repair and Replace Electrical Devices, Zone Valves, and Circulator Pumps

Here are some general suggestions and tips for replacing electrical devices on heating and air-conditioning equipment.

1. Make certain that *all* electrical power sources are de-energized. Keep in mind that some heating and air-conditioning systems are powered by more than one power source so, even if one source is off, there may still be power to some system components.

2. Make certain that all system capacitors are discharged to avoid receiving an unexpected electric shock.

3. Make certain that an exact replacement, whenever possible, for the component being replaced has been obtained.

4. In the event an exact replacement component is not available, make certain that the replacement component ratings match those of the original as closely as possible.

5. When replacing components that are directly connected to a water-carrying piping arrangement, make certain that the water from the system has been completely drained to avoid an unexpected flood.

   a. If a zone valve motor is being replaced, there is no need to drain the water system, as the valve mechanism will remain intact.

   b. If the entire zone valve is being replaced, however, the water system must be drained, as the water circuit will be accessed.

   c. If the motor on a circulator pump is being replaced and the linkage/impeller assembly is remaining in the system, the water does not need to be drained.

   d. If the entire circulator is being replaced, the system must be drained.

6. Disconnect the electrical splices one at a time, making certain to label or tag the wires so that you will be able to identify the wires when it comes to reconnecting them (Figure 10-37).

7. Carefully remove the defective component, making certain to place any screws and other small parts in a container such as a cup to prevent them from getting lost (Figure 10-38).

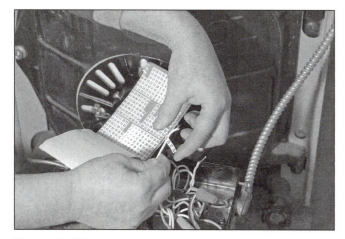

Figure 10-37: **Tag the wires**

Figure 10-38: **Place screws and small parts in a plastic cup**

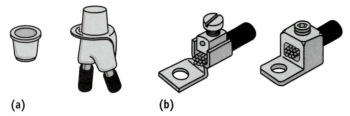

(a)   (b)   (c)   (d)

Figure 10-39: Types of wire connectors

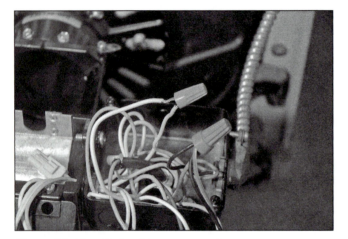

Figure 10-40: Side-by-side nut connections done correctly and incorrectly

8. Mount the new component in place before making the electrical connections.

9. Once the device has been securely mounted, begin connecting the wires one at a time, making certain that the new wiring corresponds to the wiring of the original device. Be sure to use wire nuts or other mechanical connectors (Figure 10-39).

10. Once completed, make certain that the electrical connections are tight and that no bare wire is extending from the underside of the wire nuts (Figure 10-40).

11. Make certain that no bare current-carrying conductors are making contact with the frame or casing of the device.

12. Make certain that all ground wires are properly connected (Figure 10-41).

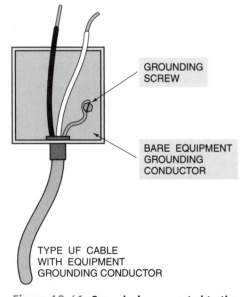

GROUNDING SCREW

BARE EQUIPMENT GROUNDING CONDUCTOR

TYPE UF CABLE WITH EQUIPMENT GROUNDING CONDUCTOR

Figure 10-41: Ground wire connected to the metal box

Lighting a Standing Pilot

When you perform any procedure on a piece of equipment, it is always recommended that the manufacturer's recommendations and procedures be used before general procedures and suggestions. Here are the steps used to light or relight a standing pilot.

1. If the gas valve is in the ON position, close the valve by turning the knob to the OFF position and allow 15 minutes for any unburned fuel to rise through the appliance.

2. Set the gas valve knob to the PILOT position and push the knob into the valve (Figure 10-42).

3. At the same time, light the pilot by using a large barbeque-type match to avoid coming in close contact with the pilot light (Figure 10-43).

4. Once the pilot light is lit, continue to depress the gas valve knob for about 1 minute.

5. After 1 minute, release the knob.

6. The knob should pop up and the pilot should remain lit. If the pilot light goes out, repeat steps 2 through 5.

7. Turn the gas valve knob to the ON position (Figure 10-44).

Figure 10-42: **Standing gas valve**

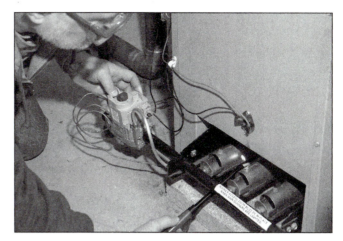

Figure 10-43: **Light the pilot**

Perform General Maintenance of a Chilled Water System

Since chilled water systems contain refrigerants, only qualified air-conditioning technicians should access the refrigeration circuits. The EPA requires that all technicians who work on the refrigeration circuits of air-conditioning and refrigeration system be certified under Section 608 of the Clean Air Act. However, a number of items can be checked by uncertified maintenance personnel.

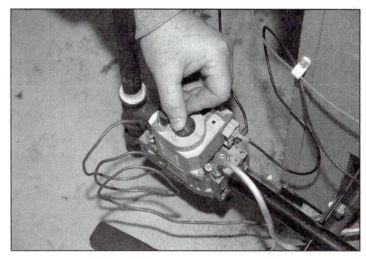

Figure 10-44: **Turn gas valve to ON**

1. Inspect the piping circuits for signs of leakage, oil, and damage.
2. Inspect the water pumps.
   a. Measure the amperage of the pumps and verify that the amperage is within an acceptable range.
   b. Listen to the pumps and make note of any unusual or abnormal noises and vibrations.
3. Check the system thermometers.
   a. Water being supplied to the chilled water coil should be in the range of 45°F.
   b. Water returning from the chilled water coil should be in the range of 55°F.
4. Measure the temperature difference between the return air temperature and the supply air temperature. This difference should be between 16°F and 20°F.
5. Make certain that all air filters on the air distribution system are clean.
6. Make certain that the blower/motor assembly is operational.
   a. Check pulley alignment.
   b. Check belts for cracks and damage.
   c. Check motor amperage.

Clean Coils

If air filters are properly installed and air is not permitted to bypass the filters, there should be very little, if any, dirt accumulation on the return air side of the evaporator coil. On occasion, though, system air filters are removed or not properly installed and air is permitted to bypass. To determine that the evaporator coil is actually clean, visually inspect the coil. This may be a difficult task given the configuration of the system and the installation practices employed. Examining the coil may be as easy as removing the access panel on the air handler or may involve having to cut an access door into the duct system if the coil is mounted on top of a furnace. In any event, once the coil has been inspected and it is determined that the coil is indeed in need of a cleaning, here are some tips and suggestions for doing so.

1. Make certain that the system has been turned off and that all rotating machinery has stopped.
2. Make certain that the area is well ventilated as some cleaning agents give off fumes that may irritate the skin or eyes.
3. Wear proper personal protection equipment such as safety glasses and gloves.
4. Using a brush, remove as much of the dirt as possible (Figure 10-45).
5. Avoid using a rigid wire brush, as the bristles may cause damage to the coil.
6. Avoid flattening the fins on the coil, as this will have a negative effect on the airflow through the coil.
7. Avoid getting cut on the evaporator coil fins. They are sharp and cuts received from them are very painful.
8. Mix the coil cleaner as directed on the product label (Figure 10-46).

Figure 10-45: **Brush the evaporator coil**

Figure 10-46: **Directions on product label**

9. Apply the water/cleaner mixture to the coil with a high-pressure sprayer and allow it to sit, allowing the chemicals to break up the accumulated dirt on the coil surface. Make certain that the strength of the high-pressure sprayer is weak enough to prevent the bending of the coil fins (Figure 10-47).
10. After the manufacturer's suggested time period, rinse the coil with high-pressure water.
11. Do not use a water hose connected to the building's water supply, as this can damage the coil fins and saturate the duct system.

Figure 10-47: **Spray the coil cleaner on the coil**

Figure 10-48: **Comparison of clean and dirty coil**

12. Depending on the amount of dirt on the coil, it may be necessary to repeat Steps 9 and 10 to ensure that the coil is as clean as possible.
13. Once the coil has been cleaned, reseal the duct if it was necessary to cut in an access door. If such is the case, be sure to avoid damaging or piercing the refrigerant lines with screws (Figure 10-48).

Lubricate Motors

Periodic motor lubrication is often required to keep the motors in good working order. Permanently lubricated motors do not need to be lubricated, but all others do. Unless otherwise specified, use a medium (20-weight) oil.

1. Remove the oil port plugs from the oil ports on the motor. Typically, there are two oil ports on a motor, so be sure to remove them both (Figure 10-49).
2. Place the oil plugs in a safe place to avoid losing them. Not replacing the oil plugs can allow dirt and dust into the motor, affecting the operation of the component.
3. Insert the oil port on the oil container into the oiling tubes (Figure 10-50).

Figure 10-49: **Remove oil plugs from motor**

Figure 10-50: **Insert oil spout into the oil port**

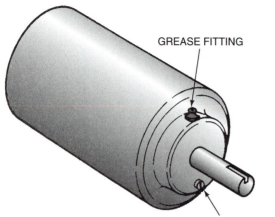

GREASE FITTING

REMOVABLE RELIEF SCREW

Figure 10-51: **Grease fitting**

4. Depending on the motor, you should add between three and six drops of motor oil into each port.
5. Make certain to replace the oil port plugs.
6. Do not over-lubricate motors.
7. If the motor is equipped with grease fittings, the lubrication process is different:
   a. Loosen and remove the relief screw on the motor. The relief screw is located on the opposite side of the motor as the grease fitting (Figure 10-51).
   b. Using a grease gun, pump grease into the motor until grease leaves through the relief screw.
   c. Replace the relief screw.
   d. Repeat this for both sides of the motor.

Follow Systematic Diagnostic and Troubleshooting Practices

When you attempt to diagnose a system problem, using a systematic approach is best. Consider these tips when you encounter a problem.

1. If the system appears to be OFF when it is turned on and operation is desired, check the main power supply to the piece of equipment. Very often, a circuit breaker or switch may have been inadvertently turned off.
2. Check for proper airflow through the air distribution system (furnace and air-conditioning applications).
   a. Reduced airflow can be responsible for a multitude of problems.
   b. Check air filters.
   c. Check belts and pulleys.
3. Check the operational controls on the system.
   a. If a zone valve is not opening, make certain that the thermostat for that zone is calling for heat.
   b. If a boiler is not operating and the water pipes are hot, the water temperature may have reached the desired temperature.
   c. If an air-conditioning system is not operating, make certain that the thermostat is set for cooling before calling for service.
   d. Check for low voltage by switching the fan switch to the ON position (furnace and cooling applications).
   e. If a boiler fails to operate, check safety devices such as high-limit controls, pressure controls, and/or low water cutoff switches.
4. Check for safety or trouble lights.
   a. Many newer controls have trouble and/or diagnostic lights.
   b. Keep all manufacturers' literature on hand. This paperwork contains valuable information regarding steps you can take to keep equipment up and running.
   c. Manufacturers' literature contains the trouble codes that often appear on the control's display.
5. Check system operating temperatures and/or pressures.
   a. Determine which parameters are within acceptable ranges.
   b. Determine which parameters are not acceptable.

6. Write down your findings.
   a. Keep logs of your findings.
   b. This may help evaluate future system problems.
   c. Keep records of what was done and when.
   d. This will also help service technicians who do the job.
7. Narrow your search.
   a. Eliminate items that are definitely operating properly.
   b. Make a list of possible system problems and examine/eliminate them as needed.
8. Ask yourself *why*?
   a. Be sure to fix the cause, not the effect.
   b. Fixing the effect does not fix the underlying problem.

Maintain and Service Condensate Systems

An integral part of the operation of an air-conditioning system is the ability to remove condensate from the structure. Quite often, condensate pumps are used to accomplish this. Depending on the location of the system, condensate pump failure can result in water damage to the structure. Even if there is no condensate pump being used, a gravity-type condensate removal system can cause damage in the event that that line becomes clogged. Here are some tips and suggestions for maintaining condensate removal systems.

1. Inspect the condensate pan under the evaporator for:
   a. signs of rust
   b. signs of damage
   c. dirt, dust, and debris accumulation
   Repair and clean as needed (Figure 10-52).
2. Test condensate lines by pouring a significant amount of water into the drain pan located under the evaporator coil.
   a. Observe the rate of water drainage.
   b. Stop pouring water into the line if water does not drain.
   c. Inspect the area around the condensate drain pan for signs of water (Figure 10-53).

Figure 10-52: Damaged condensate pan

Figure 10-53: Pour water into condensate drain pan

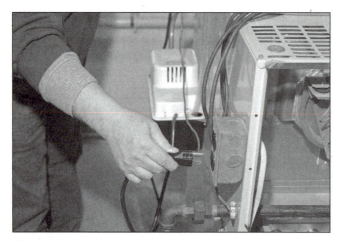

Figure 10-54: **Condensate pump**

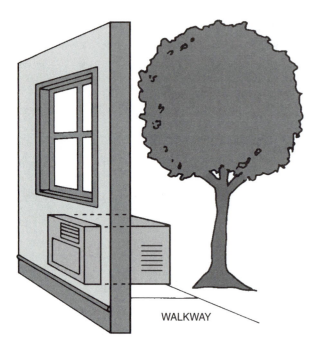

Figure 10-55: **Through-the-wall air conditioner**

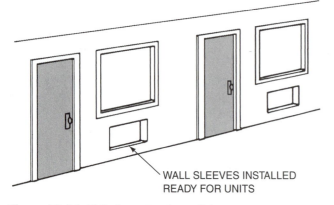

WALL SLEEVES INSTALLED
READY FOR UNITS

Figure 10-56: **Wall sleeve for air conditioner**

3. Inspect the termination point of the line.
   a. If the line terminates outdoors, observe the end of the line before introducing water to the line and again afterward.
   b. Make certain that the water is actually leaving the structure.
   c. If the line terminates in a condensate pump, make certain that the water is indeed ending up in the pump.
4. If the line is not draining, use pressurized air to blow out the line. Repeat Steps 2 and 3 to ensure that the line is now draining.
5. On systems with condensate pumps, make certain to check the operation of the pump.
6. Make certain that the pump remains plugged in or, better yet, have the pump hard-wired to ensure that there is constant power to the pump (Figure 10-54).
7. Test the pump operation by adding water to the pump and inspecting the end of the discharge pipe connected to the outlet of the pump, as in Step 3.
8. Inspect the area around the condensate pump for signs of water.

Replace Through-the-Wall Air Conditioners

If you have determined to replace a through-the-wall air conditioner, perform the following steps to replace it (Figure 10-55).

1. Remove the existing unit from the sleeve (Figure 10-56).

   Service note: Be sure to place a drop cloth or other protective barrier on the floor below the unit to protect the floor from any sharp edges on the unit.

2. Obtain all information from the unit, including the make, model, serial number, voltage rating, amperage rating, and plug type from the unit.
3. Take all unit measurements as well as the internal (daylight opening) measurements of the existing sleeve.
4. Slide the existing unit back into the sleeve.
5. With the acquired information, obtain a replacement unit, making certain that *all* measurements and specifications match those of the existing unit.
6. Uncrate and inspect the new unit.

7. Check to make certain that the sizes are correct, the voltage and amperage ratings are the same as the old unit, and the plug is the same.
8. Remove the old unit from the sleeve.
9. Clean and vacuum out the existing sleeve.
10. Slide the new unit into the existing sleeve and secure it according to the manufacturer's installation literature.
11. Make certain that all shipping materials have been removed from the unit and that the air filter is in place before putting the unit into operation.

Review Questions

1. List three common types of furnaces used in residential and commercial environments.

2. List three symptoms of a loose belt on an air distribution system.

3. List the steps for tightening the belt on an air distribution system.

4. List the steps for replacing a belt on an air distribution system.

5. Estimate the length of a belt used to connect a 6-inch pulley to a 12-inch pulley that is installed on motor and blower shafts that are 14 inches apart. Show your work.

6. How often should an air filter be replaced on a HVAC system?

7. List the steps for lubricating a motor.

8. List the steps for replacing a through-the-wall air conditioner.

Name: _____

Date: _____

HVAC

Perform General Inspection on a Furnace

Upon completion of this job sheet, you should be able to perform a general inspection on a furnace system.

Type of furnace: _____
Model: _____
Last maintenance date: _____
Maintained by: _____

1 Describe the general running condition of the furnace.

_____
_____

2 List four possible fan motor mechanical problems.

_____
_____

3 Arrange the following troubleshooting steps in the correct order.

Test the system operation

Gather information

Verify the complaint

Complete the service call

Perform the visual inspection

Isolate the problem

Correct the problem

4 List the three key indicators that pulleys are aligned properly on the blowers.

_____
_____
_____

Instructor's Response:

_____
_____
_____

Name: _____

Date: _____

HVAC

Replacing Belts

Upon completion of this job sheet, you should be able to replace belts on a furnace motor.

Type of furnace: _____

Model: _____

Last maintenance date: _____

Maintained by: _____

Check off the following tasks as they are completed:

Task Completed

Task:

____ ❶ Obtain belt information from the old belt.

____ ❷ Obtain a new, exact replacement for the old belt.

____ ❸ With a pencil, mark the position of the motor mounts on the furnace and the bolt position on the motor base.

____ ❹ Loosen the motor mounts so the motor and motor mounts can move freely.

____ ❺ Replace the belt on the pulleys.

____ ❻ Position the motor, based on the marks in Step 3, to increase belt tension and tighten the motor mounts.

____ ❼ Check the belt tension.

Instructor's Response:

_____
_____
_____

Name: _____

Date: _____

HVAC

Lighting a Standing Pilot

Upon completion of this job sheet, you should be able to light a pilot light on a gas furnace.

Type of furnace: _____
Model: _____
Last maintenance date: _____
Maintained by: _____
Check off the following tasks as they are completed:

Task Completed

Task:

____ ① Take the access cover off the furnace and look for the gas control knob.

____ ② Turn the knob to OFF and allow 15 minutes to allow unburned fuel to rise through the furnace.

____ ③ Turn the knob until the arrow is pointing to the word "Pilot."

____ ④ Push the knob to start the flow of gas.

____ ⑤ Hold a long match or large barbeque-type match up to the pilot to light the pilot.

____ ⑥ Once the pilot light is lit, continue to depress the gas valve knob for about 1 minute, then release it.

____ ⑦ Turn the knob to the ON position.

Instructor's Response:

_____
_____
_____

Chapter 11   Appliance Repair and Replacement

Introduction

One of the most common duties of a facilities maintenance technician is the replacement and repair of gas and electric appliances. Although some appliances require specialty equipment and training to repair, in most cases the facilities maintenance technician can still perform basic maintenance on these appliances.

Repairing or Replacing a Gas Stove

When repairing or replacing a gas stove, always follow the manufacturer's instruction as well as any local and state building codes. In addition, always ensure that the area is properly ventilated and free from open flame. Some of the more common problems associated with gas stoves are listed next.

Gas Burner Will Not Light

One of the most common problems associated with a gas stove is a burner that will not light. In this case, follow these steps:

1. Lift the top of the stove.
2. Remove the burner unit by lifting up the back end of the unit and sliding the front end off the gas-supply lines (Figure 11-1).
3. Use a needle or other sharp object to poke into the pilot hole and clean out any debris. Brush the remaining debris away from the tip with a toothbrush. Hold a lit match to the opening to relight the pilot. Lower the lid and turn on your burners to test them (Figure 11-2).
4. Identify a spark ignition range by a little ceramic nub located between two burners. Look for wires running to it (Figure 11-3).
5. Brush away gunk around and on the igniter with an old toothbrush. Clean the metal "ground" above the igniter wire as well. It must be clean to conduct a spark. Close the lid and turn the burner knob to "light" to test the burner.

Figure 11-1: **Burner unit**

Figure 11-2: **Cleaning pilot holes on gas burner**

Figure 11-3: **Spark ignition range on a gas burner**

Figure 11-4: **Removing oven door**

Gas Stove Will Not Heat Properly

In the case of a gas stove that will light but does not, however, heat properly, follow this procedure:

1. Remove the oven door (Figure 11-4).
2. Spread a protective covering on the bottom pan to absorb soap and water that may spill during cleaning.
3. Locate the upper burner, attached to the roof of the oven. Using a screwdriver, remove the burner cover, if there is one (Figure 11-5).
4. Using a scrub brush and warm soapy water, clean the burner flame openings to remove the debris. Use a needle or other sharp object to clean out dirt that has collected in the openings (Figure 11-6).
5. Remove the protective covering and lift out the bottom pan. The lower burner will be underneath. Clean the lower burner, following the same procedure described in Step 4.
6. Wipe up any water or dirt that collects under the burner. Let all the parts dry thoroughly; then reassemble the oven.

Figure 11-5: **Upper burner in the oven of a gas stove**

Figure 11-6: **Cleaning burner flame openings**

Replacing a Gas Stove

When a gas stove can no longer be repaired or it is no longer feasible to repair, then the gas stove must be replaced. Although replacing a gas stove is typically a simple procedure, the same safety procedure outlined in repairing a stove should be followed.

1. Shut off the gas line (Figure 11-7).
2. Lay down a piece of plywood to avoid damaging the floor and drag the stove away from the wall. Disconnect the stove from the gas lines.
3. Remove the cover plate at the base of the stove with a screwdriver.
4. If you have old copper tubing, you will first need to disconnect the line at the fitting to the stove. Loosen the nut and slide this back. This is a flared copper-type fitting (Figure 11-8).
5. Drag the unit on top of the plywood, avoiding damage to the gas line. Disconnect the old gas line at the coupling. Be careful not to crimp the copper piping (Figure 11-9).

Figure 11-7: **Turn off gas line**

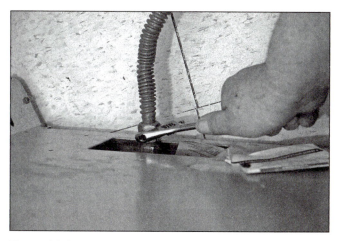

Figure 11-9: **Disconnect gas lines**

Figure 11-8: **Disconnect copper tubing**

6. If you have natural gas, skip to Step 11. If you have LP gas, you will need to replace the range orifices or spuds (Figure 11-10A and B).

7. Orifices and spuds are individually sized with marks and colors. Loosen the old spud and replace with the new one. The oven orifices will then need to be adjusted.

8. Using an appropriate size wrench, tighten down the brass fitting.

9. There may be two orifices, one for the broiler and one for the main oven.

10. Underneath the unit, you will need to reverse the plastic pin in the regulator. Remove the hex nut and flip the plastic pin and reassemble the nut (Figure 11-11).

11. To connect the gas line, use a new connector line with the threads on the pipe identical to the old connector (Figure 11-12).

12. To make the connection, coat the threads with pipe compound.

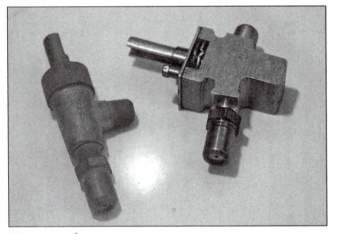

(A)

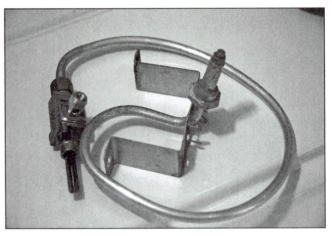

(B)

Figure 11-10: **Range orifices and spuds**

Figure 11-11: **Regulator underneath the range unit**

Figure 11-12: **Connector line connecting gas lines**

13. Tighten the fitting and test by applying soapy water. If it bubbles in a few seconds, you have a leak and will need to fix the leak (Figure 11-13).
14. The antitip bracket is mounted to the wall behind the stove as per the manufacturer's instructions (Figure 11-14).
15. Move the new stove into place.
16. Level the new stove by adjusting the legs (Figure 11-15).
17. Open the gas line. If there is any smell of gas, shut off the supply and call a qualified service technician.

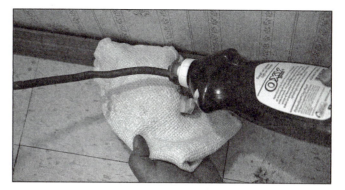

Figure 11-13: **Testing gas line**

Figure 11-14: **Antitip bracket**

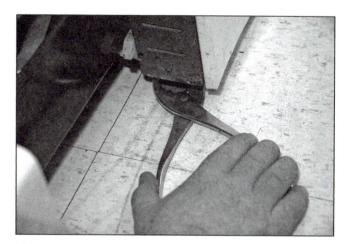

Figure 11-15: **Level stove**

Repairing and Replacing an Electric Stove

Like gas stoves, an electric stove will malfunction from time to time and therefore will require maintenance. Typically with an electric stove the **heating element** (a device used to transform electricity to heat through resistance) will be the source of trouble. In other words it eventually burns out. Sometimes, when an element burns out, you can see that the coil burns in two, or blisters and bubbles. When your heating element burns out, you have to replace it because it is not repairable.

Burned-out heating elements are one cause of a burner not working correctly, but it is not the only one. Before you replace the element, troubleshoot and identify the problem.

Troubleshooting the Problem

Before attempting to repair an electric stove, always read and follow the manufacturer's instruction. Also the technician should have a good understanding of electricity, therefore if necessary review Chapter 5 on electrical theory before attempting to troubleshoot or repair an electric stove. In addition, be sure that any replacement parts have the same electrical rating as replacement parts. Some common problems associated with electric stoves are given in Table 11-1 as well as outlined in the following steps.

1. Determine whether the element plugs into a receptacle, as most do, or is wired directly. If it plugs in, move on to Step 2. If the element is direct-wired (Figure 11-16), move on to Step 4.
2. Remove the plug-in element and inspect the prongs: Lift up the front of the element, then pull the element straight out. Check to see if the prongs are burned, pitted, or otherwise damaged. If they are, you'll need to replace the element and the receptacle (Figure 11-17).
3. If the prongs are clean, test the element: First reinstall it in the receptacle and turn on the burner—sometimes an element just needs to be reseated to work right. If it still does not heat, turn off the burner, exchange the element with

Range will not heat	
No voltage at the element (outlet)	Correct voltage.
Blown fuse or tripped breaker	Replace fuse or reset breaker. If problem persists, contact an electrician.
Broken or burnt wire in power cord	Check the continuity of the cord and replace if necessary. If the problem persists, contact an electrician.
Faulty range plug receptacle	Replace the range plug receptacle.
Burnt or oxidized prongs on range plug	Replace the range plug.
Surface burner does not heat	
Loose connections at the element	Tighten connections.
Burnt, corroded, or oxidized control switch	Clean contacts using sandpaper. If necessary replace control switch.
Burned-out element	Replace elements.
Surface burner too hot	
Switch connection reversed or incorrect	Consult the manufacturer's instructions.
Oven does not heat	
Faulty oven control	Adjust and/or replace control.
Incorrect voltage at the element	Contact an electrician.
Loose connections at the element	Tighten connections.
Burnt, corroded, or oxidized control switch	Clean contacts using sandpaper. If necessary replace control switch.
Burned-out element	Replace elements.
Faulty thermostat	Replace thermostat.
Oven over heats	
Faulty oven control	Adjust and/or replace control.
Incorrect element	Install correct element.
Stove heats unevenly	
Electric stove tilted	Level electric stove.

Table 11-1: **Common Problems Associated with an Electric Stove**

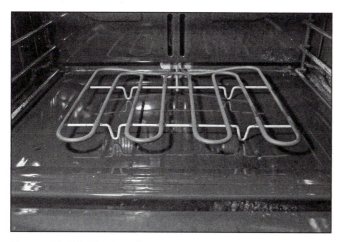

Figure 11-16: **Heating elements**

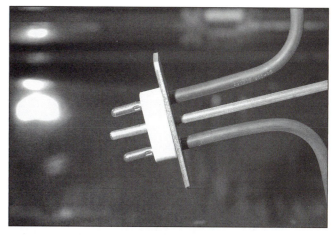

Figure 11-17: **Prongs on heating element**

another of the same size, and test. If the burner works now, the original element needs to be replaced.

4. If the element is direct-wired, lift the front of the element and pull it out until you see a white porcelain insulator with clips on each side (Figure 11-18).

5. Use a flat-head screwdriver to open the insulator and remove the clips. Then separate the two halves of the insulator.

6. Remove the screws that hold the element to its wiring. Exchange the element for another of the same size. Reassemble both elements so no bare wires are left exposed, and then turn on the burner. If the new element works, the original element needs to be replaced.

Figure 11-18: **Porcelain insulators**

Replacing an Element

As mentioned earlier one of the most common problems associated with an electric stove is the failure of the electric heating elements. To replace a heat element do the following:

1. Get a new heating element identical to the one you are replacing (Figure 11-19).

2. Install the new element in the stove. For a plug-in element, just plug it into the receptacle. For a direct-wired element, screw the new element to its wiring, reassemble the two halves of the porcelain insulator, and snap the clips in place (Figure 11-20).

3. Test the element to make sure that it's operating.

Figure 11-19: **Heating elements**

Replacing a Receptacle

Over time the contacts in a receptacle will oxidize and/or corrode and therefore will need to be replaced. When this happen, follow this procedure:

1. Disconnect the old receptacle. If it is screwed to the cooktop, use a screwdriver to disconnect it. If it is held in place by a spring steel clamp, spread the clamp and pull out the receptacle (Figure 11-21).

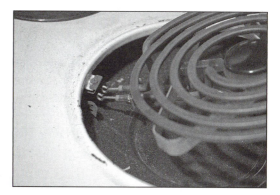

Figure 11-20: **Installing new heating elements**

2. Lift the cooktop so that you can access the receptacle wiring (Figure 11-22).

3. Remove the receptacle. Wrap the wires with masking tape and label them so that you can install the new receptacle correctly; then cut the wires (Figure 11-23).

Figure 11-21: **Disconnecting old receptacle**

Figure 11-22: **Cooktop lifted to access receptacle wiring**

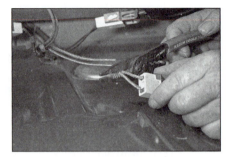

Figure 11-23: **Wires being cut**

4. Install the new receptacle. Strip the ends of the wires with a wire stripper, then twist the wires together and on wire nuts to hold them together. Reinstall the receptacle in the cooktop and install the element (Figure 11-24A and B).

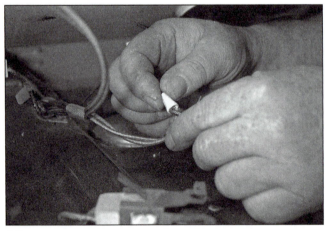

(A)

(B)

Figure 11-24: **(A) and (B) New receptacle installed**

Oven Does Not Heat Properly

As mentioned earlier besides a burnt-out heating element, a fault thermostat will cause an oven not to heat properly. The purpose of the thermostat is to regulate the temperature in the oven by permitting current to flow to the element.

Testing the Thermostat

Although some technicians believe that when in doubt about a piece of equipment or a device it is better to replace it, this is not always the right approach. Before attempting to replace a thermostat, always verify that the thermostat should be replaced. This is accomplished by doing the following:

1. Place an oven thermometer inside the oven and shut the door (Figure 11-25).
2. Turn on the oven, set it for 350°F, and let it heat for 30 minutes.
3. Check the thermometer. Most thermostats are accurate to within 25°F. If the thermostat is off by more than 50°F, the thermostat is bad and you will need to have a professional replace it. If the thermostat is off by less than 50°F, adjust the thermostat.
4. Locate the adjustment screw. On some thermostats, the adjustment screw is on the back of the thermostat knob; on others it is inside the thermostat shaft (Figure 11-26).
5. To make a temperature adjustment on the back of a knob, remove the knob and loosen the retaining screws on the back. Turn the center disk toward "hotter" or "raise" to increase the temperature or toward "cooler" or "lower" to decrease the temperature. Tighten the screws, reinstall the knob, and test the oven. Readjust the knob if necessary (Figure 11-27).
6. To make a temperature adjustment inside the shaft, remove the knob and slip a thin flat-head screwdriver into the knob until it engages the adjustment screw in the bottom. Turn the screwdriver clockwise to raise the temperature and counterclockwise to lower it. Each quarter-turn will change the temperature about 25°F.
7. Reinstall the knob and test the oven. Readjust the temperature if necessary.

Figure 11-25: **Oven thermometer**

(A)

Figure 11-26A: **Thermostat and screw**

(B)

Figure 11-26B: **Thermostat knob**

(A)

(B)

Figure 11-27: **Oven heating element**

Replacing the Oven Heating Element

As stated earlier one of the leading causes for an oven not to heat properly is a burnt-out heating element. To replace a heating element follow this procedure:

1. Remove the oven racks so that you have access to the element (Figure 11-27A and B).
2. Remove the two screws from the element mounting plate, which sits flush against the back wall of the oven.
3. Pull the element gently out as far as the wire will allow.
4. Remove the supply wires from the element terminals (Figure 11-28).
5. Replace the element with a new identical one.
6. Put the new element in place and reconnect the leads. Usually only two wires go to the element; it doesn't matter which wire attaches to which terminal, as long as they're screwed on tight.
7. Push excess wire back behind the insulation.
8. Line up holes and reinstall the mounting bracket using the same screws you removed earlier (Figure 11-29).
9. Replace the oven racks.

Figure 11-28: **Element terminals**

Figure 11-29: **Mounting brackets**

Replacing an Electric Stove

As with a gas stove, when the repair of an electric stove becomes too extensive or the replacement parts are no longer available then it might be necessary to replace it. When replacing an electric stove, always follow the manufacturer's instructions and recommendations. Also it might be necessary to update the wiring and/or breaker in the fuse box. Never purchase an oversized breaker or allow the stove to be installed with a breaker that is too small. This can cause serious electrical problems that can lead to equipment damage and/or malfunction.

1. Lay down a piece of plywood to prevent damaging the floor and drag the stove away from the wall.
2. Unplug the stove from the electrical outlet.
3. Remove the old stove and replace it with the new one.
4. Plug in the new stove.
5. Move the new stove into place.
6. Level the new stove by adjusting the legs.

Troubleshooting and Repairing an Ice Maker in a Refrigerator

If the ice maker does not make ice but you can see the arm swing into motion and you hear a buzz for about 10 seconds after it is finished, this normally means that there is a problem with the water supply line.

1. Check to make sure the water supply line is not kinked behind or beneath the refrigerator. If the ice maker has frozen up, it will need to be unthawed.
2. Unplug the refrigerator.
3. Remove the ice bin and loose ice from the ice maker.
4. Find the fill tube, the white rubber-like hose, that delivers water into the ice maker and pull the small metal cap off of the housing that holds the full tube down (Figure 11-30).
5. Warm the hose and surrounding mechanism to melt any ice blocking the mechanism. This can be done using a hair dryer or soaking the supply tubing in hot water.

Figure 11-30: **Fill tubes**

Replacing an Ice Maker

Always follow the manufacturer's instructions on replacing the ice maker. Also, whenever possible use only genuine manufacturer's replacement parts. This will ensure that the parts fit and function properly.

Repairing a Refrigerator

Major problems will require a trained refrigeration technician. However, many times the problem is simple and can be corrected. The potential problems that can easily be fixed by a facilities maintenance technician are given next.

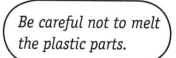

Be careful not to melt the plastic parts.

Adjusting Controls

Because each model will vary, refer to the specific refrigerator manufacturer's instructions.

Testing and Replacing Door Gaskets

1. Test the door seal in several places by closing a piece of paper in the door, and then pulling it out. There should be some resistance, indicating that the door is sealed.
2. Remove the old gasket one section at a time. Some gaskets are held on by retaining strips, others by screws or even adhesive.
3. Install an identical gasket using the retaining strips or screws or new adhesive.

> *Before doing any work on your dishwasher, turn off the power at the circuit-breaker box.*

Troubleshooting Dishwasher Problems

As with any appliance, troubleshooting is a necessary process to determine why the appliance is not operating correctly. Table 11-2 outlines some common problems associated with dishwashers.

Dishwasher doesn't work (no power)	
Blown fuse or tripped breaker	Replace fuse or reset breaker. If problem persists contact an electrician.
Dishwasher is unplugged	Reconnect the dishwasher.
Broken or burnt wire in power cord	Check the continuity of the cord and replace if necessary. If the problem persists contact an electrician.
Faulty door latch	Replace door latch.
Faulty door switch	Replace door switch.
Faulty timer motor	Replace timer motor.
Faulty selector switch	Replace selector switch.
Faulty motor	Test and replace motor relay if necessary. Test and replace motor if necessary.
Dishwasher motor does not start but the motor hums	
Check water supply	If water supply is off, turn on.
Faulty door latch	Replace door latch.
Faulty door switch	Replace door switch.
Check inlet valve filter screens	Clean inlet valve filter screens.
Check fill tube for kinks	Straighten fill line.
Dishwasher does not drain	
Faulty drain valve	Replace drain valve
Sink drain and/or drain hoses restricted	Clear restriction
Dishwasher leaks water and/or soap around the door	
Incorrect detergent	Make sure that detergent is intended for a dishwasher.
Faulty door latch	Replace door latch.
Faulty door switch	Replace door switch.
Damaged door seal	Replace door seal.
Damaged spray arm	Replace spray arm.
Dishwasher does not dry properly	
Burned-out element	Replace elements.
Faulty thermostat	Replace thermostat.

Table 11-2: **Common Problems Associated with a Dishwasher**

Water on the Floor Around the Dishwasher

Water around the dishwasher could mean that either the gasket is damaged or the sprayer is clogged. In the case of leaking or pooling water around the dishwasher, take the following steps to rectify the problem.

Damaged Gasket

1. Check your gasket for cracks or deterioration (Figure 11-31).
2. If the gasket is damaged, remove it by unscrewing it or prying it out with a screwdriver. Replace it with the same type of gasket as what was removed. Before installing the new gasket, soak it in hot water to make it more flexible.

> A dishwasher can also leak if it is not level.

Clogged Sprayer

1. Remove the sprayer and soak it in warm white vinegar for a few hours to loosen mineral deposits. Then clean out each spray hole with a pointed device such as a needle, awl, or pipe cleaner (Figure 11-32).

Figure 11-31: **Gasket around dishwasher**

Figure 11-32: **Dishwasher sprayer**

Dishwasher Overflows

If the dishwasher is overflowing and the drain and drain valve are not obstructed check the float and float switch. This is accomplished by doing the following:

1. Open the dishwasher door and locate the float switch. It should be a cylindrical-shaped piece of plastic and may be set to one side along the front of the cabinet or near the sprayer head in the middle of the machine (Figure 11-33).
2. Check the float to make sure that it moves freely up and down on its shaft. (You may have to unscrew and remove a protective cap to get to the float.) If the float sticks, you'll need to clean away any debris or mineral deposits that are causing it to jam.

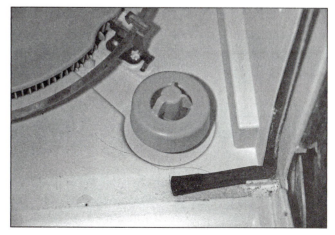

Figure 11-33: **Dishwasher float switch**

3. Pull the float off the shaft and then clean the inside of the float with a bottle brush. Clean the shaft with a scrub brush.
4. Reinstall the float and check that it moves smoothly.
5. Set the dishwasher to fill, and check to see if it overflows.

Replacing a Dishwasher

When it becomes inefficient to repair a dishwasher or the replacement parts are no longer available, then it is necessary to replace the dishwasher. When replacing a dishwasher, always follow all local and state building codes as well as all manufacturer's instructions and recommendations.

1. Turn off the power to the dishwasher circuit at the electrical service panel.
2. Shut off the hot water supply to the unit. This is typically under the sink if the supply comes from there, but it may also be under the dishwasher or in the basement if the supply comes through the wall or floor of the dishwasher opening.
3. Remove the access and lower panels at the base of both dishwashers (Figure 11-34).
4. Remove the electrical box from the old dishwasher (Figure 11-35).

Figure 11-34: **Accessing lower panels**

Figure 11-35: **Electrical box on dishwasher**

5. Unscrew the wire nuts and pull apart the wires. Start with the green wires, then white, and then black.
6. Disconnect the drain hose from the waste tee on the drain line, or the inlet on a disposer, using pliers to open a spring clamp or a screwdriver to open a screw-type clamp. Do the same where it connects to the dishwasher. If you cannot easily access that connection, you can disconnect it later (Figure 11-36).
7. Disconnect the water supply line from the water inlet on the dishwasher (Figure 11-37).
8. Lay down a piece of plywood to drag the old dishwasher onto.
9. Open the door to access and remove the screws that secure the dishwasher to the underside of the countertop. Then adjust the front leg levelers to lower the unit so that you can slide it out (Figure 11-38).
10. Take the new dishwasher out of the box and check the back to verify that all of the connections are in place.
11. Take the cap off the drain line connection at the dishwasher (Figure 11-39).

Have an old towel or drip pan handy to mop up or catch water that will spill out of drain and water lines as they are disconnected.

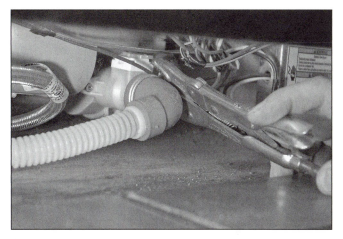

Figure 11-36: **Drain hose**

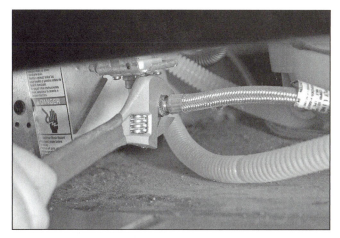

Figure 11-37: **Water supply line**

Figure 11-38: **Removing screws in counter**

Figure 11-39: **Drain line connection**

12. Attach the drain line to the dishwasher.
13. Using pliers, crimp the clamp around the hose to secure.
14. Close and lock the door; then slide/roll the machine to the opening.
15. Adjust the leveling legs as indicated by the manufacturer to raise the dishwasher (Figure 11-40). Use a level to verify that the unit is level. Install the mounting screws into the underside of the counter.
16. Reverse the removal procedures to connect water, drain, and electric lines. Cut off the exposed ends of electrical wires and use wire strippers to strip about ½ inch of insulation from the ends. Twist wires together (white-to-white, black-to-black) and on new wire connectors. Secure the ground (green) wire. Tighten the strain-relief connector.
17. Install decorative panels.
18. Follow the manufacturer's instructions to adjust the door so there is an even space on both sides. You might, for example, need to move the door spring to a new mounting hole.

Figure 11-40: **Leveling legs of dishwasher**

19. Open the water valve to check for leaks in the water line at valve and inlet connections. Restore power and operate the machine to check for drain leaks.
20. Reinstall the lower and front access panels.

Repairing a Range Hood

A range hood that does not adequately remove smoke and smells from your kitchen is usually caused by one of the following reasons:

- The grease filter or some part of the exhaust ductwork may be clogged.
- The fan may be bad.

Unclogging the Exhaust Fan

1. Remove the filter and soak it in a degreasing solution until the grease is dissolved (Figure 11-41).

(A)

(B)

Figure 11-41: Exhaust fan filter

2. Wash with warm, soapy water to remove any traces of the degreaser. Also, a filter may be put in the upper rack of the dishwasher and run it through a normal cycle.
3. Remove the exhaust fan. Unplug the fan and remove it from the hood (Figure 11-42).
4. Clean the fan blades with an old toothbrush dipped into a cleaning solution.
5. Clean the inside of the exhaust ductwork, using a plumber's snake with a heavy rag tied around the end. Push the snake through the ductwork. Soak the rag in a cleaning solution and run it through the ductwork. Rinse out the rag and repeat the operation until the duct appears to be clean (Figure 11-43).
6. Clean the exhaust hood that is attached to the outside of your house (Figure 11-44).
7. Reinstall the grease filter.

Figure 11-42: **Remove exhaust fan**

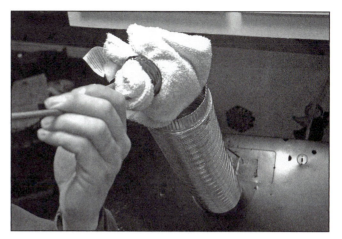

Figure 11-43: **Clean exhaust ductwork**

Figure 11-44: **Exhaust hood on outside of house**

Replacing a Range Hood

1. Remove the old range hood (Figure 11-45).
2. On the new range hood, remove the filter, fan, and electrical housing cover. Remove the knockouts for the electrical cable and the duct (Figure 11-46).

Figure 11-45: **Range hood**

Figure 11-46: **Range hood with knockouts**

Figure 11-47: **Hood ductwork**

3. Protect the surface of the cooktop with heavy cardboard and set the range hood on top of it. Then connect the house wiring to the hood. Connect the house black wire to the hood black wire and the house white wire to the hood white wire. Then connect the house ground wire under the ground screw and tighten the cable clamp onto the house wiring.
4. Using the mounting screws to install the hood, slide the hood toward the wall until the mounting screws are engaged. Tighten the screws securely with a long-handled screwdriver. Then replace the bottom cover.
5. Fasten the ductwork to the hood using duct tape to secure joints and make them airtight (Figure 11-47).
6. Install the light bulbs and replace the filters. Turn on the power at the service panel and check for proper operation.

Repairing a Microwave Oven

Because microwave leakage can be hazardous and high wattage is present, limit your microwave repairs to light bulb changes, if the light bulb is easily accessible, and checking to make sure that the oven is getting power. For other repairs, call a qualified technician to make repairs on a microwave oven.

Troubleshooting a Washer

If the washer is not getting either hot or cold water, refer to your owner's manual to ensure that the washer is not operating as it should. If the washer isn't operating as it should, there may be a problem with the water inlet valve.

Checking the Water Inlet Valve

1. Disconnect the appliance's power supply.

Figure 11-48: **Inlet valve**

2. Locate the washer's water inlet valve. It will be at the back of the washer, and it will have water hoses hooked up to its back (Figure 11-48).
3. Shut off the supply of water to your washer (Figure 11-49).
4. Disconnect both hoses at the back of the washer. Point the hoses into a bucket or a sink, and then turn on the water supply again. Do this to confirm that you are receiving adequate water pressure and that there is not some sort of blockage in the line (Figure 11-50).
5. Inspect the screens found inside the valve. Clean out any debris you find. You should be able to pop them out with a flat-head screwdriver. Do use caution when handling the screens as they are irreplaceable.

Figure 11-49: **Shutting off water supply**

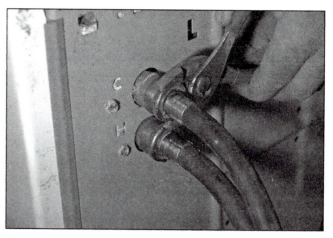

Figure 11-50: **Disconnect hoses**

Replacing Washer Inlet Valves

1. Disconnect the appliance's power supply.
2. Locate the washer's water inlet valve. It will be at the back of the washer, and it will have water hoses hooked up to its back (Figure 11-51).
3. Shut off the water supply to your washer.
4. Disconnect both hoses at the back of the washer. Point the hoses into a bucket or a sink; then turn on the water supply again. Do this to confirm that you are receiving adequate water pressure and that there is not some sort of blockage in the line.
5. Remove the screws that hold the inlet valve in place (Figure 11-52).

Figure 11-51: **Inlet valve**

Figure 11-52: **Inlet valve**

6. Remove the hose connecting the valve to the fill spout.
7. Connect your new water inlet valve to the water line.
8. Attach both water supply hoses. Turn the water on and check for leaks.
9. Reconnect the washer to the power supply.

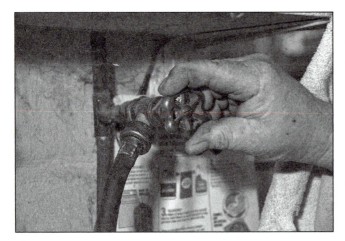

Figure 11-53: **Water faucets**

Washing Machine Fills Slowly

When a washing machine fills slowly, the problem is usually a clogged intake screen.

1. Turn off the water faucets that feed the machine (Figure 11-53).
2. Unplug the washer and pull it far enough away from the wall that you can get behind it to work.
3. Remove each water-supply hose (Figure 11-54).
4. Locate the screens and gently pry out the screens using a small flat-head screwdriver (Figure 11-55).
5. Clean the screens with an old toothbrush.
6. Reinstall the screens.
7. Reinstall the hoses and tighten the couplings securely.
8. Turn on the water to check for leaks; then plug in the machine and push it back into position.

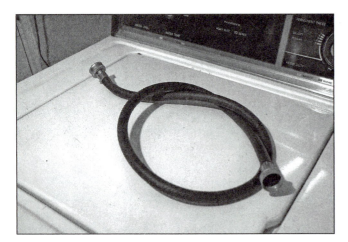

Figure 11-54: **Water-supply hose**

Figure 11-55: **Intake screen**

Installing a New Washer

The installation of a new washing machine is a simple process; however, moving the washing machine into position and out of the facility can be tricky as well as dangerous. Whenever possible, it is strongly recommended to have additional help moving the new machine into position and the old machine out of position.

1. Turn off power to the old washer.
2. Remove the old washer.
3. Clean and dry the floor and move the new washer into place to be connected.
4. Fasten the drain hose to the washer with a hose clamp. Be sure not to tighten it too much or you might strip the screw (Figure 11-56).
5. Attach the water hoses to the washer. The hot and cold on the taps and on the washer are usually clearly marked. Red indicates hot; blue indicates cold (Figure 11-57).

Figure 11-56: **Fasten drain hose**

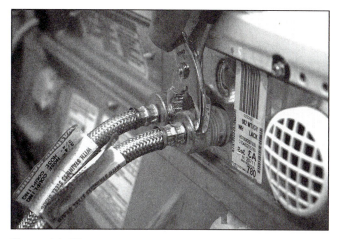

Figure 11-57: **Attach hose to the washer**

6. Plug the washing machine in and move it into place, placing the drain hose in the drainpipe when you can reach it.
7. Push the washer the rest of the way into the space, being careful not to crimp the hoses.
8. Leave about an inch and a half of space around the washer to allow room for it to vibrate.
9. Turn the water faucets on.
10. Turn the power back on.
11. Run a cycle without clothes or detergent before you use the machine to clear the water pipes and make sure that the drainage is adequate.

Troubleshooting a Dryer

Dryers typically consist of very few parts and therefore are usually simple to troubleshoot and maintain. Most dryer problems can be attributed to either a burnt-out heating element or a clogged vent.

Dryer Takes a Long Time to Dry Clothes

If a dryer is taking too long to dry clothing, it is likely that the heating element is partially or completely burned out. Follow the manufacturer's instructions for the specific dryer brand to test and replace the heating element if necessary.

The Vent Is Clogged

If the dryer feels really hot, but the clothes take forever to dry, a clogged vent could be the problem.

1. Check the vent flap or hood on the outside of the house. Make sure that you feel a strong flow of air coming out when the dryer is running. If not, try cleaning out the vent with a straightened clothes hanger (Figure 11-58).

Figure 11-58: **Vent flap**

Figure 11-59: **Dryer duct**

2. If the vent flap is not the problem, check for a kink or sag in the duct and straighten the hose if necessary.
3. If a kinked or sagging duct is not the problem, disconnect the duct from the dryer and look for blockage inside with a flashlight. To remove the blockage, shake it out or run a wadded cloth through the duct. If the duct is damaged, replace it (Figure 11-59).

Review Questions

1. List the steps for unclogging a vent.

2. What should the technician check if a dryer does not dry clothes?

3. List the steps for installing a washer.

4. List the steps for replacing a washer inlet valve.

5. List the steps for replacing a range hood.

6. List the steps for unclogging an exhaust fan.

7. List the steps for replacing a dishwasher.

8. List the steps for replacing a gas stove.

9. What should the technician check if an electric oven does not heat properly?

10. What should the technician check if a dishwasher does not dry properly?

Name: _____

Date: _____

Replacing an Electric Stove

Upon completion of this job sheet, you should be able to replace an electric stove.

In the space provided below, list the steps for replacing an electric stove.

Instructor's Response:

_____

_____

_____

Name: _____

Date: _____

Replacing a Dishwasher

Upon completion of this job sheet, you should be able to replace a dishwasher.

In the space provided below, list the steps for replacing a dishwasher.

Instructor's Response:

_____

_____

_____

Chapter 12 Trash Compactors

OBJECTIVES

By the end of this chapter, you will be able to:

Knowledge-Based

✪ Explain the purpose of interlock safety device.

Skill-Based

✪ Perform general maintenance procedures.
✪ Perform general maintenance of hydraulic devices.
✪ Perform a test of the interlock safety device.
✪ Check the general condition of a dumpster.

Glossary of a Term

Dumpster a large waste receptacle Litter is waste that is unlawfully dispose of outdoors.

Introduction

Waste and trash are usually collected in a trash can, which is then put out for trash collection. A trash compactor does exactly what its name implies. Instead of putting your trash into a trash can, you put it in the compactor, where it gets compressed to between $1/10$ and $1/12$ the space it would normally take up.

General Maintenance

A trash compactor is a relatively simple appliance with only a few components. These components include a motor, drive screws, compression ram, limit switches, door switch, exterior controls, and in some units, an odor control system (Figure 12-1).

Figure 12-1: **Trash compactor**

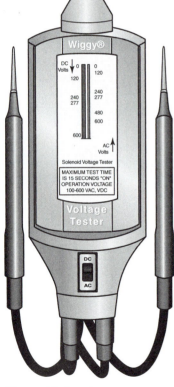

Figure 12-2: **Voltage tester**

Common Trash Compactor Problems and Solutions

As mentioned earlier, a trash compactor is a simple device with few components; however, like all equipment, malfunctions and equipment breakage can happen.

If the compactor will not start:

- Make sure that the door is completely closed.
- Make sure the compactor is plugged in securely.
- Check for a blown fuse or tripped circuit breaker.
- Inspect the electrical cord for damage.

If the motor runs but trash is not compacted:

- Drawer must be about one-third full before any compaction will take place.
- Check the outlet voltage. Measuring the voltage at an electrical outlet requires the use of a voltage tester (Figure 12-2).
- Check the power nuts for wear or obstructions. See manufacturer's specifications.
- Check the power screws for wear or obstructions. See manufacturer's specifications.
- Lubricate the power screws. See manufacturer's specifications.
- Check the drive belt/chain/gears. See manufacturer's specifications.

If the compactor starts but does not complete cycle (ram is stuck):

- Object in trash may be causing door to trigger tilt switch.
- Check for loose connections.

If the drawer is stiff or difficult to open:

- Clean the drawer tracks.
- Inspect the drawer rollers.

If the door will not open:

- Return the ram to the top position.
- Turn off the dense pack switch.
- Push the door closed while restarting compactor.
- Check the power screws for obstructions. See manufacturer's specifications.
- Check the power nuts for obstructions. See manufacturer's specifications.
- Check the ram for obstructions.

Cleaning and Deodorizing

Thoroughly clean the interior of your trash compactor regularly. It is recommended that you use a bacteria-fighting cleaner and/or degreaser to clean the ram (the platform that presses down on the garbage) and any other part of the compactor that comes into contact with the garbage.

For routine cleaning, use the following steps:

1. Always wear thick, sturdy gloves when cleaning your compactor (Figure 12-3).
2. Unplug the compactor.

3. Remove the bag and caddy, or bin, and follow the manufacturer's cleaning instructions.
4. Vacuum the interior.
5. Clean inside and outside of the compactor using warm soapy water. Rinse and dry.
6. Close the drawer and replace the caddy with a new bag.
7. Periodically check and replace the air freshener or charcoal filter.

Bacteria can grow on the inside of your trash compactor from the food waste that is put in the compactor. For temporary odor control between cleanings, spray the interior with a germ-killing deodorant/disinfectant. Also replace the filter (if there is one) once or twice a year.

General Maintenance of Hydraulic Devices

Check the compactor's preventive maintenance schedule for its most recent maintenance; also check to see if hydraulic fluid lines are adequate/intact. See the manufacturer's specifications.

Figure 12-3: **Rubber gloves**

Performing a Test of the Interlock Safety Device

The safety interlock prevents operation when the door is open. Test the safety device by opening the trash compactor door and press the buttons on the trash compactor to make sure it does not start up. If the compactor runs with the door open, follow the manufacturer's specifications to correct the problem (Figure 12-4).

Checking the General Condition of a Dumpster and Dumpster Area

In addition to maintaining the trash compactors, the facilities maintenance technician is responsible for maintaining the **dumpsters** (a large waste receptacle) and the dumpster area. This includes the following:

- Control litter (**waste** that is **unlawfully** dispose of outdoors).
- Make sure the dumpster leasing company maintains and cleans dumpster regularly.
- Return leaking dumpsters for repair immediately.
- If you must wash down a dumpster, use dry cleanup methods first, and then rinse, collect water, and discharge to appropriate drainage area.

Figure 12-4: **After disconnecting the power to the compactor, locking out the distribution panel prevents someone from turning on the power**

Review Questions

1. When troubleshooting a trash compactor, what should a contractor first check?

2. When a trash compactor starts but does not complete its cycle, what should the contractor check?

3. List the steps for cleaning a trash compactor.

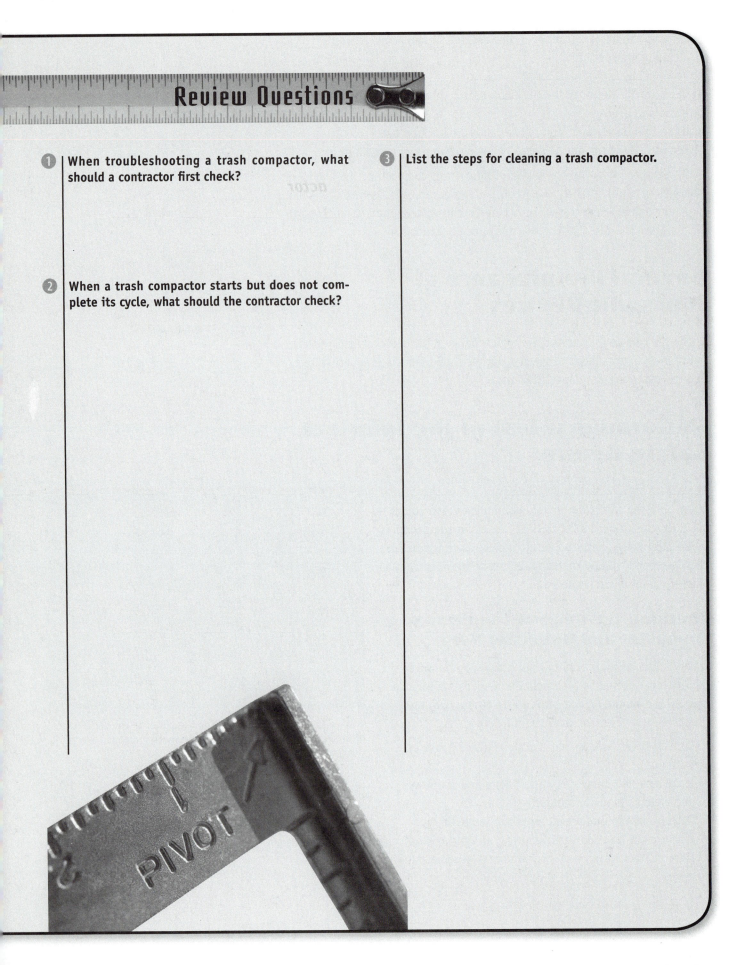

Name: _____

Date: _____

Trash Compactor

Cleaning and Deodorizing a Trash Compactor

Upon completion of this job sheet, you should be able to clean and deodorize a trash compactor.

Task Completed

Task:

____ ① Always wear thick, sturdy gloves when cleaning your compactor.

____ ② Unplug the compactor.

____ ③ Remove the bag and caddy, or bin, and follow the manufacturer's cleaning instructions.

____ ④ Vacuum the interior.

____ ⑤ Clean inside and outside of the compactor using warm soapy water. Rinse and dry.

____ ⑥ Close the drawer and replace the caddy with a new bag.

____ ⑦ Periodically check and replace the air freshener or charcoal filter.

Instructor's Response

_____

_____

_____

Chapter 13 Elevators

OBJECTIVES

By the end of this chapter, you will be able to:

Skill-Based

- ⊗ Check and inspect floor leveling.
- ⊗ Check operation of elevators.
- ⊗ Perform a test on elevator doors.

Glossary of a Term

Elevator Platform the floor of an elevator

Introduction

An elevator is defined as a device consisting of a platform used to raise and lower occupants and/or equipment from one floor to another (Figure 13-1). The objective of elevator maintenance is to ensure that the elevator system provides safe and dependable operation with maximum efficiency and minimum downtime. Failure to keep an elevator properly maintained will result in a decrease in reliability and an increase in response time. In addition, the safety of the elevator's occupants will be compromised. Full maintenance contract can and should be purchased from the manufacturer of the equipment. Purchasing a maintenance contract will assure that the manufacturer takes full responsibility for that equipment. However, even with a maintenance contract in place, the elevator should be checked on a routine basis by the facilities maintenance technician to ensure that the elevator is operating properly.

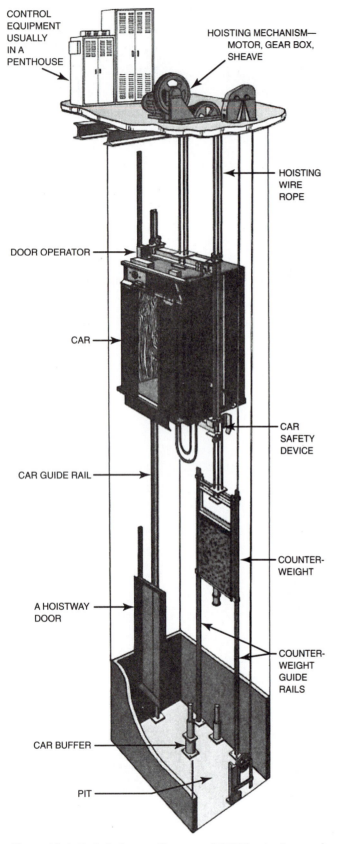

CONTROL EQUIPMENT USUALLY IN A PENTHOUSE

HOISTING MECHANISM— MOTOR, GEAR BOX, SHEAVE

HOISTING WIRE ROPE

DOOR OPERATOR

CAR

CAR SAFETY DEVICE

CAR GUIDE RAIL

COUNTER-WEIGHT

A HOISTWAY DOOR

COUNTER-WEIGHT GUIDE RAILS

CAR BUFFER

PIT

Figure 13-1: Typical elevator. (Courtesy of OTIS Elevator Company)

Checking and Inspecting Elevators

As mentioned earlier, although elevator maintenance and repair should only be performed by certified elevator mechanics, the facilities maintenance technician should still perform routine inspection of the elevator to ensure that the elevator continues to operate efficiently and safely. If any of the following situations should arise, then it is the responsibility of the facilities maintenance technician to report the problem to his or her supervisor and/or the elevator maintenance company.

- Increase in response time (waiting for the elevator)
- The platform of the elevator being not level with the floor when the door opens
- The elevator doors not opening fully at the destination floor for any reason
- Unusual noises during the operation of the elevator
- Overheating

Long Response Time (Waiting for the Elevator)

In addition to being an inconvenience to the occupants of a facility, long response time can also be an indicator that the elevator control system has or is developing a problem. On a regular basis, the time spent waiting for an elevator as well as its speed should be logged and compared to the manufacturer's specification. This is done by checking the time required for the elevator to travel from the bottom floor to the top floor during peak and non-peak times.

The Platform of the Elevator Being Not Level

When the elevator platform (the floor of the elevator) does not meet the floor of the building, it creates a tripping hazard. This can be as a result of the elevator platform either sticking above or below the structure's floor (Figures 13-2 and 13-3). According to accessibility codes, elevators must automatically level within ½ inch.

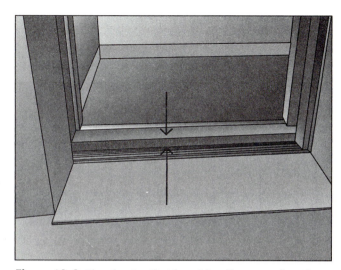

Figure 13-2: **The elevator that is not leveling properly—the elevator is above the floor**

Figure 13-3: **The elevator that is not leveling properly—the elevator is below the floor**

Overheating

Overheating is a common problem associated with electrical traction elevators because the equipment is usually located above the facility roof in a penthouse. Because the mechanical room is seldom climate controlled (heated and/or cooled instead they are ventilated—typically by louvers), they are allowed to have temperature swings depending on the outside air temperature. If the mechanical room's ventilation is located near the floor, then the outside air temperature can exceed 100°F. It should be noted that as the temperature in the mechanical room increases so does the possibility for breakdowns.

Checking the Elevator Door's Reaction Time

With a stopwatch, check the reaction time of the elevator doors (Figure 13-4).

1. Record a baseline time. This will be the time recorded the first time you measure the time the doors take to close.
2. On a regular basis, measure the time, record, and compare it against the baseline time (see Table 13-1).
3. Report the results to your supervisor.

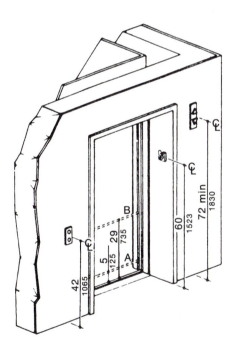

NOTE: The automatic door reopening device is activated if an object passes through either line A or line B. Line A and line B represent the vertical locations of the door reopening device not requiring contact.

Hoistway and Elevator Entrances

Figure 13-4: **Minimum entrance requirements for an elevator. (Uniform Federal Accessibility Standards)**

Elevator Door Reaction Time			
Baseline		Reaction	
Date	Time	Open	Close
Actual Reaction Time			
Date	Time	Open	Close

Table 13-1: **Elevator Door Reaction Time**

Elevators and ADA Requirements

An elevator is usually the easiest way to provide access to upper floors when access is required above the ground floor. If an elevator is installed, it must meet the ADA requirements whether the elevator was required. Every floor that the elevator serves will have to meet ADA access requirements regardless. The elevator must be automatic and have proper call buttons, car control, illumination, door timing, and reopening as shown in Figure 13-4. The elevator floor is a part of the access route so it must be similar to the floor described in walks and hallways. The shape and size of the elevator must be such that it is usable by a wheelchair, as shown in Figure 13-5.

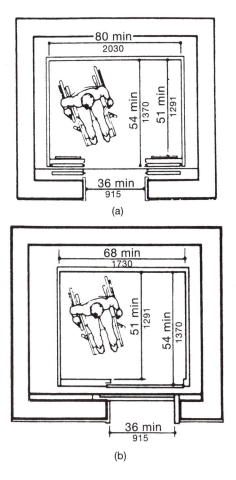

(a)

(b)

Minimum Dimensions of Elevator Cars

Figure 13-5: **Minimum dimensions of elevator cars must be considered in the initial building design. (Uniform Federal Accessibility Standards)**

Review Questions

1 True or false? If the platform of the elevator is not level with the floor when the door opens, the technician can make the required adjustments without consulting an elevator service technician.

2 True or false? If the elevator door does not open completely for any reason, the facilities maintenance technician should report it as soon as possible to a qualified elevator technician or a supervisor.

3 What should a facilities maintenance technician be looking for when inspecting an elevator?

4 What does a long response time for an elevator typically indicate?

5 Define the term elevator.

6 True or false? Only elevators that are accessible by the public are required to meet ADA requirements.

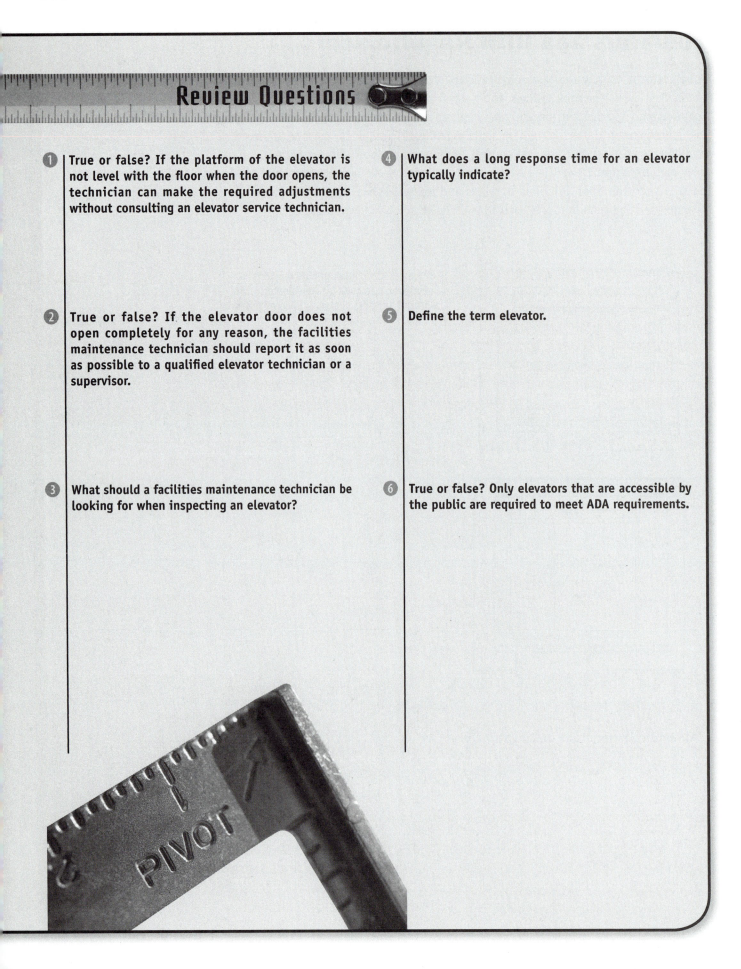

Name: _____

Date: _____

Elevators

Upon completion of this job sheet, you should be able to identify any safety requirements not being met by the elevator.

1 Is the elevator inspected and serviced on a regular schedule?   (Y/N)

2 Is the elevator capacity posted in the elevator car?   (Y/N)

3 Is there an emergency phone in the elevator?   (Y/N)

4 Is the turnaround space in the elevator 51 inches wide to conform to the Americans with Disabilities Act?   (Y/N)

Instructor's Response:

_____

_____

_____

Chapter 14 Pest Prevention

OBJECTIVES

By the end of this chapter, you will be able to:

Knowledge-Based

✪ Follow applicable safety procedures.

Skill-Based

✪ Recognize the sources of damage caused by pests.

✪ Select and apply proper techniques, chemicals, and/or materials to eradicate and/or prevent pest infiltration.

Glossary of Terms

Insecticide chemicals used to kill insects

Pesticide chemicals used to kill pests (rodents or insects)

Introduction

Pest prevention is normally not a high-priority task, until a problem occurs. It is not until a problem is reported that steps are taken to eliminate and then prevent the pest problems in the future. Pests are significant problems for people and property. The pesticides that are commonly used in pest control may pose potential health risks to the occupants as well as the environment, therefore whenever possible it is highly recommended that alternative methods should be employed.

Types of Pests and Pest Controls

The first step to controlling pests is to determine the type of pest that is infesting a facility. The following are the most common pests a facilities maintenance technician will have to deal with as well as the methods used to control them.

Figure 14-1: **Cockroach**

Figure 14-2: **Ant**

Figure 14-4: **Mouse**

Cockroaches

Controlling cockroach is seldom easy because getting the insecticide to the insect is difficult. The insecticide should have sufficient persistence to kill baby cockroaches as they hatch. If this fails, call your Environmental Health Department or pest control contractor (Figure 14-1).

Ants

Pour boiling water over the nest site and apply an **insecticide** (chemicals used to kill insects) powder. An insecticide lacquer can be applied around door thresholds or wall/floor junctions where ants run. Ant bait works so that the ant takes the bait back to the nest, killing the whole colony after a few days. Place it along where ants run (Figure 14-2).

Spiders

To control a spider, there is no need to kill it. Simply place a carton over it, and then slip a piece of thin cardboard between the carton and the surface to form a lid. Take the sealed container out of the building and let the spider go (Figure 14-3).

Figure 14-3: **Spider**

Rodents

Seal off entry points into the facilities. Ensure areas around the facilities, including the trash container areas, are clean. Poison is available as proprietary, ready-mixed bait. Serious or persistent infestations should be dealt with by a pest control contractor or the Environmental Health Department (Figure 14-4).

Flies (House and Fruit)

1. Inspection—locating the fly breeding and larval developmental sites.
2. Sanitation—the removal or elimination of the larval developmental sites. This step should eliminate the bulk of the fly problem so that mechanical and insecticidal measures will be more effective.
3. Mechanical controls—garbage receptacles with tight-fitting closures; tight windows and doors; windows securely screened if they can be opened; doors with self-closures; all holes through exterior walls for utilities, and so on, sealed; all vents securely screened; and the use of air curtains, insect light traps, sticky-surfaced traps, and others.
4. Insecticide application—using appropriately labeled pesticides.
   - Outdoors—includes the use of boric acid at the bottom of dumpsters.
   - Indoors—requires the use of automatic/metered dispensers and/or Ultra-low volume (ULV) applications on a room-by-room basis, with the low-oil formulations being more desirable (Figure 14-5).

Figure 14-5: **Fly**

Stinging/Biting Insects (Bees, Mosquitoes, Wasps, Hornets, and Ticks)

Remove the pest's food, water, and shelter by keeping the facilities clean and sanitizing outdoor areas. Keep tight-fitting lids on garbage cans and empty them regularly (Figures 14-6 through 14-10).

Figure 14-6: **Bees**

Termites

Like ants, termites are social insects that mostly feed on dead plant material (wood) and animal wastes. Currently

Figure 14-7: **Mosquito**

Figure 14-8: **Wasp**

Figure 14-9: **Hornet**

Figure 14-10: **Ticks**

Figure 14-11: **Termites queen**

there are approximately 4,000 species of termites in the world. Termites when left alone can cause serious structural damage to structures and plantation forest. Detecting and controlling termites is a job for the professionals (Figure 14-11). Although termite control should be left to the professionals, a facilities maintenance technician can do a few things to help eliminate and prevent termites. They are as follows:

- Eliminate moisture problems from leaky pipes, AC units, storm, and drainage.
- Remove food sources such as dead trees, stumps, fire wood, and damaged wood structures.

Pest Control

Before facilities maintenance technicians can effectively render pest control to a structure, they must identify the type of pest(s) that must be controlled. Once this has been accomplished, the source of food, entry, and attraction must be identified and eliminated. This includes the following:

- Check exterior doors. If you can see light under the door, this is a potential problem.
  - Install door thresholds.
  - Seal around windows.
- Check for windows that do not fit properly or have holes in the screens.
  - Seal around windows.
  - Use mesh or screens to fix holes in the screens.
- Check for openings around any objects that penetrate the building's foundation such as plumbing, electrical service, telephone wires, HVAC, and so on.
  - Seal around these objects.
- Do not store materials against the foundation of a building. This could be a nesting place for bugs.

- Do not leave outside lights on all of the time. Light attracts bugs.
  - Use motion sensor lights.
- Maintain a plant-free zone of about 12 inches around the building to deter insects from entering.
- Fix problem areas in the structure of the building that provide a nesting place for birds or rodents.
- Place outdoor garbage containers away from the building and on concrete or asphalt slabs and keep the area clean.
- Keep facilities and area around facilities clean.

If these measures do not control and/or eliminate the pest in a structure, consider selecting a **pesticide** that will control the pest and have the least possible harmful effects on the occupants and the environment.

Nonpesticidal Pest Control

Although pesticides can be successfully used to eliminate pests, they can be harmful to the occupants and pets. Therefore, it is recommended researching and understanding nonpesticidal pest control methods that can be applied to a particular type of pest. Nonpesticidal pest control methods include the following:

- Trapping
- Hoeing
- Hand weeding
- Excluding the pest with barriers
- Sanitizing the area
- Removing food, water, and/or cover for the pest
- Natural predators
- Natural insecticides (neem-tree products)
- Ultrasonic pest control (use ultrahigh frequency sound waves to repel insects and rodents)

Pest Control with Pesticides

Preventive applications of pesticides should be discouraged, and treatments should be restricted to areas of known pest activity. When pesticides are applied, the least toxic product(s) available should be used and applied in the most effective and safe manner.

Effects of Pesticides on Pests

Not all pesticides eliminate pests the same way. The most common methods of eliminating pests used by pesticides are listed as follows:

- Stomach poison—kills when swallowed
- Contact poison—sprayed directly on pest
- Fumigants—gas inhaled or absorbed
- Systemics—will kill pest when it eats the host, but does not harm host
- Protectants—prevent pest entry

Always follow manufacturer's instructions and never mix pesticides with other chemicals. Because pesticides can be harmful to the environment, never dump them into a drain or dumpster. Always follow the manufacturer's disposal instructions as well as any local, state, and federal regulations.

Pesticide Safety

Because pesticides are considered a low-tech approach to controlling and eliminating pests, they do not discriminate against whom they will kill and/or affect. Therefore, when working with pesticides, it is important to use the following safety procedures:

- Choose the lowest toxicity pesticide that can be used legally on the target area, crop, or plant and that will safely and effectively control the pest.
- Plan ahead and buy no more pesticide than you need.
- Keep pesticides separate from other items.
- Make sure you have the proper safety and application equipment available and know how to use it.
- Read, understand, and follow the pesticide's label directions.
- Examine the area to be treated and the surrounding area. If there are plants or animals that could be harmed by the pesticide, don't spray if you cannot guarantee they will not be injured.
- Never put potentially hazardous wastes, such as pesticides, directly in the garbage or pour remaining chemicals down the drain. Check if your community has a household hazardous waste collection program.
- Store pesticides out of reach of children—in locked cabinets or in cabinets with childproof latches.
- Store pesticides only in their original containers with labels visible and intact.

Applying Pesticides

The pesticide application equipment you use is important to the success of your pest control job. First, you must select the right kind of application equipment; then you must use it correctly and take good care of it. Remember, pests of any kind are a nuisance to tenants. When working on a complaint about pests, be as nonintrusive as possible, but do whatever is necessary to eliminate the pests. Once the elimination or preventive measures have been performed, check back with the tenant to make sure the pests have been eliminated.

Sprayers

Sprayers are the most common pesticide application equipment. They are standard equipment for nearly every pesticide applicator and are used in every type of pest control operation. Sprayers range in size and complexity from simple, handheld models to intricate machines weighing several tons.

Hand Sprayers

Hand sprayers are often used to apply small quantities of pesticides. They can be used in structures and can be used outside for spot treatments or in hard-to-reach areas. Most operate on compressed air supplied by a hand pump.

- Pressurized can (aerosol sprayer)—consists of a sealed container of compressed gas and pesticides (Figure 14-12).
- Trigger pump sprayer—forces the pesticide and diluents through the nozzle by pressure created when the trigger is squeezed. The capacity of trigger pump sprayers ranges from 1 pint to 1 gallon (Figure 14-13).

Figure 14-12: **Pesticide aerosol can**

Figure 14-13: **Trigger pump pesticide sprayer**

- Hose-end sprayer—causes a fixed rate of pesticide to mix with the water flowing through the hose to which it is attached (Figure 14-14).
- Push–pull hand pump sprayer—works with a hand-operated plunger that forces air out of a cylinder, creating a vacuum at the top of a siphon tube. The suction draws pesticide from a small tank and forces it out with the air flow. Capacity is usually 1 quart or less (Figure 14-15).
- Compressed air sprayer—usually a hand-carried sprayer that operates under pressure created by a self-contained manual pump (Figure 14-16).

Figure 14-14: **Spraying with hose-end pesticide sprayer (Image copyright Alistair Scott, 2009. Used under license from Shutterstock.com)**

Figure 14-15: **Push–pull hand sprayer pump**

Figure 14-16: **Compressed air pesticide sprayer**

Figure 14-17: **Estate pesticide sprayer**

Figure 14-18: **Power backpack sprayer**

Small Motorized Sprayers

Small motorized sprayers are usually not self-propelled. They may be mounted on wheels, so they can be pulled manually, mounted on a small trailer for pulling behind a small tractor, or skid-mounted for carrying on a small truck. They may be low-pressure or high-pressure sprayers, according to the pump and other components with which they are equipped.

- Estate sprayers—mounted on a two-wheel cart with handles for pushing (Figure 14-17).
- Power backpack sprayer—usually a backpack-type sprayer that has a small gasoline-powered engine. This model can generate high pressure and is best suited for low-volume applications of diluted or concentrated pesticides (Figure 14-18).

Review Questions

1. Describe the difference between pest control with pesticides and nonpesticidal pest control.

2. Describe how to control the following pests:

   Cockroaches

   Ants

   Spiders

   Termites

3. List the way in which pesticides kill pests.

4. What is the difference between a pesticide and an insecticide?

5. How do ultrasonic pest controls manage pest?

Name: _____

Date: _____

Pest Prevention

Identify and Remove Pests

Upon completion of this job sheet, you should be able to identify pests that may be causing problems at your facility and develop a plan to remove them.

1 Inspect your facility or training area and make a list of the pests that you observe or pests that the tenants report to you. If possible, use a digital camera to photograph the pests for easier identification.

2 Document where the pests were observed and what they were doing at the time of observation.

3 Estimate the level of infestation to determine if there needs to be a plan in place to remove the pests.

4 Research the appropriate method for removing the pests if deemed necessary and develop a plan.

5 Implement your plan if applicable.

Instructor's Response:

_____
_____
_____

Chapter 15 Landscaping and Groundskeeping

OBJECTIVES

By the end of this chapter, you will be able to:

Knowledge-Based

- ✪ Identify the various parts of a plant.
- ✪ Discuss the proper procedure for installing a retaining wall.
- ✪ Discuss the proper procedure for removing snow.

Skill-Based

- ✪ Maintain and police grounds including mowing, edging, planting, mulching, leaf removal, and other assigned tasks.
- ✪ Perform basic small engine repair and preventive maintenance according to the manufacturer's specifications.
- ✪ Perform basic swimming pool maintenance that does not require certification.
- ✪ Remove refuse and snow as required.
- ✪ Maintain public areas including hallways, kitchens, and lobbies.
- ✪ Repair asphalt using cold-patch material.

Glossary of Terms

Groundskeeping the activity of tending an area for aesthetic or functional purposes. It includes mowing grass, trimming hedges, pulling weeds, planting flowers, and so on

Landscaper an individual that modifies the feature of an area (land) such as construct stone walls, wooden fences, brick pathways, statuary, fountains, benches, trees, flowers, shrubs, and grasses

Belly deck a mower that is built like a car and has a blade deck underneath the driver and engine

Organic mulch constitutes natural substances such as bark, wood chips, leaves, pine needles, or grass clippings

Inorganic mulch constitutes gravel, pebbles, black plastic, and landscape fabrics

Introduction

*Groundskeeping is the activity of tending an area for aesthetic or functional purposes. It includes mowing grass, trimming hedges, pulling weeds, planting flowers, and so on. A **groundskeeper** is a person who maintains the functionality and appearance of landscaped grounds and gardens, whereas a **landscaper** is an individual that modifies the feature of an area (land) such as construct stone walls, wooden fences, brick pathways, statuary, fountains, benches, trees, flowers, shrubs, and grasses.*

Landscaping and groundskeeping workers maintain grounds using hand or power tools or equipment. Workers typically perform various tasks, which may include sod laying, mowing, trimming, planting, watering, fertilizing, digging, raking, sprinkler installation, and installation of mortarless segmental concrete masonry wall units.

Type of Grass	Desired Height (Inches)
Common Bermuda	1.5
Hybrid Bermuda	1
Kentucky Blue	2
St. Augustine	2.5
Zoysia japonica	1.5
Zoysia matrella	1

Table 15-1: **Desired Grass Height for Common Grasses**

Mowing

Mowing is the process of maintaining the height of the lawn using a lawnmower to cut the grass. Proper lawn maintenance is more involved than simply starting a power mower and making a few laps around the grounds. Although it is not rocket science, there are a few basic rules that should be followed:

1. Never cut grass too short (see Table 15-1).
2. Never remove more than one-third of the grass height in one session.
3. Never mow the grass when it is wet.
4. Change the direction in which the grass is cut with each mowing.
5. On average a lawn should be mowed once a week.
6. Growing season is usually from May to October.

Lawn Mowing

Depending on the size of the lawn to be maintained, most facilities maintenance technicians use either a push mower (which may be self-propelled) or a riding mower (Figure 15-1).

Riding mowers are generally available in different sizes and types. The most common riding mower for residential use is one with a "belly" deck. A **belly deck** is a mower that is built like a car (four wheels, a driver's seat, a steering wheel, engine located in the front) and has a blade deck underneath the driver and engine (Figure 15-2).

Edging and Trimming

Traditionally edging and/or trimming was done as the final step in mowing a lawn. However, depending on the size of the groundskeeping crew it should be

Figure 15-1: **Push mower (Image copyright Margo Harrison, 2009. Used under license from Shutterstock.com)**

Figure 15-2: **Riding mower**

done before or while the lawn is being mowed. This allows the clipping generated to be either mulched and left on the lawn or mulched and bagged by the mower. Trimming and edging is a simple but all-important practice that gives a lawn a finished, manicured look. When edging:

- Use a handheld edger between pavement and grass. Place the wheel on the pavement with the blade over the edge and push and pull. For large lawns, use a power-driven model (Figure 15-3).
- Edge along walkway to prevent grass from growing out past the borders.
- Edge along the edges of turf and landscape beds to prevent grass from growing into the beds.

When trimming:

- Use grass shears around delicate plants and trees that could be damaged using string trimmers (Figure 15-4).
- Use a string trimmer in areas that are difficult or impossible to reach with a lawn mower (Figure 15-5).

Figure 15-3: **Handheld edger**

Figure 15-4: **Grass shears (Image copyright Mark Studio, 2009. Used under license from Shutterstock.com)**

Figure 15-5: **String trimmer**

Introduction to Botany

Botany is a branch of science that is concerned with the study of plant life and its development. Plants are vital to all life forms on this planet by providing food, fuel, medicine, and the generation of oxygen. Therefore the study of plants is critical for:

- The production of food
- The production of medicine
- Environmental changes due to changes in plant life

Parts of a Plant

To understand how to take care of plants, understand the different parts of plants and their function. Although there are thousands of different plant varieties, all plants produce seeds and most produce their own food using photosynthesis. The basic parts of a plant are roots, stems, leaves, flowers, and fruits (see Figure 15-6).

Roots

Roots of a plant are used to absorb water and minerals from the soil as well as help stabilize and anchor the plants to the ground.

Stems

Stems are the plant's transportation system that helps transport the water and minerals taken in by the roots. In addition, stems are used to transport the food produced by the leaves to the remaining portions of the plant.

Leaves

Leaves are the tiny plant food factories, in which sunlight is captured and combined with the minerals and water to produce sugar that the plant uses to produce food.

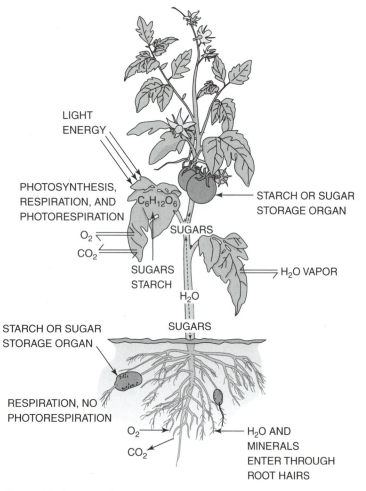

Figure 15-6: **Parts of a plant**

Flowers

Flowers are the reproductive portion of the plant. Flowers contain both pollen and tiny eggs called ovules, which when fertilized (pollinated) turn into fruit.

Fruits

Fruit is a protective covering for the plant's seed. In other words, fruit is a ripened ovary.

Seeds

Seeds are small embryonic plants that contain stored food enclosed in a covering.

Mulching

Mulch is defined as any loose material placed over soil to control weeds and conserve soil moisture by providing 10 to 25% reduction in soil moisture loss from evaporation. In addition, mulch helps keep the soil well-aerated by reducing soil compaction that results when raindrops hit the soil. It also reduces water runoff and soil erosion. Mulching trees and shrubs is a good method to reduce landscape maintenance and keep plants healthy.

Nature produces large quantities of mulch all the time with fallen leaves, needles, twigs, pieces of bark, spent flower blossoms, fallen fruit, and other organic material. There are basically two types of mulches: organic and inorganic. Both types may have their place in the landscaping. **Organic mulch** constitutes natural substances such as bark, wood chips, leaves, pine needles, or grass clippings. Organic mulches attract insects, slugs, cutworms, and the birds that eat them. They decompose over time and need to be replaced after several years. **Inorganic mulches**, such as gravel, pebbles, black plastic, and landscape fabrics, do not attract pests and they do not decompose. A 2–4-inch layer (after settling) is adequate to prevent most weed seeds from germinating. Mulch should be applied to a weed-free soil surface (Figure 15-7).

Figure 15-7: **Mulch around plants (Image copyright Lijuan Guo, 2009. Used under license from Shutterstock.com)**

Aeration

Lawn aeration involves the removal of small soil plugs or cores out of the lawn. Although hand aerators are available, most aeration is done mechanically with a machine having hollow tines or spoons mounted on a disk or drum. Known as a core aerator, it extracts ½- to ¾-inch diameter cores of soil and deposits them on your lawn. Aeration holes are typically 1–6 inches deep and 2–6 inches apart. Other types of aerators push solid spikes or tines into the soil without removing a plug (spiking). These aerators are not as effective because they can contribute to compaction. Core aeration is a recommended lawn care practice on compacted, heavily used turf and to control thatch buildup (Figure 15-8).

Aeration should be done twice a year, spring and fall. Mark sprinkler heads with flag or paint before aeration.

Figure 15-8: **Aeration machine**

1. The soil should be moist but not wet.
2. Lawns should be thoroughly watered 2 days prior to aerating, so that tines can penetrate deeper into the soil and soil cores easily fall out of the tines. If you are aerating after prolonged rainfall, wait until the soil has dried somewhat so that soil cores do not stick in the hollow tines.
3. Aerate the lawn in at least two different directions to ensure good coverage. Be careful on slopes, especially steep ones, as well as near buildings and landscape beds.

Amending Soil Prior to Installation

A common misconception of most people is the distinction between dirt and soil. Basically dirt is the soil that has been depleted of the nutrients needed to support and maintain plant life (its fertility). The primary nutrients required to support life are nitrogen, phosphorus, and potassium. Amending the soil is the process of changing its ability to sustain plant life, that is, adding necessary nutrients and/or changing the texture (the ratio of sand, silt, and clay). The type of plant life that is being planted will determine the type and amount of nutrients as well as the soil texture needed.

Nitrogen

Plants use nitrogen to produce protein (in the form of enzymes) and nucleic acids. Nitrogen is present in chlorophyll, a green pigment responsible for photosynthesis. It is distributed throughout a plant from older to younger tissue. A plant that is deficient in nitrogen will have a yellowing in the older leaves.

Phosphorus

Phosphorus is another essential element in the process of photosynthesis. It is also used in the development of the plant's root system as well as the production of the plant's flowers.

Potassium

Potassium is essential for the overall health of the plant by aiding in the production of proteins, photosynthesis, fruit quality, and reduction of diseases.

Using Bushes and Shrubs in Landscapes

Both flowering shrubs and evergreen bushes have their place in landscaping. Many flowering shrubs attract birds with their berries and provide both fall foliage and winter interest. Evergreen bushes can be pruned into hedges or can function as privacy screens. Either flowering shrubs or evergreen bushes can stand alone as specimens used for focal points (see Figures 15-9 and 15-10).

Planting Bushes and Shrubs

When planting bushes and shrubs, always follow the instruction provided with the plant or the retailer in which the plant was purchased. To plant a shrub or a bush, follow these guidelines:

1. Select a location for the plant where the following are kept in mind:
   - Growing condition (temperature, water conditions, sunlight, etc.)
   - Potential hazards (to the plant)
2. When digging the receiving hole for the plant, be sure that it is twice as wide and deep as the pot or rootball of the plant.
3. Add a layer of mulch that is approximately one-fourth the depth of the receiving hole.
4. Add a cup of slow release fertilizer to the receiving hole.

Mulching the beds is done mid-spring, usually with bark or woodchips 2–3 inches in depth to help keep the weeds to a minimum.

Figure 15-9: **Flowering shrubs (Image copyright P. Phillips, 2009. Used under license from Shutterstock.com)**

Figure 15-10: **Evergreen bushes (Image copyright zhuda, 2009. Used under license from Shutterstock.com)**

5. Add two shovelful of dirt to cover the fertilizer.
6. Fill the hole with water and allow it to absorb into the compost and surrounding soil.
7. Remove the plant roots from the pot or burlap wrap.
8. Gently spread out its roots.
9. Place the root ball centrally in the hole and backfill to the soil line on the plant.
10. Gently tamp the soil to hold the bush upright.

Retaining Walls

A retaining wall is a structure that is designed to hold back soil and/or rock from a facility or an area. Retaining walls are used to control erosion, extend usable grounds, and control changes in grade (the slope of the land). When built correctly, they can add beauty and value to any property.

Installing a Cinder Block Retaining Wall

Concrete block retaining walls, also known as cinderblock retaining walls, are more extensive to construct and are usually longer lasting than other forms of retaining walls (landscaping timbers, see Figure 15-11). However, with a little planning, these types of retaining walls can be easily constructed. Before attempting

Figure 15-11: **Cinder block retain wall (Image copyright Steven Frame, 2009. Used under license from Shutterstock.com)**

to construct a cinderblock retaining wall, always consult local and state building codes. To construct a cinderblock retaining wall:

1. Determine where the retaining wall must be located.
2. Excavate the footing. The size of the footing will be determined by the height of the wall. Normally for a retaining wall 3-foot high, the footing should be 24 inches wide and 8 inches deep.
3. Place three rows of ½ inch steel rebar 6 inches from the outside edges, and one in the center should be placed in the footer about ⅓ inch the depth of the footer from the bottom.
4. Using 4,000 psi concrete, pour the footer.
5. Place ½ inch rebar perpendicular to the footer where the row of blocks will be installed. Ensure that the position of the rebar will match the cavities of the blocks!
6. Snap a chalk line to position the first row of cinder blocks. Lay a ½ inch thick mortar bed and apply mortar to the blocks, tap the blocks into position, and level as required.
7. For the second course, alternate the joints as it is a standard practice in masonry for maximum strength.

Figure 15-12: Castlerock retaining wall (Image copyright Denise Kappa, 2009. Used under license from Shutterstock.com)

8. For retaining walls over 3-foot tall, deadmen should be installed every 4 feet.
9. Attach galvanized deadmen cables parallel to the retaining wall and buried in undisturbed soil.
10. On the uphill side of the footing, place preformed weeping pipe in a layer of gravel or crushed stone.
11. Backfill the first two courses adjacent to the wall with gravel and pack into place using a tamper.
12. Fill the first two courses with concrete.
13. Install all additional courses filling all of them with concrete and deadmen supports as required.
14. Install retaining wall facing (flagstone, brick, stucco) to give the wall a finish look.
15. Cap the final course as desired and complete the backfilling.

Installing a Castlerock Retaining Wall

Castlerock or mortarless block retaining walls are much simpler to install and maintain (see Figure 15-12). In addition, if you make a mistake and most likely you will, castlerock retaining walls can be easily dismantled and adjusted or corrected. Finally the skill level required to construct a cinderblock retaining wall is much greater than that of a castlerock retaining wall. To construct a castlerock retaining wall:

1. Determine where the retaining wall must be located.
2. Excavate the footing. Normally for a castlerock wall 3-foot high, the footing should be 18 inches wide

and 6 inches deep. The foundation course should be below ground level. Add a layer of leveling sand or paver base (1–2 inches) and tamp.

3. Set the first castle rock into place checking it to ensure that it is level. If the block is not level then using a hammer, tap the block to adjust it.

4. Continue this until the entire foundation course has been set.

5. After the foundation course has been set, starting with the end of the retaining wall, offset the second block (like you would for cinder blocks) and set into position.

6. Attaching a line between the two blocks and a line level, check to see if the blocks are level with one another. If the blocks are not level adjust them until they are level.

7. Once the blocks are level continue laying the course.

8. For each course repeat Steps 6 and 7 until the desired height has been reached.

9. To prevent soil from washing through the spaces in the retaining wall, backfill using crushed rock or gravel from the retaining wall to 8–12 inches out from the retaining wall.

Installing a Landscaping Timber

Like castlerock, landscaping timber retaining walls are much simpler to install and maintain than cinder blocks (see Figure 15-13). To construct a landscaping timber retaining wall:

1. Determine where the retaining wall must be located.

2. Excavate the footing. Normally for a landscaping timber retaining wall 3-foot high, the footing should be 18 inches wide and 6 inches deep. The footer should now be filled with crushed stone or gravel and tamped flat.

3. Using a chainsaw, cut the landscaping timber to size as needed.

4. Using a drill, drill holes through the timbers for the rebar. The holes should be spaced about 4-foot apart.

5. Place the first course of landscaping timber into position and drive the 24–36 inch rebar through the timber into the ground.

6. Before adding subsequent courses, check each course to ensure that it is level. If the course is not level adjust the course using a sledge hammer.

7. Lay the next course into position and connect it to the previous course using 12-inch galvanized spikes about every 24 inches.

8. To every third course add a deadmen anchored to the course using 12-inch galvanized spikes.

9. Once the desired height has been achieved, backfill the retaining wall with crushed rock or gravel from the retaining wall to 8–12 inches out from the retaining wall.

Figure 15-13: **Landscaping retaining wall (Image copyright Bonnie Watton, 2009. Used under license from Shutterstock.com)**

Paving Stone Walks

Paving stones can offer an attractive, quick, and easy alternative to concrete in constructing a patio or walkway for a fraction of the cost.

Figure 15-14: Paving stone walk (Image copyright Elena Elisseeva, 2009. Used under license from Shutterstock.com)

Installing Paving Stone Walks

The process for installing a paving stone walk is similar to that of a castlerock retaining wall (see Figure 15-14). To install a paving stone walk:

1. Determine where the walk will be located.
2. Excavate the walk area. If the walk is designed for pedestrian traffic then the area should be excavated 7–9 inches with a 4–6-inch gravel base. For vehicular traffic the area should be excavated 9–11 inches with a 6–8-inch gravel base. The base should be tamped flat and level.
3. Install the edge restrains in their desired location and shape. Edge restraints are used to prevent the pavers from moving.
4. Install the bedding sand. The bedding sand is installed by laying 1-inch conduit 6–8-feet apart in the area to install the pavers followed by screening the sand into place using a 2 × 4. Once the bedding sand is screened into place, remove the 1-inch conduit.
5. Starting in one of the corners lay the pavers in the desired pattern.
6. Sweep the surface to remove any debris. Spread masonry sand over the surface, sweeping it into the joints leaving surplus sand on the pavers. Tamp the pavers down using the plate compactor.
7. Sweep the remaining excess dry sand over the surface filling the joints.

Irrigation Systems

An irrigation system is a network of piping, valves, and controls used to artificially maintain the moisture level in the soil. The most common types of irrigation systems currently in use today are ditch irrigation, terraced irrigation, drip irrigation, sprinkler system, rotary systems, and center pivot irrigation.

Winterizing the Irrigation System

Every year, before the first freeze, winterization becomes the priority for all irrigation systems that are in parts of the country where the frost level extends below the depth of the installed piping. Winterization is important because it helps minimize the risk of freeze damage to the irrigation system.

Manual Drain Method

This method is used when the manual valves are located at the end and low points of the irrigation piping (Figure 15-15).

To drain these systems:

1. Shutoff the irrigation water supply. The shutoff will be located in the basement and will be a gate/globe valve, ball valve, or stop and waste valves (see Figures 15-16 through 15-18).
2. Open all the manual drain valves.

Figure 15-15: **Valves on the irrigation piping**

Figure 15-16: **Gate valve**

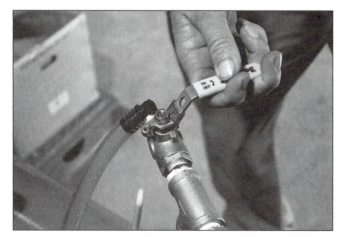

Figure 15-17: **Ball valve**

Figure 15-18: **Stop/waste valve**

3. Once the water has drained out of the mainline, open the boiler drain valve or the drain cap on the stop and waste valve and drain all the remaining water that is present between the irrigation water shutoff valve and the backflow device (Figure 15-19).

4. Open the test cocks on the backflow device. If your sprinklers have check valves, you will need to pull up on the sprinklers to allow the water to drain out the bottom of the sprinkler body (Figure 15-20).

Figure 15-19: **Boiler drain valve/drain cap**

Figure 15-20: **Test cocks**

Figure 15-21: **Valves at the end of irrigation piping**

Automatic Drain Method

This method is used when the automatic drain valves are located at the end and low points of the irrigation piping (Figure 15-21). These valves will automatically open and drain water if the pressure in the piping is less than 10 psi.

1. To activate the automatic drain valves, shut off the irrigation water supply and activate a station to relieve the system pressure. The shutoff will be located in the basement and will be a gate/globe valve (Figure 15-18), ball valve, or stop and waste valves (Figures 15-22 and 15-23).

Figure 15-22: **Gate valve**

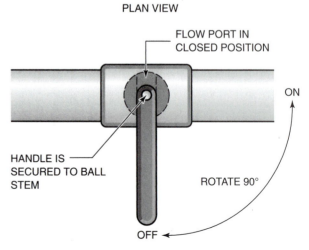

PLAN VIEW

FLOW PORT IN
CLOSED POSITION

ON

HANDLE IS
SECURED TO BALL
STEM

ROTATE 90°

OFF

Figure 15-23: **Ball valve**

2. Once the water has drained out of the mainline, open the boiler drain valve or the drain cap on the stop and waste valve and drain the remaining water that is present between the irrigation water shutoff valve and the backflow device (Figure 15-19).

3. Open the test cocks on the backflow device. If your sprinklers have check valves, you will need to pull up on the sprinklers to allow the water to drain out the bottom of the sprinkler body (Figure 15-20).

"Blow-Out" Method

It is recommended that a qualified licensed contractor perform this type of winterization method. The blow-out method utilizes an air compressor with a cubic foot per minute (CFM) rating of 125–185 for any mainline of 2 inches or less and a PSI of 50–80.

1. Open the test cocks on the vacuum breaker (Figure 15-20).
2. Shutoff the irrigation water supply and open the drain on the supply line.
3. Once the line is drained, close the drain.
4. Attach the compressor to the mainline via a quick coupler, hose bib, or other type of connection, which is located before the backflow device (Figure 15-24).
5. Activate the station on the controller that is the zone or sprinklers highest in elevation and the farthest from the compressor (Figure 15-25).
6. Do not close the backflow isolation or test cock valves. Slowly open the valve on the compressor; this should gradually introduce air into the irrigation system. The air pressure should be constant at 50 psi. If the sprinkler heads do not pop up and seal, increase the air until the heads do pop up and seal. The air pressure should NEVER exceed 80 psi.
7. Activate each station/zone starting from the farthest station/zone from the compressor, slowly working your way to the closest station/zone to the compressor. Each station/zone should be activated until no water can be seen exiting the heads. This should take approximately 2–4 minutes per station/zone.

Spring Irrigation Startup

If the system was correctly winterized in the fall, the chances of cracks and breaks due to freezing were greatly reduced. But even properly winterized systems are

Figure 15-24: **Compressor attached to the mainline**

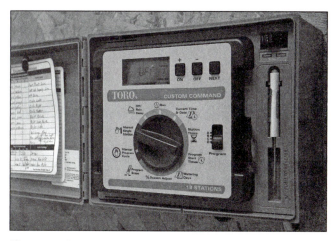

Figure 15-25: **Station controller**

subject to damage from extreme conditions. Therefore the spring startup procedure should be done in four stages:

1. Safely reintroducing water to each zone
2. Checking winter damage, making repairs as needed, and cleaning or replacing nozzles
3. Examining the entire system to see that it is still operating the way it is supposed to and providing even coverage
4. Resetting the controls

Starting Up the Irrigation System in the Spring

Follow this procedure when first starting an irrigation system in the spring:

Figure 15-26: **Manual drain valve**

1. Before turning on any water to the system, make sure that all manual drain valves are returned to the CLOSED position (Figure 15-26).
2. Open the system main water valve slowly to allow pipes to fill with water gradually. If these valves are opened too quickly, sprinkler mainlines are subjected to high surge pressures, uncontrolled flow, and water hammer.
3. Verify the proper operation of each zone valve by manually activating it from the controller.
4. Activate each station on the controller, checking for proper operation of the zone. Check for proper operating pressure (low pressure indicates a line break or missing sprinkler), proper rotation and adjustment of sprinkler heads, and adequate coverage. Check and clean filters on poorly performing sprinklers. Adjust heads to grade as necessary.
5. Reprogram the controller for automatic watering. Replace the controller backup battery if necessary.
6. Uncover and clean the system rain sensor, if applicable (Figure 15-27).
7. Finish and clean any in-line filters for drip irrigation zones.

Pool Maintenance

Routine pool maintenance is an unavoidable fact of life if you want to keep your pool looking clean, sparkling, and inviting day after day.

Cleaning the Pool Deck

In addition to maintaining the pool, most facilities maintenance technicians are responsible for the upkeep of the pool deck. When maintaining the pool deck area:

1. Remove as much debris as possible from the pool.
2. Sweep or use a hose to remove the debris near the pool. Cover pool if necessary (Figure 15-28).

Figure 15-27: **Rain sensor**

Cleaning the Surface of the Pool

Dirt floating on the surface of the water is easier to remove than it is from the bottom. Remove floating debris off the surface using a leaf rake and telepole. As the net fills, empty it into a trash can or plastic garbage bag. Do not empty the skimming debris into the garden or on the lawn for the debris is likely to blow right back into the pool as soon as it dries out.

There is no particular method to skim, but as you do, scrape the tile line, which acts as a magnet for small bits of leaves and dirt. The rubber–plastic edge gasket on the professional leaf rake will prevent scratching the tile (Figure 15-29).

If there is scum or general dirt on the water surface, squirt a quick shot of tile soap over the length of the pool. The soap will spread the scum toward the edges of the pool, making it more concentrated and easier to skim off.

Figure 15-28: **Pool deck**

Maintaining the Pool

In addition to maintaining the appearance of the pool (removing trash, dirt, etc.) the quality of the water must also be maintained. Maintaining the integrity of the water in the pool can be accomplished by:

1. Sanitize your pool with a stabilized chlorine product to provide protection against bacteria. These products are generally available in stick or tablet form and are fed into a distribution container near the pump and filter system.

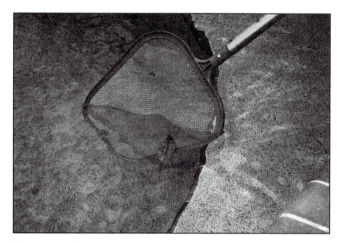

Figure 15-29: **Skim debris from the pool**

2. Use an algae preventive product or inhibitor to help keep the more than 15,000 kinds of algae from ever getting started. This liquid product is simply poured into the water near the skimmer intake so that the pump system can distribute it to all areas of the pool.
3. Find a pool professional or supply dealership that has a water test facility, or access to one, and that offers computer analysis of samples you bring in.
4. Check for chlorine and pH levels daily. To check pH in pool water, use a kit from your local pool store. It contains tubes that hold about 4 ounces of pool water. You then put in drops of liquid that comes with the test kit, shake for about 30 seconds, and then match the colors with the accompanying chart to check the chlorine and pH levels (yellow for chlorine and red for pH).

Snow Plowing

Unfortunately, as with rainfall, you cannot always predict when or how much snow will fall in any one spot. When it does fall, it is the responsibility of the facilities maintenance technician to remove the snow. This does not mean that the technician has to remove the snow himself or herself; the snow removal service can be contracted out.

Recommended Snow Removal Procedures

The following procedure should be followed when removing snow:

Figure 15-30: **Clear walkways**

Figure 15-31: **Snow on side of parking lot**

- Steps, large walk-through areas, and large entry ways shall be partially shoveled with a path along the railings for initial opening of these areas. Handicap areas must be fully accessible (Figure 15-30).
- Cleanup operations after a storm shall involve completion of opening all walks and entry ways and deicing.
- Snow shall be pushed back from sides of roadways, walks, and parking lots (Figure 15-31).
- Stairs and entry ways shall have all remaining snow removed.
- Ice choppers and ice-melt applications will then be employed on remaining ice.
- Application of ice-control products will follow plowing, based on current weather conditions, or when freezing occurs, as determined by the grounds supervisor.
- Return trips to the surface to remove melted ice and slush will complete the cleaning of all surfaces.
- Return trips for sanding and salting equipment shall be made as often as necessary on roads, as determined by the grounds supervisor.

When parking lots are plowed, snow should be piled so as not to block thoroughfares and sidewalk areas. If snow has to be pushed up over a curbed area, it should be piled so that it will not fall back into the lot and will still be clear of any adjacent sidewalks.

Small Engine Repair

The facilities maintenance technician should be able to perform basic small engine repair and preventive maintenance according to manufacturer's specifications. However, to avoid voiding the warranty of a lawnmower, trimmer, chain saw, etc., never work on any equipment that is still under warranty.

Common Problems with Small Engines

Some common problems associated with small engines in which a facilities maintenance technician should be able to address are engines that will not start, engines that smoke, and engines that sputter.

Engine Will Not Start

One of the most common problems that a facilities maintenance technician will have to troubleshoot is engines that will not start.

- Check for proper fuel level—Fill gas tank to just below the fill neck so there is room for the gas to expand and slosh around. If you fill the tank all the way to the top, then gas will leak out as the engine shakes around.
- Check to see if the fuel valve is in the ON position.
- Check the spark plug—If it is "carbon shorted," carbon present between the electrode gap, clean or replace it. If the plug is pitted, burned, or has cracked porcelain, replace it with an identical replacement spark plug (Figure 15-32).
- Check for spark—With a commercially available spark tester, test for spark by putting the spark plug wire (high-tension lead) on one side of the tester and clip the other side of the tester to the shroud, fins, or head bolts (anything metallic). Or ground the plug, with high tension lead on it, to the head of the engine and crank on the engine. Spark is present when you see blue sparks jump the electrode gap.
- Prime or choke the engine—If the engine has a primer or choke, use them! Prime the engine three times and then start. Choke engine until it starts to fire and then open the choke.

If the engine still won't start after going through these steps, take it in to a repair center to be repaired.

Engine Is Smoking

The second common problem that a facilities maintenance technician will have to troubleshoot is engines that smoke.

- Dirty or plugged air filter—Replace or clean air filter (Figure 15-33).
- Wrong grade of oil—Oil that is less than 30w HD will vaporize if used for a long time. Use only a good grade of 30w HD oil for all four-cycle lawn equipment.
- Worn valve guides—A valve guide that is worn out due to excessive wear on the engine must be taken to the shop for a replacement guide bushing. Common causes are mowing on a hillside for excessive amounts of time and debris buildup on the cooling fins (Figure 15-34).

Figure 15-32: **Spark plug (Image copyright Dewayne Flowers, 2009. Used under license from Shutterstock. com)**

Figure 15-33: **Dirty air filter**

Figure 15-34: **Valve guide**

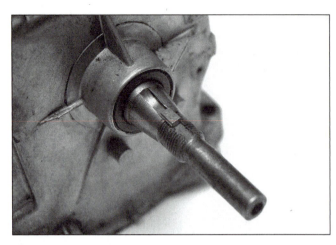

Figure 15-35: Flywheel key

Figure 15-36: Flywheel

- Choke is still on—Open the choke as soon as the engine fires and continues to run without faltering.
- Too much oil—Correct the amount of oil in the crankcase. Fill only to the FULL mark.

If the engine still will not start after going through these steps, take it in to a repair center to be repaired.

Engine That Sputters

Finally, engines that sputter are something that a facilities maintenance person will have to troubleshoot.

- Some water is in the gas—Remove this old fuel and replace with fresh midgrade gasoline.
- Electrical system is grounding out on the equipment (not engine)—Check for cracks for exposed wires. Repair as necessary.
- Flywheel key sheared—Replace key (Figure 15-35).
- Stop wire is rubbing on flywheel—Remove flywheel, tape up wire if not severed, and reroute it so that it will not touch flywheel (Figure 15-36).
- Bad spark plug (Figure 15-37)—Replace the spark plug with an identical new one (Figure 15-38).

If the engine still will not start after going through these steps, take it in to a repair center to be repaired.

Service Recommendations

Up to this point the focus has been on repairing small engines. However, preventive maintenance is the most critical task that a facilities maintenance technician will perform on a small engine. Proper maintenance will extend the life of the equipment and reduce the repair cost.

- Change the oil—This is the most important thing you can do to make your engine live forever. Change your oil with a good grade of SAE 30w HD for Service SC and higher. Do not overfill. This will cause the engine to smoke or blow oil out of the breather into the air filter. On an engine with a dipstick, fill to the top line of the crosshatch mark. On engines without dipsticks, fill to the top of the hole. On horizontal engines without dipsticks, fill until oil comes out of the fill hole.
- Clean or replace the air filter—This is also a very important part of engine maintenance. Clean the foam filters with soap and hot water. Reoil and squeeze out excess. Paper elements must be replaced if extremely dirty (Figure 15-39).
- Change spark plugs—Replace them once a season instead of trying to clean and reuse them (Figure 15-40).
- Remove fuel before storing engine—Doing this will prevent unwanted maintenance in the spring time.

Figure 15-37: **Bad spark plug**

Figure 15-38: **Good spark plug**

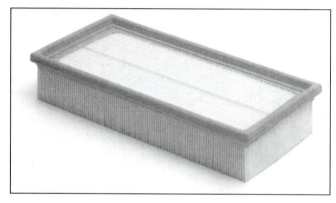

Figure 15-39: **Air filter**

If the engine still will not start after you have completed all of these steps, take it in to a repair center to be repaired.

Figure 15-40: **Spark plug**

Maintaining Public Areas

Public areas are defined as areas that are not assigned to individuals. This includes kitchens, hallways, lobby areas, and stairwells—areas open to the public.

Recommended Maintenance Procedures for Public Areas

All facility must be maintained for both cosmetic and safety reasons. In some cases dirty facilities can be a health violation that could result in a fine and even the closing of the facility. The following are recommended maintenance procedures for common public areas.

Bathroom

The maintenance of restrooms is mostly a safety concern. In other words improperly maintained restrooms can spread disease.

- Disinfected on a daily basis
- Paper products restocked on a daily basis
- Thoroughly cleaned once a week
- Hand soap replaced as needed

Stairways

The maintenance of stairways is a safety issue. In North America the accident rate of stairs is rather high, that is, thousands of people are injured each year from falls on stairways.

- Thoroughly cleaned at least once a week
- Swept on a daily basis

Hallways

Although hallways are not as dangerous as stairways they should also be maintained on a regular basis.

- Buffed to a shine at least once a week, twice time permitting
- General cleaning on a daily basis
- Walls cleaned weekly and spot cleaned as needed
- Alcoves and shelves thoroughly cleaned
- Water fountains disinfected daily

Trash Cans

If trash cans are not properly maintained, then they can be used as a food source for insects and pests. Therefore, it is imperative that they be properly maintained.

- Emptied on a daily basis
- Disinfected and thoroughly cleaned once a week

Lobbies

Lobbies are often the first exposure that a person will have with a facility; therefore it is critical that they be properly maintained.

- Thoroughly cleaned weekly
- General cleaning on a daily basis

Repairing Asphalt Using Cold-Patch Material

Although asphalt driveways and parking lots are fairly durable, they do require some maintenance, especially in cold areas where freeze/thaw cycles are the norm. Minor damage caused by water getting into small cracks and then freezing can quickly become major problems. Periodic sealing will help keep these cracks from starting, but they will not prevent damage caused by settling of the ground under the driveway or improper installation.

Cold patch is a fast, permanent, easy-to-use repair material for asphalt and concrete surfaces.

Repairing Small Cracks

Repairing small cracks will usually postpone and/or prevent major cracks and problems from developing later. Therefore it is important to repair cracks as they are found regardless of their size.

1. Fill any cracks in a blacktop drive as soon as possible to keep water from getting under the slab and causing more serious problems. Cracks that are ½ inch and wider are filled with asphalt cold patch, which is sold in bags and cans. Narrow cracks are treated with crack filler, which is available in cans, plastic pour bottles, and handy caulking cartridges.
2. Use a masonry chisel, wire brush, or similar tool to dig away chunks of loose and broken material from the crack (Figure 15-41).
3. Sweep out the crack with a stiff-bristled broom (Figure 15-42).

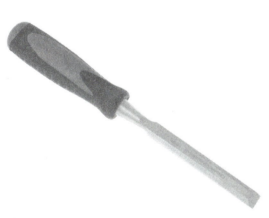

Figure 15-41: **Use chisel or sharp object to clean out the crack**

Figure 15-42: **Sweep out the crack with stiff-bristled broom**

4. Use a garden hose with a pressure nozzle to clean off all dust.
5. Apply the crack filler. For repairs deeper than 2 inches, apply and tamp cold patch in 1–2-inch layers. Add enough material so that tamping or tire rolling leaves a slight crown. If over time a patch settles below the road surface, clean the surface of the patch, add another layer, and compact (Figure 15-43).

Figure 15-43: **Apply crackfiller to the crack**

Repairing Large Holes with Cold Patch

Large holes or potholes require asphalt cold patch, which has larger aggregate than paste patches. For this application, purchase 60–70-pound bags. Cold patch resembles driveway asphalt but is treated with chemicals to keep it workable. If the temperature stays above 50°F, you don't have to heat it the way highway crews heat fresh asphalt—but you do have to roll it.

1. Clear loose debris, and remove jagged edges around the hole with a hammer and cold chisel (Figures 15-44 and 15-45).
2. Pile on enough cold patch to leave a slight mound after tamping (Figure 15-46).

Figure 15-44: **Clear loose debris**

Figure 15-45: **Use hammer or cold chisel to remove jagged edges around a hole**

3. Fill deep holes in two stages, tamping in between, to avoid leaving a water-collecting depression.
4. Apply the weight of your car, driving it slowly over a piece of ¾-inch plywood or a layer of sand spread over the cold patch (Figure 15-47).

Figure 15-46: **Cold patch piled in a hole**

Figure 15-47: **Drive a car over the cold patch to spread it**

Review Questions

1. List the steps for planting brushes and shrubs.

2. List three methods for winterizing an irrigation system.

3. List the steps for starting up the irrigation system in the spring.

4. List the steps for maintaining a swimming pool.

5. If the lawnmower's engine does not start, what should the technician check?

6. What should the technician check if the engine on the lawnmower starts to sputter?

7. What are the three primary nutrients needed by plants and how do they benefit the plant?

8. What are the steps for creating a 12' × 24' paving stone patio?

Name: _____

Date: _____

Groundskeeping

Lawnmower Maintenance

Upon completion of this job sheet, you should be able to identify a need for mower maintenance.

Task Completed

____ ❶ Use the gas and oil recommended by the manufacturer.

____ ❷ Blades are sharp.

____ ❸ Blades and crankshaft are tight.

____ ❹ Underside of mower is cleaned after each use.

____ ❺ Check grass-catcher bag for wear and tear or deterioration.

____ ❻ Check mower wheels, bearings, and axles for wear and lubrication.

____ ❼ Mower is thoroughly inspected every year.

Instructor's Response:

_____

_____

_____

Name: _____

Date: _____

Botany

Plant Identification

Upon completion of this job sheet, you should be able to research the web to determine the growing conditions of a plant.

Task Completed

Using the Internet determine the growing conditions of the following plants.

Little Jamie Whitecedar

Dwarf Fothergilla

Blaauw's Juniper

Rosegold Pussy Willow

Ginny Bruner Holly

Mary Nell Holly

Yoshino Cryptomeria

Swamp White Oak

Instructor's Response:

_____
_____
_____

Chapter 16

Basic Math for Facilities Maintenance Technicians

OBJECTIVES

By the end of this chapter, you will be able to:

Knowledge-Based

- ✪ State the difference between a real number and a whole number.
- ✪ State the difference between an integer and a whole number.

Skill-Based

- ✪ Add whole and Real numbers and Fractions.
- ✪ Subtract whole and Real numbers and Fractions.
- ✪ Multiply whole and Real numbers and Fractions.
- ✪ Divide whole and Real numbers and Fractions.
- ✪ Solve problems involving multiple operations with whole numbers.

Glossary of Terms

Integer any whole number including zero that is positive or negative, factorable or nonfactorable

Real number any number including zero that is positive or negative, rational or irrational

Whole numbers counting numbers including zero

Fraction a number that can represent part of a whole

Introduction

Regardless of the assigned task, it is imperative that the facilities maintenance technician have a good grasp of basic math principles and techniques dealing with whole numbers, decimals, fractions, percent and percentage, and area and volume. In the following sections we will be discussing some of the principles and techniques associated with them.

Integers

An **integer** is defined as any whole number including zero that is positive or negative, factorable or nonfactorable. Examples of integers are 0, 1, 4, −5, and −9. In math, integers range from negative to positive infinity.

Real or Real Numbers

A **real number** is any number including zero that is positive or negative, rational (a number that can be expressed as an integer or a ratio of two integers, ½, 2, −5) or irrational (square root of 2). Examples of real numbers are −4, −4.25, −3, 0, 2, and 6.5.

Whole Numbers

By definition **whole numbers** are counting numbers including zero. Whole numbers are an important part of our everyday life. They are used in every aspect of modern living. This includes everything from ordering lunch to filling out an employment application. To be an effective facilities maintenance technician you should have a good understanding of the basic operation or whole numbers. In other words you must be able to add, subtract, multiple, and divide to properly order supplies, maintain inventory, and so on.

Basic Principles of Working with Whole Numbers

When working with whole numbers it is important to have a good working knowledge of multiplication as well as addition tables. Most importantly it is necessary to correctly line up columns of numbers properly. The numerals on the right should line up over one another for addition, subtraction, and multiplication. Also keep long involved problems simple by breaking them down and solving them one step at a time. Finally, when working with whole numbers that have units always remember to carry the units to the answer.

Addition and Subtraction of Whole Numbers

When performing addition and/or subtraction of whole numbers, it is easy to arrange the numbers in a table format, and then starting from the right-hand side of the table add or subtract each column while moving to the left.

Examples:

		270 pounds	2,768	112 ft
		814 pounds	814	96 ft
	13	58 pounds	644	40 ft
	+ 8	+ 9 pounds	+ 555	+ 57 ft
	21	1,151 pounds	4,781	305 ft

	29	57 ft	28 cu yd
	− 5	− 18 ft	− 13 cu yd
	24	39 ft	15 cu yd

2,768	114 lb	120 in
− 814	− 102 lb	− 106 in
1,954	12 lb	14 in

Multiplication of Whole Numbers

Multiplication can be thought of as repeated addition, that is, 2×4 is the same as saying $2 + 2 + 2 + 2$ or $4 + 4$.

Examples:

9	22	427 ft	377 gal
× 5	× 3	× 23	× 14
45	66	9,821 ft	5,278 gal

54	24	$12.00
× 4	× 7	× 37
216	168	$444.00

Division of Whole Numbers

When a whole number division problem is arranged in a tabular format, the bottom number is the divisor and the top number is the dividend.

Examples:

12	144	96 ft	5,012 lb
÷ 4	÷ 6	÷ 8	÷ 14
3	24	12 ft	358 lb

14,400 V	352 oz	$720.00
÷ 120	÷ 16	÷ 14
120 V	22 oz	$51.43

Decimals

As mentioned earlier whole numbers are all the counting numbers including zero. However, in reality not all numbers are considered to be whole numbers. For example, how do you express quantities that are less than one? In these situations we can express the quantity as either a fraction (which we will be discussing later in this chapter) or a decimal number. In the decimal system, numbers are placed on either side of a decimal point. The numbers on the left-hand side of the decimal point represent whole numbers whereas those on the right-hand side represent the portion of the whole numbers that is less than one (Figure 16-1). When a number does not have a whole number portion associated with it (i.e., the whole number portion is equal to zero), it is said to be a decimal fraction.

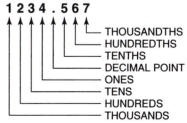

Figure 16-1: **Decimal number**

Addition and Subtraction of Decimals

To add or subtract decimal numbers, line up the decimal points and perform the operation exactly as you would do for whole numbers. Be careful to line up the decimal points and put the decimal point in the same place for the answer.

Examples:

41.12	56.92		
5.25	7.80	9.4	24.3
+ 2.60	+ 61.23	− 5.2	− 5.6
48.97	125.95	4.2	18.7

13.52	47.58
+ 8.41	− 5.31
21.93	42.27

42.4
× .482
848
3392
1696
20.4368

(Decimal point is 4 places from the right)

Figure 16-2: Multiplication of decimals

Multiplication of Decimals

When multiplying decimals, don't line up the decimal points. Just line up the digits on the right, as you would do for whole numbers. Next multiply the top row by each digit in the bottom row (see Figure 16-2). Finally the location of the decimal in the answer is determined by counting the numbers to the right of the decimal of the numbers being multiplied and moving the decimal in the answer to the left the total number of places.

Division of Decimals

The division of decimal numbers is similar to that of whole numbers. If the divisor is not a whole number, move the decimal point to the right until it is a whole number. Next move the decimal point in the dividend the exact same number of spaces (to the right). Finally put the decimal point directly above that in the dividend and divide the problem as you would do for any other whole number problem (see Figure 16-3).

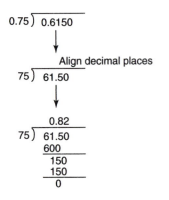

Figure 16-3: Division of decimals

Fractions

By definition a **fraction** is a number that can represent part of a whole. Fraction problems are best worked longhand. Fractions can be converted to decimal numbers and therefore worked on a calculator and/or by hand using the methods described for decimals. Common fractions, frequently just called fractions, are used for many other things, such as parts of volume measure (½ cup) or parts of a dollar (a quarter). A whole inch may be divided into equal parts in many ways. For instance, an inch may be divided into eight equal parts. Each part is an eighth of an inch (⅛ inch). If five of these parts were needed in measuring a length, the quantity would be five-eighths of an inch and would be written as ⅝ inch. Each fraction is made up of two numbers:

Numerator
‾‾‾‾‾‾‾‾‾
Denominator

Sometimes it is necessary to change fractions to equivalent fractions. Equivalent fractions are fractions that have the same value. The value of a fraction is not changed when both the numerator and the denominator are multiplied or divided by the same number. Often when working with two or more fractions, it is necessary to find the lowest common denominator. The lowest common denominator is the smallest denominator that is evenly divisible by each of the denominators of the fractions.

Addition and Subtraction of Fractions

To add fractions, express them with their least common denominators, add the numerators, and then write the sum of the numerators over the common denominator. For subtraction, fractions must have a common denominator. To subtract a fraction from another fraction, express the fractions as equivalent fractions having a common denominator. Subtract the numerators. Write their difference over the common denominator.

Examples:

$$\frac{3}{1} - \frac{17}{5} = -\frac{2}{5} \qquad \frac{1}{17} - \frac{3}{7} = -\frac{44}{119} \qquad \frac{6}{13} - \frac{8}{9} = -\frac{50}{117}$$

$$\frac{5}{20} - \frac{19}{17} = -\frac{59}{68} \qquad \frac{18}{17} + \frac{6}{14} = 1\frac{59}{119}$$

Multiplication of Fractions

To multiply two or more fractions, multiply the numerators and multiply the denominators. Write as a fraction with the product of the numerators over that of the denominators. Express the answer in the lowest terms.

Examples:

$$\frac{2}{20} \times \frac{16}{4} = \frac{2}{5} \qquad \frac{19}{19} \times \frac{14}{19} = 1\frac{5}{9} \qquad \frac{4}{19} \times \frac{18}{16} = \frac{9}{38}$$

Division of Fractions

Division is the inverse of multiplication. Dividing by 4 is the same as multiplying by 1/4. So 4 is the inverse of 1/4, and vice versa. (Remember, 4 can be written as 4/1.) The inverse of 5/16 is 16/5. To divide fractions, invert the divisor (the part the other number is being divided by) and multiply. Express the answer in the lowest terms.

Examples:

$$\frac{4}{3} \div \frac{18}{6} = \frac{4}{9} \qquad \frac{2}{2} \div \frac{17}{11} = \frac{11}{17} \qquad \frac{14}{20} \div \frac{15}{17} = \frac{119}{150}$$

Percent and Percentages

Percentages is a way of expressing a number as a fraction of 100. In other words, percentage means per hundred, just as miles per hour means miles each hour. Percentages may be expressed as either a percentage or a decimal. To express a percentage as a decimal, first change the number to a fraction and then to a decimal. For example, 25% can be converted to a decimal as follows:

25%   Start with a percentage. The "%" sign may be considered a magnet that pulls the decimal point two places to the right. Removing the "%" sign moves the decimal point two places to the left.

25/100 Drop the percentage symbol and divide the percentage by 100.

0.25   Decimal equivalent.

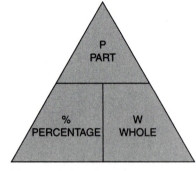

Figure 16-4: Relationship between percentages and whole numbers

Working with Percentages

When working with percentages, remember to specify the relationship. Figure 16-4 illustrates the relationship between percentages and whole numbers. For example, when determining percentage (%), cover the percentage section of the diagram; the fractional parts P (part) and W (whole) remain with P above W. Therefore, the formula for percentage is % = P/W.

Example:

60 is 80% of _____.

60/80% = whole number

60/0.80 = whole number

60/0.80 = 75

Areas and Volumes

The area of a shape is the size of the inside of the shape. Area is always measured in square units—for instance, square inches, square feet, and square yards. Volume is the space enclosed by a three-dimensional figure. Volume can be calculated by first determining the area of an object and then multiplying the area by the object's depth (Figures 16-5 through 16-12).

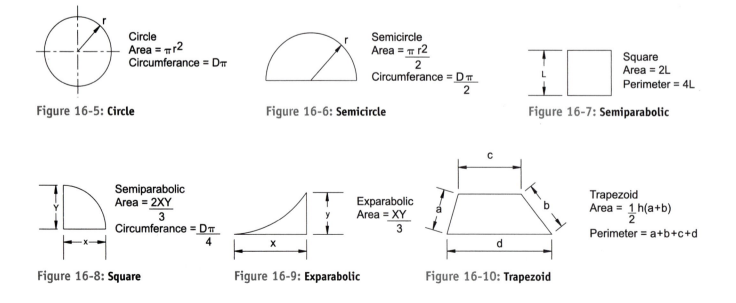

Circle
Area = πr^2
Circumferance = $D\pi$

Figure 16-5: Circle

Semicircle
Area = $\dfrac{\pi r^2}{2}$
Circumferance = $\dfrac{D\pi}{2}$

Figure 16-6: Semicircle

Square
Area = 2L
Perimeter = 4L

Figure 16-7: Semiparabolic

Semiparabolic
Area = $\dfrac{2XY}{3}$
Circumferance = $\dfrac{D\pi}{4}$

Figure 16-8: Square

Exparabolic
Area $\dfrac{XY}{3}$

Figure 16-9: Exparabolic

Trapezoid
Area = $\dfrac{1}{2}h(a+b)$
Perimeter = a+b+c+d

Figure 16-10: Trapezoid

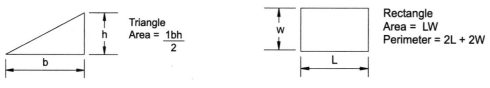

Figure 16-11: Triangle   **Figure 16-12: Rectangle**

Combined Operations

There are two keys to solving a combined-operations problem: First, analyze the problem to determine what information is given and what you must calculate. Second, write down the steps you will go through. One of the steps very often will be to get everything into the same units. When solving problems that combine operations, remember the following rules:

Rule 1: Perform any calculations inside parentheses.

Rule 2: Perform all multiplications and divisions, working from left to right.

Rule 3: Perform all additions and subtractions, working from left to right.

   For example:

   Evaluate $4 + 5 \times (6 + 2) \div 3 - 5$ using the order of operations.

   Solution:

Step 1: $4 + 5 \times (6 + 2) \div 4 - 5 = 4 + 5 \times 8 / 4 - 5$ Parentheses

Step 2: $4 + 5 \times 8 \div 4 - 5 = 4 + 40 / 4 - 5$ Multiplication

Step 3: $4 + 40 \div 4 - 5 = 4 + 10 - 5$ Division

Step 4: $4 + 10 - 5 = 14 - 5$ Addition

Step 5: $14 - 5 = 9$ Subtraction

Review Questions

Complete the following:

1 $58 + \dfrac{13 + 32 - 9}{3} - 17 =$ _____

2 The lowest common denominator for $\dfrac{7}{12}$, $\dfrac{2}{3}$, and $\dfrac{1}{2}$ is _____.

3 The lowest common denominator for $\dfrac{3}{4}$, $\dfrac{7}{10}$, $\dfrac{1}{3}$, and $\dfrac{8}{15}$ is _____.

4 Using the prime factors, the lowest common denominator for $\dfrac{2}{3}$, $\dfrac{7}{10}$, $\dfrac{5}{9}$, and $\dfrac{11}{12}$ is _____.

5 Add the following decimal fractions:

```
  1478.35
  4362.31
+  116.96
```

a. 5,960.62   c. 5,957.62
b. 6,191.54   d. 5,879.65

6 Add the following decimal fractions:

```
  3,425.26
  1,334.63
+ 1,606.53
```

a. 6,368.42   c. 6,364.42
b. 6,366.42   d. 5,563.16

7 Add the following decimal fractions:

```
  1,270.54
  1,798.54
+ 3,400.99
```

a. 6,473.33   c. 6,470.07
b. 13,272.31   d. 4,203.00

8 Add the following decimal fractions:

```
  1,014.19
  2,567.85
+ 1,093.83
```

a. 4,677.86   c. 4,673.86
b. 4,675.87   d. 4,128.95

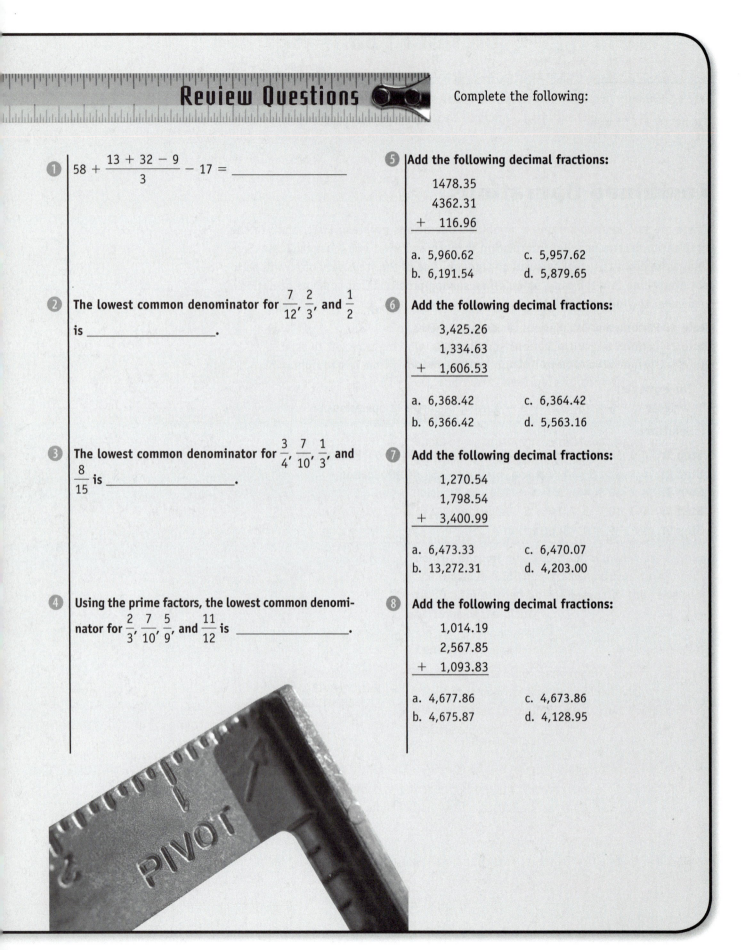

Review Questions

9 | **Add the following decimal fractions:**

$$
\begin{array}{r}
2,656.21 \\
5,235.27 \\
+\quad 5,091.55 \\
\end{array}
$$

a. 12,986.03   c. 12,983.03
b. 23,166.13   d. 9,588.67

10 | **Subtract the following decimal fractions:**

4,722.57 − 3,889.43 = _____

a. 837.14   c. 829.14
b. 3332.56   d. 833.14

11 | **Subtract the following decimal fractions:**

3,721.43 − 3,633.047 = _____

a. 88.38   c. 79.36
b. 795.24   d. 09.82

12 | **Multiply the following decimal fractions:**

$$
\begin{array}{r}
4,228.43 \\
\times\quad 3,883.72 \\
\end{array}
$$

a. 16,422,038.16   c. 16,422,025.47
b. 39,181.90   d. 1,824,670.50

13 | **Multiply the following decimal fractions:**

$$
\begin{array}{r}
3,035.03 \\
\times\quad 5,870.45 \\
\end{array}
$$

a. 17,816,975.90   c. 17,816,991.86
b. 55,869.05   d. 1,979,663.99

14 | **Multiply the following decimal fractions:**

$$
\begin{array}{r}
102.53 \\
\times\quad 1,448.38 \\
\end{array}
$$

a. 5,896.03   c. 148,495.03
b. 148,502.40   d. 148,499.03

15 | **Multiply the following decimal fractions:**

$$
\begin{array}{r}
3,523.46 \\
\times\quad 5,998.62 \\
\end{array}
$$

a. 27,517.95   c. 21,135,897.62
b. 27,519.42   d. 21,135,919.74

16 | **Divide the following decimal fractions:**

216.53 ÷ 06.07

Review Questions

⑰ **Divide the following decimal fractions:**

397.27 ÷ 33.14

⑱ **Add the following fractions:**

$\frac{3}{4} + \frac{7}{8} + \frac{5}{16}$

⑲ **Subtract the following fractions:**

$9\frac{3}{8} - \frac{3}{4}$

⑳ **Multiple the following:**

$\frac{3}{4} \times 12 \times 1\frac{5}{8}$

㉑ **Divide the following:**

$5\frac{3}{5} \div \frac{15}{16}$

㉒ **Add the following:**

362 + 1,491 + 73 + 29,248

㉓ **Subtract the following:**

4,793 − 404

㉔ **Multiply the following:**

189 × 9

㉕ **Add the following:**

13,328 + 238

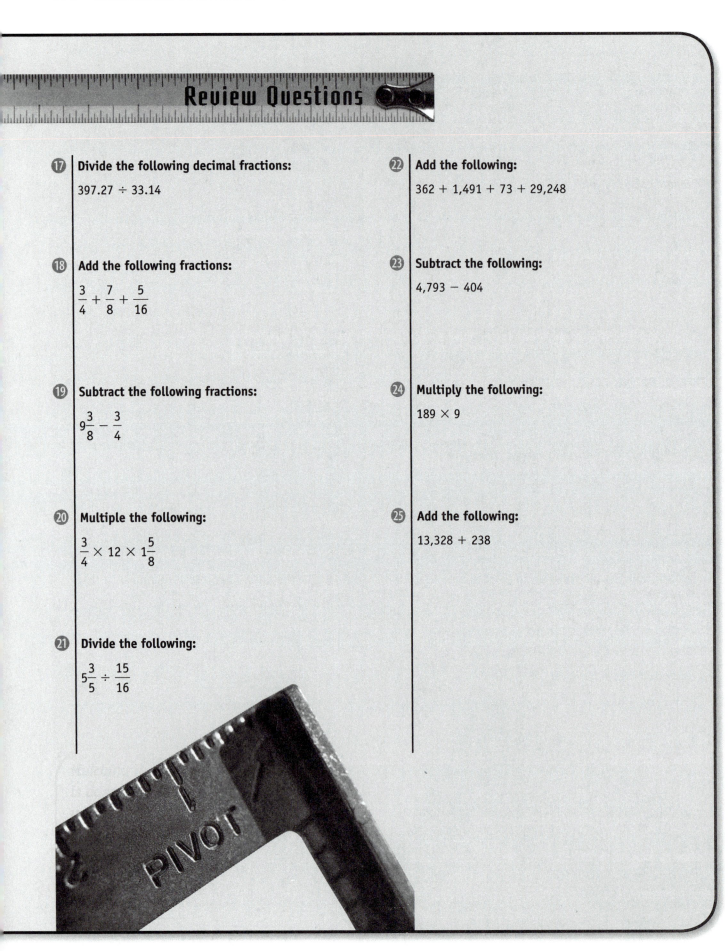

Name: _____

Date: _____

Whole Number Addition

Upon completion of this job sheet, you should be able to add whole numbers.

Procedure

Add the following whole numbers:

3,887	1,308	5,041
5,233	3,783	179
+ 5,602	+ 3,951	+ 5,578

2,974	3,270	3,615
4,784	5,489	1,268
+ 948	+ 4,833	+ 2,309

2,373	1,027	4,552
2,181	218	50
+ 2,758	+ 1,052	+ 5,962

1,821
2,245
+ 2,565

Name: _____

Date: _____

Whole Number Subtraction

Upon completion of this job sheet, you should be able to subtract whole numbers.

Procedure

Subtract the following whole numbers:

2,291 − 585   = _____

3,341 − 3,173 = _____

4,430 − 1,162 = _____

3,957 − 561   = _____

4,174 − 1,931 = _____

4,879 − 2,660 = _____

5,816 − 4,598 = _____

2,251 − 1,699 = _____

1,763 − 1,571 = _____

2,147 − 417   = _____

Name: _____

Date: _____

Whole Number Multiplication

Upon completion of this job sheet, you should be able to multiply whole numbers.

Procedure

Choose the correct answer for the following multiplication:

793 × 449	a. 2,139 b. 356,057	c. 355,954 d. 118,651
5,600 × 2,435	a. 12,906 b. 13,636,000	c. 13,638,102 d. 4,546,033
2,667 × 4,013	a. 10,694 b. 10,705,401	c. 10,705,399 d. 10,702,671
88 × 2,564	a. 225,632 b. 226,628	c. 226,626 d. 113,313
3,821 × 4,959	a. 23,657 b. 18,948,339	c. 18,948,944 d. 18,948,948
1,755 × 3,565	a. 6,254,950 b. 6,256,575	c. 6,254,941 d. 694,994
2,051 × 4,270	a. 19,130 b. 8,757,770	c. 8,756,491 d. 8,756,495
2,396 × 49	a. 117,404 b. 2,541	c. 116,392 d. 38,797
1,154 × 961	a. 4,038 b. 1,108,994	c. 1,109,057 d. 369,685
5,808 × 4,999	a. 29,034,192 b. 20,805	c. 29,032,560 d. 9,677,519

Name: _____

Date: _____

Whole Number Division

Upon completion of this job sheet, you should be able to divide whole numbers.

Procedure

Divide the following whole numbers:

$931 \div 19$ = _____

$1{,}150 \div 46$ = _____

$306 \div 34$ = _____

$2{,}597 \div 49$ = _____

$858 \div 39$ = _____

$1{,}232 \div 56$ = _____

$2{,}065 \div 59$ = _____

$28 \div 2$ = _____

$2{,}450 \div 50$ = _____

Name: _____

Date: _____

Adding Fractions

Upon completion of this job sheet, you should be able to add fractions.

Procedure

Add the following common fractions:

$\dfrac{6}{13} + \dfrac{8}{9}$ = _____

$\dfrac{19}{9} + \dfrac{6}{14}$ = _____

$\dfrac{18}{17} + \dfrac{6}{14}$ = _____

$\dfrac{4}{1} + \dfrac{19}{4}$ = _____

$\dfrac{10}{16} + \dfrac{4}{19}$ = _____

$\dfrac{3}{12} + \dfrac{10}{1}$ = _____

$\dfrac{8}{4} + \dfrac{2}{5}$ = _____

$\dfrac{8}{4} + \dfrac{16}{19}$ = _____

$\dfrac{19}{19} + \dfrac{1}{10}$ = _____

$\dfrac{12}{15} + \dfrac{7}{12}$ = _____

Name: _____

Date: _____

Subtracting Fractions

Upon completion of this job sheet, you should be able to subtract fractions.

Procedure

Subtract the following common fractions:

$$\frac{3}{1} - \frac{17}{5} = \underline{\hspace{2cm}}$$

$$\frac{1}{17} - \frac{3}{7} = \underline{\hspace{2cm}}$$

$$\frac{20}{20} - \frac{15}{10} = \underline{\hspace{2cm}}$$

$$\frac{20}{9} - \frac{14}{19} = \underline{\hspace{2cm}}$$

$$\frac{12}{19} - \frac{13}{15} = \underline{\hspace{2cm}}$$

$$\frac{5}{20} - \frac{19}{17} = \underline{\hspace{2cm}}$$

$$\frac{4}{3} - \frac{12}{20} = \underline{\hspace{2cm}}$$

$$\frac{9}{12} - \frac{10}{15} = \underline{\hspace{2cm}}$$

$$\frac{16}{2} - \frac{10}{10} = \underline{\hspace{2cm}}$$

$$\frac{14}{18} - \frac{16}{7} = \underline{\hspace{2cm}}$$

Name: _____

Date: _____

Multiplying Fractions

Upon completion of this job sheet, you should be able to multiply fractions.

Procedure

Multiply the following common fractions:

$$\frac{2}{20} \times \frac{16}{4} = \underline{\hspace{2cm}}$$

$$\frac{20}{16} \times \frac{2}{6} = \underline{\hspace{2cm}}$$

$$\frac{19}{19} \times \frac{14}{9} = \underline{\hspace{2cm}}$$

$$\frac{19}{8} \times \frac{13}{18} = \underline{\hspace{2cm}}$$

$$\frac{11}{18} \times \frac{12}{14} = \underline{\hspace{2cm}}$$

$$\frac{4}{19} \times \frac{18}{16} = \underline{\hspace{2cm}}$$

$$\frac{3}{2} \times \frac{11}{19} = \underline{\hspace{2cm}}$$

$$\frac{9}{11} \times \frac{9}{14} = \underline{\hspace{2cm}}$$

$$\frac{15}{1} \times \frac{9}{10} = \underline{\hspace{2cm}}$$

$$\frac{9}{12} \times \frac{10}{2} = \underline{\hspace{2cm}}$$

Name: _____

Date: _____

Dividing Fractions

Upon completion of this job sheet, you should be able to divide fractions.

Procedure

Divide the following common fractions:

$$\frac{4}{3} \div \frac{18}{6} = \underline{\hspace{2cm}}$$

$$\frac{3}{19} \div \frac{4}{8} = \underline{\hspace{2cm}}$$

$$\frac{2}{2} \div \frac{17}{11} = \underline{\hspace{2cm}}$$

$$\frac{2}{10} \div \frac{15}{1} = \underline{\hspace{2cm}}$$

$$\frac{14}{20} \div \frac{15}{17} = \underline{\hspace{2cm}}$$

$$\frac{7}{2} \div \frac{1}{19} = \underline{\hspace{2cm}}$$

$$\frac{6}{5} \div \frac{13}{2} = \underline{\hspace{2cm}}$$

$$\frac{11}{13} \div \frac{12}{16} = \underline{\hspace{2cm}}$$

$$\frac{18}{13} \div \frac{12}{12} = \underline{\hspace{2cm}}$$

$$\frac{16}{20} \div \frac{18}{9} = \underline{\hspace{2cm}}$$

Chapter 17 Blueprint Reading for Facility Maintenance Technicians

Introduction

Because technology is evolving at an ever-increasing rate and systems are becoming more and more complex, it is becoming ever more important that the technician have a basic understanding of blueprint reading. This is especially true when working on Hydronic systems (water-based heating systems), electrical systems, and alarm systems. Blueprints are drawn plans of homes and buildings that are used to build the structure. The contractor and/or technician doing the work must have a good understanding of the portion of the prints that apply to their individual trade, electrical, piping, duct work, and location of equipment. In addition understanding the basics of blueprint reading can save a technician numerous hours troubleshooting a problem with a particular piece of equipment or a system.

Linear Measurement

Linear measurement is defined as the measurement of two points along a straight line (Figure 17-1). All objects, whether they are man-made or the result of natural conditions and/or forces, consist of points and lines. A point can be the center location of a gas line (Figure 17-2) or one of the edges of a rectangular duct (Figure 17-3). Whatever the application of points and/or lines, the fact remains the same; any person entering a technical field must understand how to locate them.

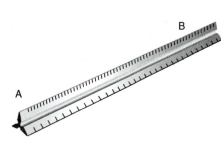

Figure 17-1: **Linear measurement**

Figure 17-2: **The proposed center line of a gas line**

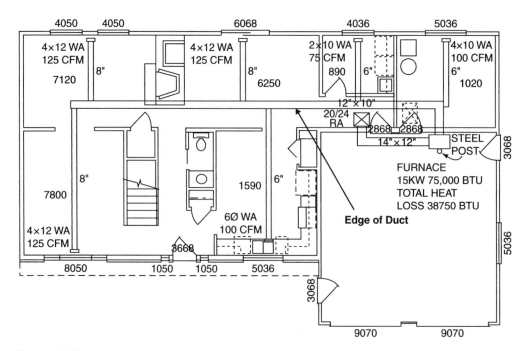

Figure 17-3: **Edge of a rectangular duct supplying air to a room in a facility**

Currently two basic systems, **English system of measure** and **metric system of measure,** are used to make measurements in the world today. The English system of measure is the unit of measure currently used in the United States by most technicians (see Table 17-1). Its base units are inch (in), foot (ft), yard (yd), and mile (mi).

The base unit for the metric system is the meter. This base unit is divided into larger and smaller units (in multiple of 10) by adding prefixes. Common metric prefixes are deci (10), centi (100), and milli (1,000) (see Table 17-2). For example a meter contains 10 decimeters.

Unit	Divisions
English	
1 inch (in)	
1 foot (ft)	12 in
1 yard (yd)	3 ft
1 mile (mi)	5,280 ft

Table 17-1: **English Conversion Factors**

Unit	Divisions
Units of Length	
10 millimeters (mm)	= 1 centimeter (cm)
10 centimeters	= 1 decimeter (dm) = 100 millimeters
10 decimeters	= 1 meter (m) = 1,000 millimeters
10 meters	= 1 dekameter (dam)
10 dekameters	= 1 hectometer (hm) = 100 meters
10 hectometers	= 1 kilometer (km) = 1,000 meters

Table 17-2: **Metric Conversion Factors**

Although the English system of measure is most commonly used in the United States, a facility maintenance technician should still be able to recognize and convert from one unit to another in the metric system of measure. Table 17-3 provides the conversion factors for converting from one system of units to another.

Starting with	Multiply	To Find
Inches	2.5	centimeters
Feet	30	centimeters
Yards	0.9	meters
Miles	1.6	kilometers
centimeters	0.3937	inches
Meters	1.1111	yards
kilometers	0.625	miles

Table 17-3: **Converting from Metric to English and English to Metric**

For example, to convert 12'6" to centimeters the technician would perform the following steps:

Convert from Inches to Centimeters

Step #1 Converting 12' into inches

12 ft × 12 in/ft = 144 in

Step #2 Next add the inch portion of the original measurement

144 in + 6 in = 150 in

Step #3 Find the conversion factor using Table 17-3

120 in × 2.5 cm/in = 300 cm

Convert from Centimeters to Inches

Step #1 Find the conversion factor using Table 17-3

300 cm × 0.4 in/cm = 150 in

Step #2 Convert from inches to feet and inches by dividing by 12

120 in / 12 ft/in = 10 ft

Therefore the measure would be written as 10 ft.

Reading Linear Measurement on a Blueprint

When a mechanical drawing is created on the computer, typically it is drawn at a scale of 1:1 (also known as full scale). However, in most cases when the file is printed, it is printed at a reduced scale so that it can fit onto a single sheet of paper (typically 8 1/2 × 11, 24 × 36, or 30 × 42). When a drawing is printed at full scale, determining the length of a line is easy to accomplish. This is done be simply placing a ruler along the length of the line and reading the dimensions (Figure 17-4). However, if the drawing is printed at a reduced scale (a scale smaller then 1:1), then a scale factor must be applied to determine the length of a line not dimensioned (Figures 17-5, 17-6, and 17-7)

When a drawing is printed to a scale ¼ times smaller than its actual size, then that drawing is said to be drawn at ¼ scale and is therefore read as every ¼" is

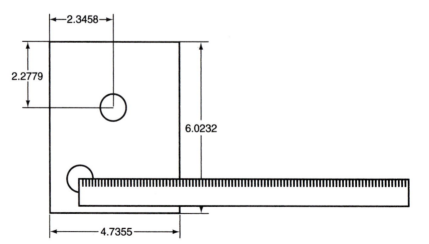

Figure 17-4: **Reading the length of a line drawn at full scale**

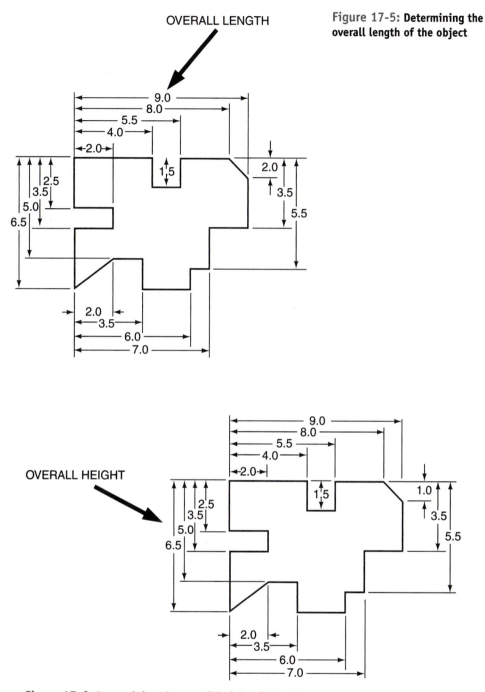

Figure 17-5: **Determining the overall length of the object**

Figure 17-6: **Determining the overall height of the object**

equal to 1'-0". For example, if a portion of a building footing plan measures 2" using a ruler, then the actual length of the footing can be determined by multiplying the measured length by the inverse (meaning swap the top and bottom numbers) of the drawings scale factor. For a scale factor of ¼ the inverse or multiplier would be 4 and therefore the true length of the portion of the footing would be 8ft (Figure 17-8).

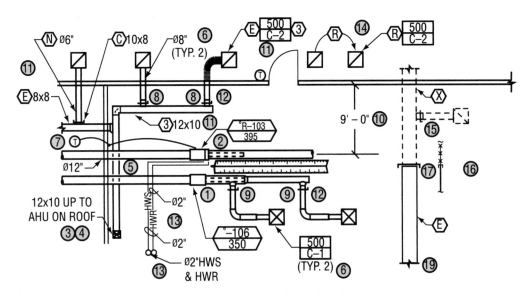

Figure 17-7: **Reading the length of a line that is not dimensioned**

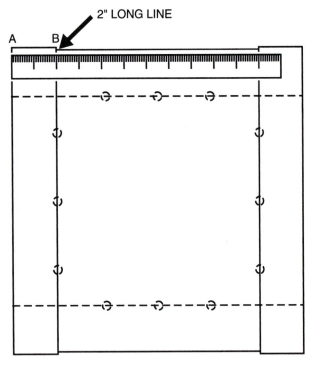

Figure 17-8: **Determining the length of a portion of a footing using a ruler**

Table 17-4 gives the conversion factors for various scales commonly used on engineering drawing.

Drawing Scale	Multiplier
1/16"=1'-0"	16
3/32"=1'-0"	10.66
1/8"=1'-0"	8
3/16"=1'-0"	5.33
1/4"=1'-0"	4
3/8"=1'-0"	2.66
1/2"=1-0"	2
3/4"=1'-0"	1.33
1"=1'-0"	1
1 1/2"=1'-0"	1.5
3"=1'-0"	0.33
Half size	2
Full size	1

Table 17-4: **Drawing Scale Conversion Factors**

Determining Linear Measure Using a Scale

When a drawing is created, one of the responsibilities of the draftsman is to note the scale at which the drawing was created. On average this information is listed in the drawing title block region. The title block region is usually located along the bottom edge or lower right-hand corner of the drawing (Figure 17-9). The length and position of items not dimensioned on a drawing can be determined using a ruler and the scale factor as discussed in the previous section. However, an easier method of determining the length and position of objects not dimensioned is to measure them on the drawing using a scale (Figures 17-10 and 17-11). The scale most commonly used in residential and light commercial is the architect's scale.

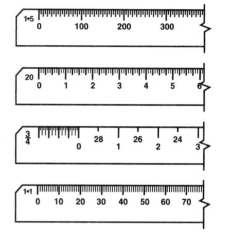

Figure 17-9: Typical engineering drawing containing a title block

Figure 17-10: Examples of engineers', architect's, and metric scales

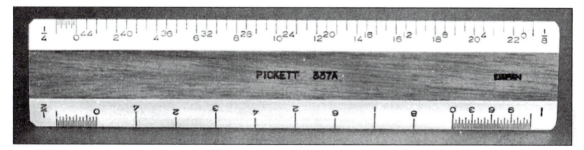

Figure 17-11: Six-inch architectural scale

The Architect's Scale

The architect's scale is a type of ruler in which a range of precalibrated ratios are illustrated. These scales can be made of a variety of materials and contain as few as two scales (ratio) and as many as eleven (on ten of them, each 1" represents a foot and is subdivided into multiples of 12). The most commonly used architect's scale used is the triangular scale. This scale receives it name because its cross-section is in the shape of a triangle (Figures 17-11 and 17-12). Architect's scales are standardized into feet and inches ranging from 1/16" = 1' 0" through 3" = 1' 0".

Figure 17-12: Twelve-inch triangular architect's scale

Reading the Architect's Scale

If you understand how to read one of the ratios (scales) on an architect's scale, then reading the remaining will be simple. Because these scales are used to represent feet and inches, the scale is divided into two sections. The first section represents feet whereas the second section represents inches and fractions of an inch graduated in 1/16", 1/8", ¼", and ½" increments. This will be different from an engineer's scale that is graduated in 1/1,000", 1/100", and 1/10 of an inch measured with caliper type devices and is used for manufacturing machine products. For example, using the scale shown in (Figure 17-13), the marks on the right-hand side of the zero line indicate 1 foot increments, whereas the marks on the left-hand side represent inches and fraction of an inch.

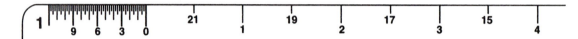

Figure 17-13: **Architect's scale 1"= 1"-0"**

To find the length of a line using the 1" scale, the following procedure should be used:

Step #1 Position the scale zero mark indicator on the beginning of the line or object to be measured (see Figure 17-14).

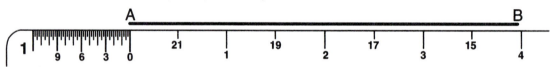

Figure 17-14: **Placing the architect's scale on the beginning of the line to be measured**

Step #2 Moving from left to right, count the number of feet marks until you reach the end of the line or the foot mark nearest to the end (Figure 17-15).

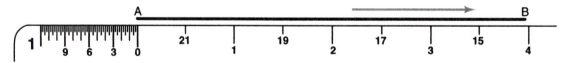

Figure 17-15: **Reading the architect's scale moving from left to right**

Step #3 Slide the scale to the right until the end of the line is on the foot mark indicated in **Step #2** (Figure 17-16).

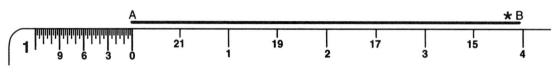

Figure 17-16: **Finding the foot mark nearest to the end of the line to be measured**

Step #4 Starting at the zero, count the number of inch marks to the end of the line. See Figure 17-17.

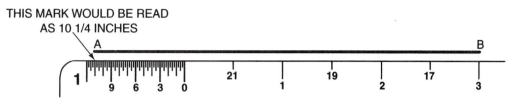

Figure 17-17: **Determining the inches and fraction of an inch**

Angular Measurement

Understanding and reading linear measurements are only a part of the skills needed to interrupt a blueprint; the facility maintenance technician must also be able to read and understand angular measurements (measuring angles) as well.

Angles

Three types of units are used to express an angle: angular degrees, radians, and gradients, but typically only angular degrees are used on blueprints. Since the circumference of any circle contains 360 degrees, an angular degree is equal to 1/360th of the circumference of a circle. This means that by drawing a circle (of any size) and dividing its circumference into 360 equal segments (called arc lengths), the angle formed by constructing a line from the center of the circle to the endpoints of one arc length would produce a wedge equal to 1 degree (Figure 17-18).

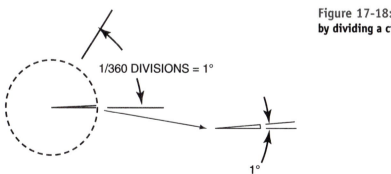

Figure 17-18: **Wedge produced by dividing a circle 360 times**

Reading and Measuring Angles

Angles are usually given on a blueprint with either a dimension or a leader; however, from time to time it may be necessary to determine the angle on a blueprint using a protractor. A protractor is an instrument consisting of a half circle with a midpoint (center) marked on the horizontal position (base) of the protractor. This midpoint, also called the reference point, is marked on the protractor. The half circle is divided into degrees (180). The degrees are labeled from right to left and vice versa, allowing for angles to be measured from either direction (Figure 17-19).

To measure an angle using a protractor, first place the end of the protractor base angle on the vertex of the angle to be measured (Figure 17-20). Next align the baseline with one of the sides of the angle to be measured (Figure 17-21). Finally, count the number of degrees of the side adjacent to the baseline.

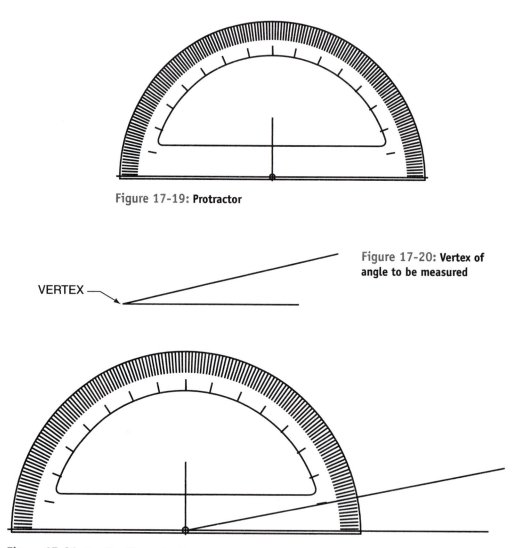

Figure 17-19: **Protractor**

Figure 17-20: **Vertex of angle to be measured**

VERTEX —

Figure 17-21: **Reading the protractor**

Standard Abbreviations and Symbols

Standard abbreviations and symbols have been developed for the engineering and architectural community that not only facilitate the development of blueprints but also ensure consistency in their interpretation. Therefore, a good understanding of what these symbols are is critical in developing blueprint reading skills. An example of some of the abbreviations used in the HVAC industry is shown in Table 17-5.

Abbreviations	Definition
A/D	Analog to digital
AC	Air conditioning–alternating current
ACH	Air changes per hour
ACM	Asbestos -containing material
BACnet	A data communication protocol for building automation and control networks
CFC	Chlorofluorocarbon
CFM	Cubic feet per minute
CFU	Colony forming units
CHWP	Chilled water pump

Table 17–5: **HVAC Abbreviations**

Like abbreviations, symbols are used to speed up the drawing process by using a shorthand method of representing commonly used equipment. Typically, an engineering or architectural firm will use a standard set of symbols commonly used and accepted in the building trade industry. However, if a nonstandard symbol is used then it is typically identified in the drawing legend. A sample of commonly used symbols is illustrated in Figure 17-22.

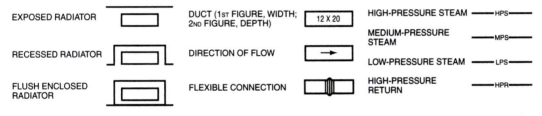

Figure 17-22: **HVAC symbols**

Two-Dimensional Views

The way in which an object is presented or viewed is extremely important. If a part is to be manufactured and the angle or view shown does not provide the necessary information, then the part could never be manufactured to design expectations. The following section examines the most common method of representing objects on engineering and architectural drawings.

Multiview drawings

When something is to be constructed or created using a blueprint, it is the job of the drafting person to supply all the necessary information regarding the objects sizes and the location of its features. All this can easily be accomplished when the object is relatively simple, and all the essential data can be provided in one view. For example, suppose that a 2-inch square piece of steel plating that is ½" thick is to be fabricated with a ½" diameter hole drilled at its center. All the required information about this part can be contained in one drawing consisting of a single view (Figure 17-23). When a more complex part is produced, a single view is not

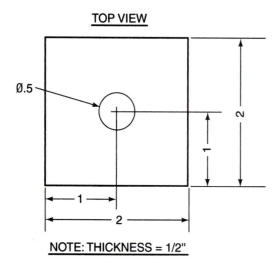

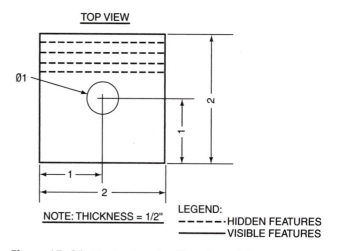

Figure 17-23: Single view of a 2" × 2" × 1/2" steel plate with a 1/2" hole drilled in the center

Figure 17-24: Single view of a 2" × 2" × 1/2" steel plate with a 1/2" hole drilled in the center and containing a series of grooves as hidden features.

sufficient to clearly show all its features. In other words, a series of grooves on the opposite side of this same part would be shown as hidden features, resulting in a drawing that is difficult to interpret (Figure 17-24). To solve this problem, additional views must be created that will reveal all the hidden attributes of the part. In this instance, an additional view of the side is required (Figure 17-25A).

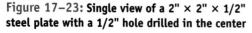

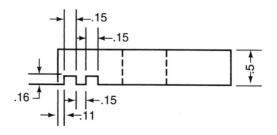

Figure 17-25A: Orthographic projection of a 2" × 2" × 1/2" steel plate with a 1/2" hole drilled in the center and containing a series of grooves

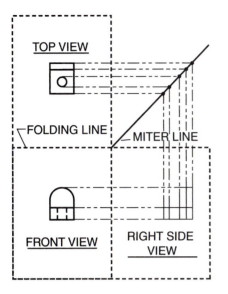

Figure 17-25B: **Orthographic projection showing has the right side view was created**

Very complex parts might contain views showing the front, sides, back, bottom, and top of the object. A drawing containing two or more of these views is called a multiview drawing. The different views in an orthographic projection are created by projecting the lines from one view to another. In other words, by projecting the edge of the object down from the top view, the width of the object can be established in the front view (Figure 17-25B). Another way to look at this is to suppose that the object is placed in a glass box (Figure 17-26), with the object's surfaces and features being projected onto the surface of the glass (Figure 17-27). Once the object's features have been transferred to the glass box, the cube is then unfolded to reveal six different views of the object (Figure 17-28). The primary views used on an engineering drawing are front view, right side view, left side view, top view, bottom view, and rear view.

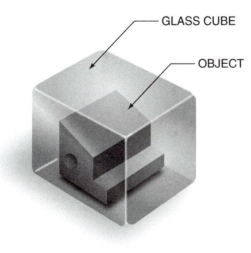

Figure 17-26: **Object encased in a glass cube**

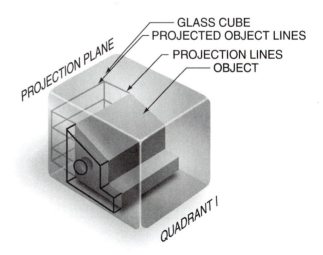

Figure 17-27: **Object encased in glass cube with object lines projected onto surface of glass cube**

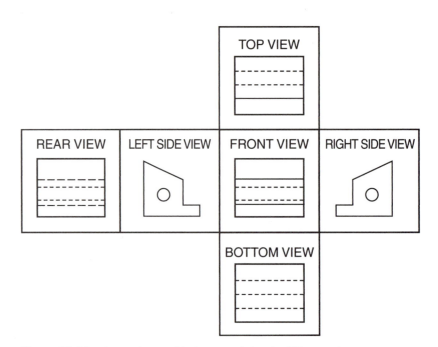

Figure 17-28: **Glass cube unfolded to reveal the six different views**

Orthographic Projection and Architectural Drawings

Architectural drawings are typically not labeled the same as engineering drawings; however, in reality they are still created based on the same principles. In an architectural drawing the top view is not referred to as a top view but instead it is called a plan. In addition, architectural drawings do not refer to left and right side views as left and right side views, instead they are referred to as elevations.

Floor Plans

The floor plan provides a representation of where to locate the major items of a home. The plan shows the location of walls, doors, windows, cabinets, appliances, and plumbing fixtures. The drawing allows the owner to evaluate the project to assure that the building will meet the current and future needs of the business once constructed. The floor plan also serves as a key tool in the communication process between the design team and the building team.

Types of Plans

The floor plan is the skeleton of framework for the development of other drawings required to complete the construction of the structure. These drawings include electrical plans, fire protection plans, framing plans, plumbing plans, and HVAC or mechanical plans.

Plumbing Plans

The plumbing plan contains the size and location of each piping system contained in the proposed structure. Typically piping on a plumbing plan is represented using a single line in which the different functions of that piping system are represented using different line types (Figure 17-29). For example, drain lines are typically shown on plumbing plans using a heavier line, whereas the domestic cold water lines are presented as light lines having long dashes and short dashes. Domestic hot water, on the other hand, is shown as a series of long dashes followed by two short dashes. All vent piping is shown as a series of dashes.

In addition to recognizing the types of plumbing lines used in a facility, the technician must also be able to recognize various plumbing fittings and valves used in plumbing drawings. The most commonly used symbols are shown in Figures 17-30 and 17-31.

HVAC Plans

The HVAC plan contains the size and location of each HVAC system contained in the proposed structure. Typically piping on a HVAC plan is represented using a single line in which the different functions

Soil and Waste, Above Grade...	_____
Soil and Waste, Below Grade.........................	__ ___ ___
Vent...	_ _ _ _ _ _ _
Cold Water..	__ _ _ __ _
Hot Water..	__ _ _ _ __
Hot Water Return..............................	__ _ _ __
Fire Line...	—F— F—
Gas Line...	—G— G—
Acid Waste..	—— ACID ——
Drinking Water Supply..............	__ _ __ _ __
Drinking Water Return..............	__ _ _ _ __
Vacuum Cleaning......................	—V— V—
Compressed Air.........................	——— A ———

Figure 17-29: **Plumbing linetypes**

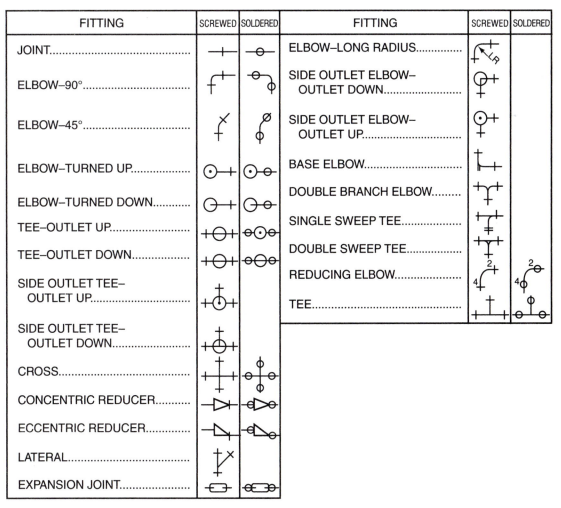

Figure 17-30: **Common plumbing fittings**

VALVE	SCREWED	SOLDERED
GATE VALVE..................................		
GLOBE VALVE..............................		
ANGLE GLOBE VALVE.....................		
ANGLE GATE VALVE........................		
CHECK VALVE...............................		
ANGLE CHECK VALVE.....................		
STOP COCK.................................		
SAFETY VALVE.............................		
QUICK-OPENING VALVE..................		
FLOAT VALVE..............................		
MOTOR-OPERATED GATE VALVE...		

Figure 17-31: **Common plumbing valves**

REFRIGERANT LIQUID	—— RL ——	
REFRIGERANT DISCHARGE	—— RD ——	
REFRIGERANT SUCTION	—— RS ——	
CONDENSER WATER SUPPLY	—— CWS ——	
CONDENSER WATER RETURN	—— CWR ——	
CHILLED WATER SUPPLY	—— CHWS ——	
CHILLED WATER RETURN	—— CHWR ——	
MAKEUP WATER	—— MU ——	
HUMIDIFICATION LINE	—— H ——	
DRAIN	—— D ——	

Figure 17-33: **HVAC air-conditioning linetypes**

HIGH-PRESSURE STEAM	—— HPS ——
MEDIUM-PRESSURE STEAM	—— MPS ——
LOW-PRESSURE STEAM	—— LPS ——
HIGH-PRESSURE RETURN	—— HPR ——
MEDIUM-PRESSURE RETURN	—— MPR ——
LOW-PRESSURE RETURN	—— LPR ——
BOILER BLOW OFF	—— BO ——
CONDENSATE OR VACCUUM PUMP DISCHARGE	—— VPD ——
FEEDWATER PUMP DISCHARGE	—— FPD ——
MAKEUP WATER	—— MU ——
AIR RELIEF LINE	—— V ——
FUEL OIL SUCTION	—— FOS ——
FUEL OIL RETURN	—— FOR ——
FUEL OIL VENT	—— FOV ——
COMPRESSED AIR	—— A ——
HOT WATER HEATING SUPPLY	—— HW ——
HOT WATER HEATING RETURN	—— HWR ——

Figure 17-32: **HVAC heating linetypes**

of that piping system are represented using different line types (Figures 17-32 and 17-33). For example, drain lines are typically shown on HVAC plans using a single line with the letter "D" inserted evenly throughout the line.

In addition to recognizing the types of HVAC lines used in a facility, the technician must also be able to recognize equipment and duct symbols used in HVAC drawings. The most commonly used symbols are shown in Figures 17-34 and 17-35.

Elevations

An elevation is an orthographic drawing that shows one side of a building. In true orthographic projection, the elevations would be displayed as shown in Figure 17-36. The true projection is typically modified as shown in Figure 17-37 to ease viewing.

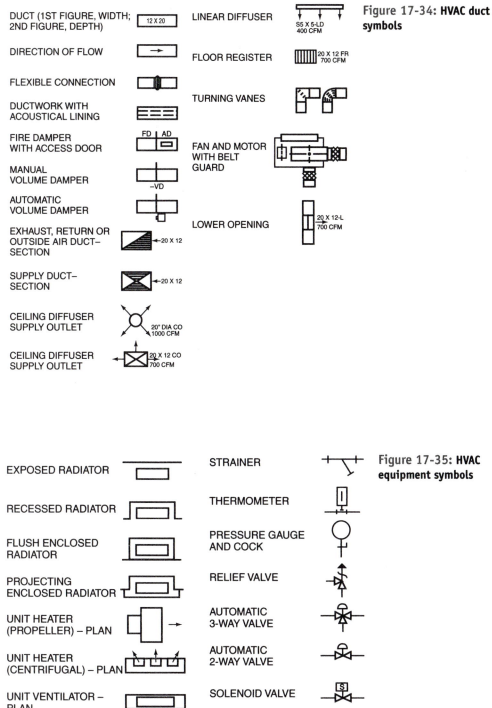

DUCT (1ST FIGURE, WIDTH; 2ND FIGURE, DEPTH)

DIRECTION OF FLOW

FLEXIBLE CONNECTION

DUCTWORK WITH ACOUSTICAL LINING

FIRE DAMPER WITH ACCESS DOOR

MANUAL VOLUME DAMPER

AUTOMATIC VOLUME DAMPER

EXHAUST, RETURN OR OUTSIDE AIR DUCT– SECTION

SUPPLY DUCT– SECTION

CEILING DIFFUSER SUPPLY OUTLET

CEILING DIFFUSER SUPPLY OUTLET

LINEAR DIFFUSER

FLOOR REGISTER

TURNING VANES

FAN AND MOTOR WITH BELT GUARD

LOWER OPENING

Figure 17-34: HVAC duct symbols

EXPOSED RADIATOR

RECESSED RADIATOR

FLUSH ENCLOSED RADIATOR

PROJECTING ENCLOSED RADIATOR

UNIT HEATER (PROPELLER) – PLAN

UNIT HEATER (CENTRIFUGAL) – PLAN

UNIT VENTILATOR – PLAN

STEAM

DUPLEX STRAINER

PRESSURE-REDUCING VALVE

AIR LINE VALVE

STRAINER

THERMOMETER

PRESSURE GAUGE AND COCK

RELIEF VALVE

AUTOMATIC 3-WAY VALVE

AUTOMATIC 2-WAY VALVE

SOLENOID VALVE

Figure 17-35: HVAC equipment symbols

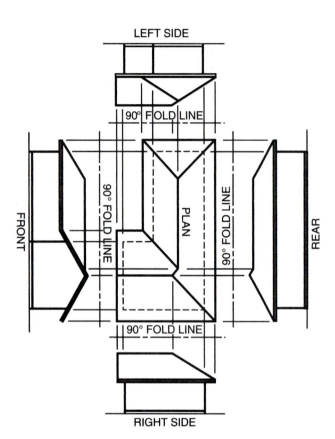

Figure 17-36: **Elevations are orthographic projections showing each side of a structure**

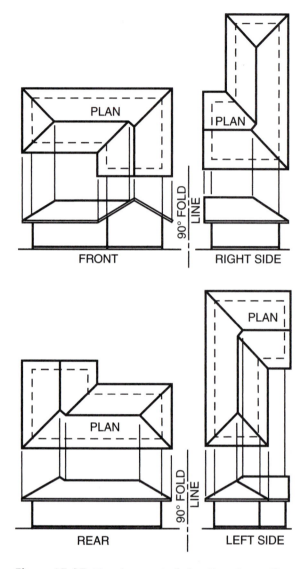

Figure 17-37: **The placement of elevations is usually altered to ease viewing. They are grouped so that a 90° rotation exists between views.**

No matter how they are displayed, it is important to realize that between each elevation, projection and the plan view is an imaginary 90° fold line. An imaginary 90° fold line also exists between elevations in Figure 17-37. Elevations are drawn to show exterior shapes and finishes, as well as the vertical relationships of the building levels. By using the elevations, sections, and floor plans, the exterior shape of a building can be determined.

Review Questions

1 _____ is defined as the measurement of two points along a straight line.

2 All objects, whether they are man-made or the result of natural conditions and/or forces, consist of _____ and _____.

3 Currently there are two basic systems used to make measurements in the world today: _____ and _____.

4 The _____ base units are inch (in), foot (ft), yard (yd), and mile (mi).

5 The _____ is based on the meter.

6 When a mechanical drawing is created on the computer, it is typically drawn at a scale of 1:1 (also known as _____ scale).

7 The title block region is typically located along the _____ edge or _____ of the drawing.

8 The length and position of items not dimensioned on a drawing can be determined using a _____.

Review Questions

9 Three types of units are used to express an angle: _____, _____, and _____.

10 Angles are typically given on a blueprint with either a _____ or a leader.

11 When these three angles are added, their sum is equal to _____ degrees.

12 Each side of a triangle has a name: the _____ _____.

13 True or False, In a right triangle the hypotenuse is always the longest side.

14 The primary views used on an engineering drawing are _____ _____.

Name: _____

Date: _____

Unit Conversion

Upon completion of this job sheet, you should be able to convert from one unit to another for both metric and English.

Procedure

Convert the following:

144 feet	= _____	inches
234 inches	= _____	feet
234,539 inches	= _____	miles
5 miles	= _____	inches
23 miles	= _____	yards
100 feet	= _____	meters
245.9 centimeters	= _____	feet
123 millimeters	= _____	feet
256 yards	= _____	meters
452.1 inches	= _____	centimeters

Instructor's Response:

_____
_____
_____

Name: _____

Date: _____

Reading a Scale

Upon completion of this job sheet, you should be able to read a scale.

Procedure

Using the following scale, determine the length of lines "A," "B," "C," "G," and "H."

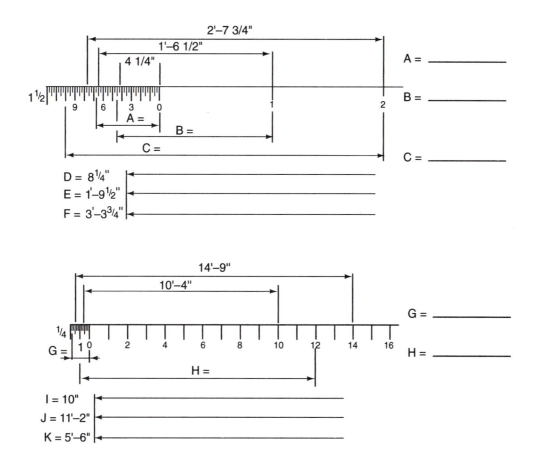

A = _____

B = _____

C = _____

D = $8\frac{1}{4}$"

E = $1'-9\frac{1}{2}$"

F = $3'-3\frac{3}{4}$"

G = _____

H = _____

I = 10"

J = 11'–2"

K = 5'–6"

Instructor's Response:

_____

_____

_____

Name: _____

Date: _____

Determine Angles

Upon completion of this job sheet, you should be able to identify angles.

Procedure

Determine the angles of the following:

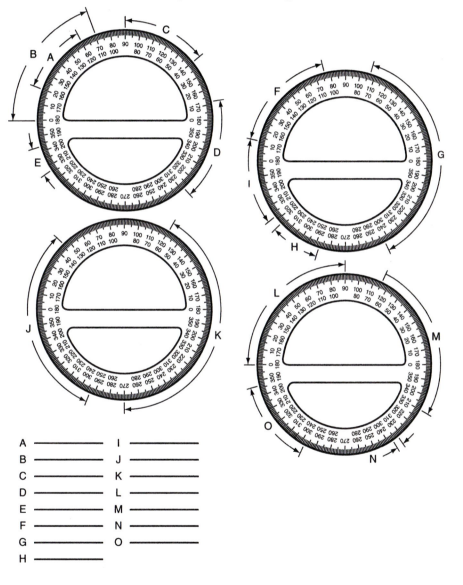

A —————		I —————	
B —————		J —————	
C —————		K —————	
D —————		L —————	
E —————		M —————	
F —————		N —————	
G —————		O —————	
H —————			

Instructor's Response:

_____

_____

_____

Name: _____

Date: _____

Valve Identification

Upon completion of this job sheet, you should be able to identify various valves on a blueprint.

Procedure

Identify the following valve:

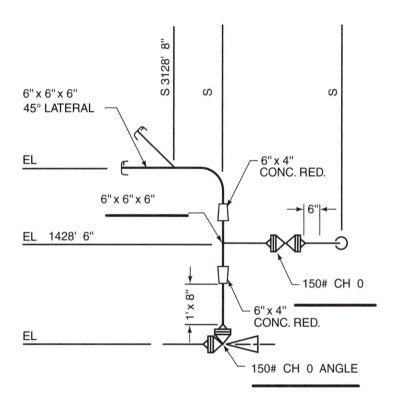

Instructor's Response:

_____

_____

_____

Name: _____

Date: _____

Line Type Identification

Upon completion of this job sheet, you should be able to identify various line types used on blueprints.

Procedure

Identity the following line types:

————————————   ——— RL ———

————————————   ——— RD ———

————————————   ——— RS ———

————————————   ——— CWS ———

————————————   ——— CWR ———

————————————   ——— CHWR ———

————————————   ——— MU ———

————————————   ——— H ———

————————————   ——— D ———

Instructor's Response:

_____

_____

_____

Fasteners

Screws and Bolts

Wood screws

Phillips flat head
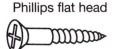

Slotted oval head

Slotted flat head

Slotted round head

Hex bolts

Carriage bolt

Full thread tap bolt

Standard bolt

Socket bolts

Socket head

Socket button head

Socket flat head

Socket set screw
with cup point

Machine screws

Phillips flat head

Slotted oval head

Phillips pan head

Combination round head

Slotted flat head

Combination truss head

Phillips oval head

Slotted round head

Sheet metal screws

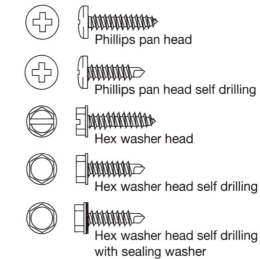

Phillips flat head

Phillips oval head

Phillips truss head

Phillips pan head

Phillips pan head self drilling

Hex washer head

Hex washer head self drilling

Hex washer head self drilling
with sealing washer

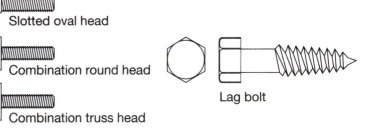

Lag bolt

Washers and Nuts

Washers

Flat washer
(USS and SAE)

Lock washer
internal tooth

Lock washer

Finishing washer

Lock washer
external tooth

Dock washer

Nuts

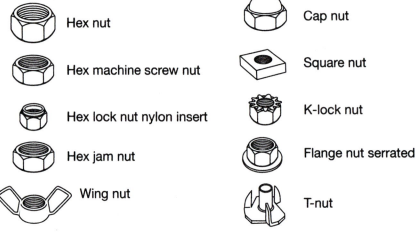

Hex nut   Cap nut

Hex machine screw nut   Square nut

Hex lock nut nylon insert K-lock nut

Hex jam nut   Flange nut serrated

Wing nut   T-nut

Nails and Other Fasteners

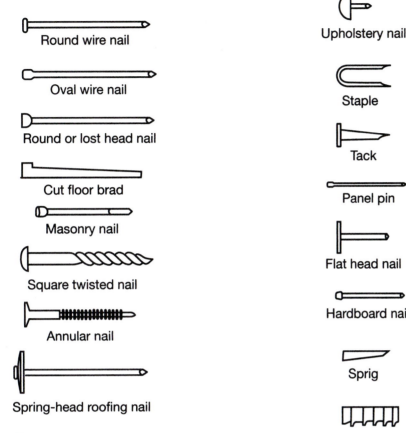

Round wire nail

Oval wire nail

Round or lost head nail

Cut floor brad

Masonry nail

Square twisted nail

Annular nail

Spring-head roofing nail

Cut clasp nail

Upholstery nail

Staple

Tack

Panel pin

Flat head nail

Hardboard nail

Sprig

Corrugated fastener

Wire Connectors

CRIMP CONNECTORS USED TO SPLICE AND TERMINATE 20 AWG TO 500 KCMILS ALUMINUM-TO-ALUMINUM, ALUMINUM-TO-COPPER, OR COPPER-TO-COPPER CONDUCTORS.	**A**	PROPERLY CRIMP THEN TAPE
CONNECTORS USED TO CONNECT WIRES TOGETHER ON COMBINATIONS OF 18 AWG THROUGH 6 AWG CONDUCTORS. THEY ARE TWIST-ON, SOLDERLESS, AND TAPELESS. *WIRE-NUT® and WING-NUT® are registered trademarks of IDEAL INDUSTRIES, INC. Scotchlok® is a registered trademark of 3M.	**B**	WIRE CONNECTORS VARIOUSLY KNOWN AS WIRE-NUT,® WING-NUT,® AND SCOTCHLOK.®
CONNECTORS USED TO CONNECT WIRES TOGETHER IN COMBINATIONS OF 16, 14, AND 12 AWG CONDUCTORS. THEY ARE CRIMPED ON WITH A SPECIAL TOOL, THEN COVERED WITH A SNAP-ON INSULATING CAP.	**C**	CRIMP-TYPE WIRE CONNECTOR AND INSULATING CAP
SOLDERLESS CONNECTORS ARE AVAILABLE IN SIZES 14 AWG THROUGH 500 KCMIL CONDUCTORS. THEY ARE USED FOR ONE SOLID OR ONE STRANDED CONDUCTOR ONLY, UNLESS OTHERWISE NOTED ON THE CONNECTOR OR ON ITS SHIPPING CARTON. THE SCREW MAY BE OF THE STANDARD SCREWDRIVER SLOT TYPE, OR IT MAY BE FOR USE WITH AN ALLEN WRENCH OR SOCKET WRENCH.	**D**	SOLDERLESS CONNECTORS
COMPRESSION CONNECTORS ARE USED FOR 8 AWG THROUGH 1,000 KCMIL CONDUCTORS. THE WIRE IS INSERTED INTO THE END OF THE CONNECTOR, THEN CRIMPED ON WITH A SPECIAL COMPRESSION TOOL.	**E**	COMPRESSION CONNECTOR
SPLIT-BOLT CONNECTORS ARE USED FOR CONNECTING TWO CONDUCTORS TOGETHER, OR FOR TAPPING ONE CONDUCTOR TO ANOTHER. THEY ARE AVAILABLE IN SIZES 10 AWG THROUGH 1,000 KCMIL. THEY ARE USED FOR TWO SOLID AND/OR TWO STRANDED CONDUCTORS ONLY, UNLESS OTHERWISE NOTED ON THE CONNECTOR OR ON ITS SHIPPING CARTON.	**F**	SPLIT-BOLT CONNECTOR

Appendix C

Conversion Tables

Fraction	Decimal	Fraction	Decimal	Fraction	Decimal
1/64	0.0156	11/32	0.3438	23/32	0.7188
1/32	0.0313	23/64	0.3594	47/64	0.7344
3/64	0.0469	**3/8**	**0.3750**	**3/4**	**0.7500**
1/16	0.0625	25/64	0.3906	49/64	0.7656
5/64	0.0781	13/32	0.4063	25/32	0.7813
3/32	0.0938	27/64	0.4219	51/64	0.7969
7/64	0.1094	7/16	0.4375	13/16	0.8125
1/8	**0.1250**	29/64	0.4531	53/64	0.8281
9/64	0.1406	15/32	0.4688	27/32	0.8438
5/32	0.1563	31/64	0.4844	55/64	0.8594
11/64	0.1719	**1/2**	**0.5000**	**7/8**	**0.8750**
3/16	0.1875	33/64	0.5156	57/64	0.8906
13/64	0.2031	17/32	0.5313	29/32	0.9063
7/32	0.2188	35/64	0.5469	59/64	0.9219
15/64	0.2344	9/16	0.5625	15/16	0.9375
1/4	**0.2500**	37/64	0.5781	61/64	0.9531
17/64	0.2656	19/32	0.5938	31/32	0.9688
9/32	0.2813	39/64	0.6094	63/64	0.9844
19/64	0.2969	**5/8**	**0.6250**		

Conversion Factors

Two basic systems, **English** and **metric systems**, are used to make measurements in the world today. The English system of measure is the unit of measure currently used in the United States today by most technicians. Its base units are inch (in), foot (ft), yard (yd), and mile (mi).

Unit	Divisions
English	
1 inch (in)	6 picas
1 foot (ft)	12 inches
1 yard (yd)	3 feet
1 mile (mi)	5,280 feet

English Conversion Factors

The base unit for the metric system is meter. This base unit is further divided into larger and smaller units (in multiples of 10) by adding prefixes. Common metric prefixes are deci- (10), centi- (100), and milli- (1000). For example, a meter contains 10 decimeters.

Unit	Divisions
	Units of Length
10 millimeters (mm)	1 centimeter (cm)
10 centimeters	1 decimeter (dm) = 100 millimeters
10 decimeters	1 meter (m) = 1,000 millimeters
10 meters	1 dekameter (dam)
10 dekameters	1 hectometer (hm) = 100 meters
10 hectometers	5.1 kilometer (km) 51,000 meters

Metric Conversion Factors

The metric system of measure is occasionally used in the United States and therefore it is important that the technician be able to recognize and convert from one unit to another. The following table gives conversion factors for converting from one system to another.

Starting with	Multiply	To Find
inches	2.5	centimeters
feet	30	centimeters
yards	0.9	meters
miles	1.6	kilometers
centimeters	0.4	inches
centimeters	0.0333	inches
meters	1.1111	yards
kilometers	0.625	miles

Converting from Metric to English and English to Metric

For example, to convert 12 feet 6 inches to centimeters, perform the following steps.

Convert from Inches to Centimeters

Step 1   Converting 12 feet into inches.

$$12 \text{ ft} \times 12 \text{ in/ft} = \textbf{144 in}$$

Step 2   Next add the inch portion of the original measurement.

$$144 \text{ in} + 6 \text{ in} = \textbf{150 in}$$

Step 3   Find the conversion factor using the previous table.

$$150 \text{ in} \times 2.5 \text{ cm/in} = \textbf{\textit{375 cm}}$$

Convert from Centimeters to Inches

Step 1   Find the conversion factor using the previous table.

$$375 \text{ cm} \times 0.4 \text{ in/cm} = \textbf{140 in}$$

Step 2   Convert from inches to feet and inches by dividing by 12.

$$150 \text{ in}/12 \text{ ft/in} = 12 \text{ ft with a remainder of } 6$$

Therefore, the measure would be written as **_12 ft 6 in or 12′ 6″._**

Electrical Wire Gauge and Current Chart

AWG Gauge	Diameter (inches)	Ohms/1000 ft	Maximum Amps
0000	0.46	0.049	302
000	0.4096	0.0618	239
00	0.3648	0.0779	190
0	0.3249	0.0983	150
1	0.2893	0.1239	119
2	0.2576	0.1563	94
3	0.2294	0.197	75
4	0.2043	0.2485	60
5	0.1819	0.3133	47
6	0.162	0.3951	37
7	0.1443	0.4982	30
8	0.1285	0.6282	24
9	0.1144	0.7921	19
10	0.1019	0.9989	15
11	0.0907	1.26	12
12	0.0808	1.588	9.3
13	0.072	2.003	7.4
14	0.0641	2.525	5.9
15	0.0571	3.184	4.7
16	0.0508	4.016	3.7
17	0.0453	5.064	2.9
18	0.0403	6.385	2.3

Small Engine Recommended Preventive Maintenance Charts

Recommended Chainsaw Preventive Maintenance				
	Daily	Weekly	Monthly	As Needed
Sprocket	Inspect			Replace
Fuel filter		Clean		
Muffler			Clean	
Muffler screen	Clean			Replace
Fuel tank			Clean	
Spark plug		Clean and adjust		Replace
Fuel, oil, and hoses	Check			
Air filter	Clean	Replace		
Screws, nuts, and bolts	Inspect and tighten			
Chain	Inspect and sharpen			

Recommended Lawnmower Preventive Maintenance				
	Daily	Weekly	Monthly	As Needed
Fuel filter		Clean		
Muffler			Clean	
Fuel tank			Clean	
Spark plug		Clean and adjust		Replace
Fuel, oil, and hoses	Check			
Air filter	Clean	Replace		
Screws, nuts, and bolts	Inspect and tighten			

Recommended String Trimmer Preventive Maintenance				
	Daily	Weekly	Monthly	As Needed
Fuel filter		Clean		
Muffler			Clean	
Fuel tank			Clean	
Spark plug		Clean and adjust		Replace
Fuel, oil, and hoses	Check			
Air filter	Clean	Replace		
Screws, nuts, and bolts	Inspect and tighten			

Appendix D

Using Sine, Cosine, and Tangent to Determine Angles and Lengths of Lines of a Triangle

As indicated by its name, a triangle contains three angles. When these three angles are added, their sum is equal to 180°, or angle A + angle B + angle C = 180°. Using this fact, a missing angle can be calculated if the other two angles are given. For example, the missing angle of a triangle containing a 70° and a 55° angle is found by adding the two known angles, and then subtracting their product from 180° (180° − [70° + 55°]). After performing the calculation, the missing angle is found to be 55°.

Each side of a triangle has a name: the hypotenuse, opposite, and adjacent sides. In a right triangle the hypotenuse is always the longest side. The other two sides, opposite and adjacent, are labeled relative to the acute angle (any angle less than 90°) being focused on in a given calculation (Figures D-1 and D-2). The adjacent side is

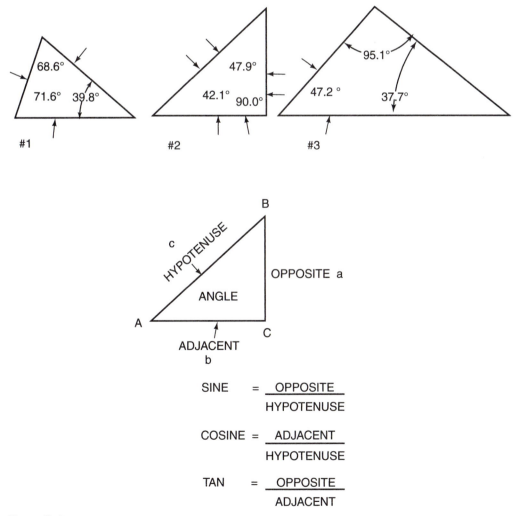

Figure D-1

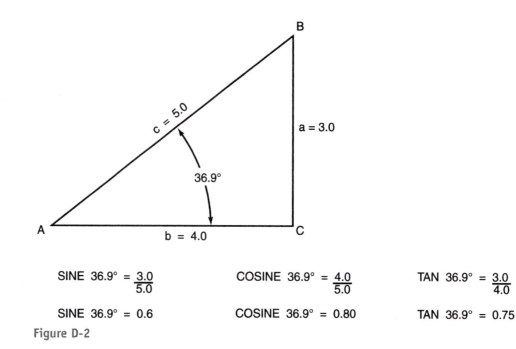

$$\text{SINE } 36.9° = \frac{3.0}{5.0} \qquad \text{COSINE } 36.9° = \frac{4.0}{5.0} \qquad \text{TAN } 36.9° = \frac{3.0}{4.0}$$

$$\text{SINE } 36.9° = 0.6 \qquad \text{COSINE } 36.9° = 0.80 \qquad \text{TAN } 36.9° = 0.75$$

Figure D-2

the side adjacent to the angle in question. The opposite is the side opposite to that angle.

In a right triangle, there is a direct relationship between the angles and the lengths of the sides. This relationship can be summed up in three fundamental trigonometric functions: sine, cosine, and tangent. The sine function is defined as the ratio of the side opposite to an acute angle divided by the hypotenuse or (Sine A = a/c), as shown in Figures D-1 and D-2. The cosine function is defined as the ratio of the side adjacent to an acute angle divided by the hypotenuse or (Cosine A = b/c), and the tangent function is defined as the ratio of the side opposite to an acute angle divided by the side adjacent (Tan A = a/b). Another important concept in trigonometry is the Pythagorean Theorem. It states that the square of the hypotenuse of a right triangle is equal to the sum of the square of the other two sides or $R^2 = X^2 + Y^2$. In this equation R is equal to the hypotenuse, and X and Y are equal to the adjacent and opposite sides of the triangle. With this equation, the missing side of a right triangle can be determined if the other two sides are given by applying the Pythagorean Theorem. To calculate the hypotenuse, algebra would have to be used to isolate the variable R. In other words, the hypotenuse is found by solving for R, not R^2. To isolate R, the square root of both sides of the equation must be taken, and doing so would yield the equation $R = \sqrt{(X^2 + Y^2)}$. For example, given a right triangle that contains an adjacent side (b) equal to 3 inches and an opposite side (a) equal to 4 inches, the hypotenuse can be found by using the formula: ($R = \sqrt{3^2 + 4^2}$, $R = \sqrt{25}$, R = 5), as shown in Figure D-3.

#1 Acute triangle—all angles are less than 90°.
#2 Right triangle—one angle is 90°.
#3 Obtuse triangle—one angle is greater than 90°.

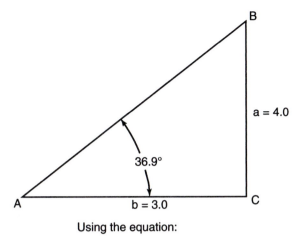

Using the equation:

$R^2 = A^2 + B^2$

Solving for R not R^2

the equation becomes

$$R = \sqrt{B^2 + A^2}$$

$$R = \sqrt{9 + 16}$$

$$R = \sqrt{25}$$

$$R = 5$$

Figure D-3

Glossary

Alternating current electron flow that flows in one direction and then reverses at regular intervals.

Ampere (amp) unit of current flow.

Appreciation the expression of gratitude toward your customers.

Asphyxiation loss of consciousness that is caused by a lack of oxygen or excessive carbon dioxide in the blood.

Assigning tasks giving a task to someone to complete.

Atom the smallest particle of an element.

Back miter an angle cut starting from the end and coming back on the face of the stock.

Belly deck a mower that is built like a car and has a blade deck underneath the driver and engine.

Booking the process used to activate the paste.

Boring jig a tool frequently used to guide bits when boring holes for locksets.

Box nail a thin nail with a head, usually coated with a material to increase its holding power.

Brad a thin, short, finishing nail.

Carbon dioxide by-product of natural gas combustion that is not harmful.

Carbon monoxide a poisonous, colorless, odorless, tasteless gas generated by incomplete combustion.

Cardiopulmonary resuscitation (CPR) an emergency first-aid procedure used to maintain circulation of blood to the brain.

Class A fire extinguishers fire extinguishers used on fires that result from burning wood, paper, or other ordinary combustibles.

Class B fire extinguishers fire extinguishers used on fires that involve flammable liquids such as grease, gasoline, or oil.

Class C fire extinguishers fire extinguishers used on electrically energized fires.

Class D fire extinguishers fire extinguishers typically used on flammable metals.

Competence having the skills, knowledge, ability, or qualifications to complete a task.

Confidence having a belief in yourself and your abilities.

Continuous load a load in which the maximum current is expected to continue for 3 hours or more.

Conventional current flow theory states that current flows from positive to negative in a circuit.

Coulomb one coulomb is the amount of electric charge transported in one second by one ampere of current.

Courtesy acting respectful toward your customers.

Current (or amperage) the flow of electrons through a given circuit, which is measured in *amps*.

Diagnostics the process of determining a malfunction.

Direct current electron flow that flows in only one direction; used in the industry only for special applications such as solid-state modules and electronic air filters.

Dumpster a large waste receptacle.

Duplex nail a double-headed nail used for temporary fastening such as in the construction if wood scaffolds.

Electron theory states that current flows from negative to positive in a circuit.

Element any of the known substances (of which ninety-two occur naturally) that cannot be separated into simpler compounds.

Elevator platform the floor of an elevator.

Empathy the capacity to understand your customers' state of mind or emotion.

Engineered panels human-made products in the form of large reconstituted wood sheets.

English system of measure a system of measurement primarily used in the United States based on the dimensions of the human body.

Faceplate markers a tool used to lay out the mortise for the latch faceplate.

Finishing nail a thin nail with a small head designed for setting below the surface of finished materials.

Fraction a number that can represent part of a whole.

Frostbite injury to the skin resulting from prolonged exposure to freezing temperatures.

Frostnip the first stage of exposure, which causes whitening of the skin, itching, tingling, and loss of feeling.

Green lumber lumber that has just been cut from a log.

Ground fault circuit interrupter (GFCI) electrical device designed to sense small current leaks to ground and de-energize the circuit before injury can result.

Groundskeeping the activity of tending an area for aesthetic or functional purposes. It includes mowing grass, trimming hedges, pulling weeds, planting flowers, and so on.

Hardwood deciduous trees.

Heating element a device used to transform electricity into heat through electrical resistance.

Honesty acting truthfully with your customers.

Inorganic mulch constitutes gravel, pebbles, black plastic, and landscape fabrics.

Insecticides chemicals used to kill insects.

Integer any whole number, including zero that is positive or negative, factorable or nonfactorable.

Kill spot an inconspicuous area used to start and end wallpaper in a room.

Landscaper an individual that modifies the feature of an area (land) such as construct stone walls, wooden fences, brick pathways, statuary, fountains, benches, trees, flowers, shrubs, and grasses.

Law of centrifugal force states that spinning object has a tendency to pull away from its center point and that the faster it spins, the greater the centrifugal force will be.

Law of charges states that like charges repel and unlike charges attract.

Linear measurement the measurement of two points along a straight line.

Mask a faux finish used to apply a new color over a dry base coat to create an image or shape.

Matter a substance that takes up space and has weight.

Metric system of measure a system of measurement that uses a single unit for any physical quantity.

Multispur bits a power-driven bit, guided by a boring jig, that is used to make a hole in a door for the lockset.

Mural a faux finish used to give the illusion of scenery or architectural elements.

Occupational Safety and Health Administration (OSHA) branch of the U.S. Department of Labor that strives to reduce injuries and deaths in the workplace.

Ohm's law relationship between voltage and current and a material's ability to conduct electricity.

Organic mulch constitutes natural substances such as bark, wood chips, leaves, pine needles, or grass clippings.

Personal protective equipment (PPE) any equipment that will provide personal protection from a possible injury.

Pesticides chemicals used to kill pests (as rodents or insects).

Plies layers of wood used to build up a product such as plywood.

Power the electrical work that is being done in a given circuit, which is measured in wattage (watts) for a purely resistive circuit or volt-amps (VA) for an inductive/capacitive circuit.

Priority giving a task precedence over others.

Real number any number including zero that is positive or negative, rational or irrational.

Reliability the quality of being dependable.

Resistance the opposition to current flow in a given electrical circuit, which is measured in *ohms*.

Self-talk what you say silently to yourself as you go through the day or when you are faced with difficult situations.

Softboard a low-density fiberboard.

Softwood coniferous, or cone-bearing, trees.

Striker plate a plate installed on the door jamb against which the latch on the door engages when the door is closed.

Task an activity that needs to be performed to complete a project.

Thermostat a device used to control the temperature of water by controlling the heat source.

Troubleshooting the process of performing a systematic search for a resolution to a technical problem.

Valence shell the outermost shell of an atom.

Voltage the potential electrical difference for electron flow from one line to another in an electrical circuit.

Watt a unit of power applied to electron flow. One watt equals 3.414 Btu.

Whole numbers counting numbers including zero.

Index